Manual of Clinical Virology

Manual of Clinical Virology

Danny L. Wiedbrauk, Ph.D.
*Chief, Virology and Serology
William Beaumont Hospital
Royal Oak, Michigan*

Sheryl L. G. Johnston, M.S.
*Supervisor, Virology and Seroimmunology
St. Vincent Hospital
Green Bay, Wisconsin*

RAVEN PRESS ● NEW YORK

Raven Press, Ltd., 1185 Avenue of the Americas, New York, New York 10036

© 1993 by Raven Press, Ltd. All rights reserved. This book is protected by copyright. No part of it may be reproduced, stored in a retrieval system, or transmitted, in any form or by any means, electronic, mechanical, photocopying, or recording, or otherwise, without the prior written permission of the publisher.

Made in the United States of America

Library of Congress Cataloging-in-Publication Data

Wiedbrauk, Danny L.
    Manual of clinical virology / Danny L. Wiedbrauk, Sheryl L.G. Johnston.
      p. cm.
  Includes bibliographical references and index.
  ISBN 0-88167-984-4
  1. Diagnostic virology—Handbooks, manuals, etc. I. Johnston, Sheryl L. G. II. Title.
  [DNLM: 1. Virus Cultivation—methods—laboratory manuals. 2. Viruses—isolation and purification—laboratory manuals. QW 25 W642m]
QR387.W54 1992
616'.0194—dc20
DNLM/DLC
for Library of Congress   92-49313
                                                                                                            CIP

    The material contained in this volume was submitted as previously unpublished material, except in the instances in which some of the illustrative material was derived.
    Great care has been taken to maintain the accuracy of the information contained in the volume. However, neither Raven Press nor the authors can be held responsible for errors or for any consequences arising from the use of information contained herein.
    Some of the names of products referred to in this book may be registered trademarks or proprietary names, although specific reference to this fact may not be made, however, the use of a name without designation is not to be construed as a representation by the publisher or author that it is in the public domain. In addition, the mention of specific companies or of their products or proprietary names does not imply any endorsement or recommendation of the part of the publisher or author.
    Authors were themselves responsible for obtaining the necessary permission to reproduce copyright materials from other sources.

Contents

Preface, *vii*

Acknowledgments, *ix*

Part I. General Laboratory Procedures

1. Laboratory Design and Equipment 1
2. Laboratory Safety 6
3. General Methodologies 11
4. Virus Testing Protocols 18
5. Specimen Collection and Processing 22
6. Mammalian Cell Culture Procedures 33
7. Quality Assurance 45

Part II. Specific Detection Methods

8. Adenoviruses 54
   At a Glance, 55
9. Chlamydiae 64
   At a Glance, 65
10. *Clostridium difficile* Toxin Assay 77
    At a Glance, 78
11. Cytomegalovirus 82
    At a Glance, 83
12. Enteroviruses 92
    At a Glance, 93
13. Epstein-Barr Virus 98
    At a Glance, 99
14. Hepatitis A Virus 105
    At a Glance, 106
15. Herpes Simplex Virus 109
    At a Glance, 110
16. Human Herpesvirus 6 121
    At a Glance, 122

17. Influenza Virus 127
    At a Glance, 128

18. Measles Virus 141
    At a Glance, 142

19. Mumps Virus 152
    At a Glance, 153

20. Parainfluenza Virus 161
    At a Glance, 162

21. Human Papillomavirus 172
    At a Glance, 173

22. Polyomavirus 175
    At a Glance, 176

23. Human Parvovirus 181
    At a Glance, 182

24. Respiratory Syncytial Virus 184
    At a Glance, 185

25. Human Retroviruses 196
    At a Glance, 197, 199, 201, 203

26. Rhinovirus 208
    At a Glance, 209

27. Rotavirus 214
    At a Glance, 215

28. Rubella Virus 222
    At a Glance, 223

29. Varicella-Zoster Virus 229
    At a Glance, 230

Part III. Appendices

A. Glossary of Commonly Used Terms 241
B. Reagent Resources 248
C. Reagent Formulations 255

Subject Index 267

Color section appears following page 52.

Preface

This manual provides a stand-alone virus isolation guide that can be used by clinical laboratory personnel, physicians, graduate students, and clinical researchers. The original concept for this book arose from our efforts to modernize our procedure manuals and to incorporate the essential elements described in the NCCLS guidelines for laboratory procedures. This manual is a collection of procedures that contain the essential NCCLS elements and quality assurance protocols. We have also provided standard methods for producing quality control slides and positive control cultures. Wherever possible, we have included tube culture, centrifugation-enhanced (shell vial), and direct fluorescent antibody procedures even though most laboratories do not perform all these procedures on every virus.

Our laboratories often get telephone calls from physicians, residents, nursing personnel, and hospital administrators requesting information on appropriate specimen selection, turnaround times, immune responses during virus infection, and the appropriateness of disinfection procedures. The *At a Glance* sections were written in response to these often harried requests for "no frills" information about virus isolations, viral epidemiology, and viral serologies. Because they are concise enough to be conveyed over the telephone, the *At a Glance* information has allowed our laboratories to quickly respond to these basic information requests.

The *Manual of Clinical Virology* has been a significant undertaking for both of us. We hope you find it as useful in your laboratory as it has been in ours.

D.L. Wiedbrauk
Sheryl L.G. Johnston

Acknowledgments

No work of this type can be written alone. We would therefore like to acknowledge the work and assistance of our friends and colleagues who helped make this book possible. John Gibson took all the photomicrographs, and we are extremely grateful for his help. Elizabeth Ostler, Ruth Bollinger, Cindy Podgorski, and John Gibson reviewed a number of chapters and set up numerous cultures and DFA slides for the photography sessions. Curt Gleaves reviewed early versions of the book and made a number of valuable suggestions. Harlan Bloy's review of the final draft was sincerely appreciated. We would also like to thank Dr. Kerry Willis of Raven Press for sticking by us as the book evolved. Her encouragement and helpful criticism made the book better and more succinct. Finally, we are grateful for the support provided by William Beaumont Hospital and St. Vincent Hospital. This book could not have been written without the assistance and support of these institutions.

CHAPTER 1

Laboratory Design and Equipment

INTRODUCTION

The virology and cell culture laboratory can be established in a variety of rooms and situations. Cell cultures must be protected from bacterial and fungal contamination. Therefore, the ideal cell culture area should be maintained under positive pressure by infusion of HEPA-filtered air. In addition, the amount of equipment and furniture in the cell culture area should be kept to a minimum to provide for ease of cleaning and to minimize areas where dust will collect. When the ideal situation is not available, a laminar airflow hood can provide a clean environment for cell culture. If the hood must also be used for virus isolation and cell culture, the culture procedures should be performed first, and the hood should be thoroughly decontaminated before and after each use.

In contrast with the cell culture area, the ideal virology area is maintained under negative pressure and the airflow out of the room is passed through a HEPA filter. The purpose of this procedure is to contain any viral agents that might be liberated from spilled cultures or leaking specimens. Because many laboratories cannot adequately control the airflow to maintain these conditions, Class 2 and 3 biological containment hoods are often used as surrogates for good environmental engineering. However, high-tech equipment cannot replace good aseptic technique. Proper specimen handling and virus isolation procedures must be rigorously followed in order to protect laboratory workers and the environment from unnecessary contamination.

The ideal virology laboratory with multiple rooms and separate ventilation systems is often unattainable, especially in crowded laboratories or when establishing a virology laboratory in one corner of a clinical microbiology laboratory. Small single-room laboratories can work so long as key design elements are incorporated into the plan. Clean areas should be established at one end of the room and the washup and sterilization areas should be located at the other end. Transitional areas between the "clean" and "dirty" portions of the room can be used for storage, incubators, freezers, media preparation, and microscopes.

In general, the laboratory layout should include adequate space for storage and for the operation of equipment. Furniture and equipment should be placed to facilitate cleaning. All surfaces should be non-porous and easily disinfected. Disruptive airflows should be avoided in order to minimize dust contamination. The laboratory should have sufficient air exchanges to minimize the buildup of toxic chemicals and to maintain even temperature and humidity levels. A properly designed laboratory can significantly reduce cell culture contamination problems.

The Clean Area

All sterile procedures should take place in the clean area. Clean areas should be situated away from high traffic areas that may cause drafts and stir dust. The clean area must be designed for easy cleaning and should be dusted and damp mopped daily to control dust. The laminar flow cabinet is the most important piece of equipment in the clean area. This cabinet should stand far enough away from the wall to allow for cleaning and dusting. Alternatively the cabinet can be sealed into the wall with a plastic or taped strip. The laboratory design should allow the laminar airflow cabinet to remain clean and uncluttered. Storage space should be readily available to accommodate instruments, pipettes, media, pumps and other objects. The interior of the cabinet should be able to withstand daily disinfection and germicidal UV lighting.

The Work Area

The working area of the laboratory should be designed so that there is adequate ventilation and temperature control. Ideally, the work area should be arranged to minimize the amount of movement necessary to accomplish the work. Traffic through the work area should be kept to a minimum because traffic creates air disturbances and dust contamination problems.

The workbench should be situated at a convenient height (either standing or sitting height) to minimize fatigue. The surface of the workbench should be impermeable to water, acetone, alcohols, stains, and bleach. The surface should also be smooth to facilitate easy disinfection. The working area should also contain adequate storage for glassware, pipettes, syringes, office supplies, etc. These supplies should be placed in closed cabinets and drawers to minimize dust contamination.

The floor of the laboratory should be impervious to bleach and acetone and constructed in a manner that allows easy decontamination and cleaning. Floors should be kept clear of boxes and supplies to facilitate cleaning and minimize dust accumulation.

Laboratory personnel should be discouraged from bringing houseplants and flowers into the laboratory. Although aesthetically pleasing, plants and flowers can be a source of insects, microorganisms, pollen, and dust contamination.

The Washup Area

The washup area should be located near the entrance of the laboratory, where the traffic flow will be the greatest. Placing refrigerators for specimen storage and autoclaves in this area will decrease the traffic in the rest of the laboratory. In addition, routine equipment maintenance will produce fewer disruptions of the clean area of the laboratory. The washup area should have plenty of countertop space for soaking, cleaning, and processing glassware and a designated specimen receiving area. The entrance into this area of the laboratory should also have sufficient space for laboratory personnel to store coats and shoes, without having to bring them further into the laboratory. The washup area should also have an area for hanging lab coats and a place where laboratory personnel can wash their hands before leaving the laboratory.

Laminar Airflow Hoods (Biological Safety Cabinets)

The laminar flow hood can be of two types. Class I cabinets have an open front with negative pressure ventilation and a HEPA-filtered air exhaust system. While these cabinets are designed to protect the user from infectious agents, Class I cabinets are not suitable for cell culture operations. In contrast, the horizontal flow (clean bench) cabinet works well for cell culture but these units must not be used for handling infectious agents. The Class IIA cabinet is the best all-round biological safety cabinet for general virology usage. Class II cabinets provide protection of the user, the environment, and the cultures by means of a recirculating HEPA-filtered vertical laminar airflow, and HEPA-filtered exhaust air. Class II cabinets can be purchased as freestanding or bench top units. These units must be installed so that laboratory personnel can sit comfortably. Adjustable chairs are preferred so that different technicians can clearly view the work space without bending and so that the operator's face is fully protected by the glass panel in the front of the cabinet.

Biological safety cabinets should be large enough (typically 4-6 feet wide by 2 feet deep) to adequately handle the work and equipment needed. Safety cabinets should be easy to dismantle to allow for cleaning and decontamination. Cabinets should be as quiet as possible so that laboratory personnel working in the cabinet can hear others talking in the room. Loud cabinets can mask fire alarm signals and emergency announcements and can cause noise fatigue in laboratory personnel. All biological safety cabinets should be inspected and certified by a qualified inspector when the cabinets are installed or moved. Cabinets should be certified at least annually thereafter.

Laboratory workers must be trained to use and decontaminate biological safety cabinets. They also must be aware that the cabinet does not provide absolute protection against contaminants. Interruption of the air flow by arms, equipment, or supplies may allow aerosols to escape from the cabinet. Laboratory personnel must be able to work in the center of the work space with only a minimum amount of equipment interrupting the air flow. In addition, the hands and arms of laboratory personnel should be protected against contamination when necessary.

When operating, biological safety cabinets generate a significant amount of heat. Therefore, any laboratory contemplating installation of biological safety cabinets should consult their building engineers to determine if the additional heat load can be handled by the environmental control equipment. In general, warm laboratories will be much warmer after installation of biological safety cabinets and laboratory workers often refuse to wear gloves and lab coats under these conditions.

Microscopes

Microscopes are an essential part of any virology laboratory. Microscopes are used to examine tube cultures, perform cell counts, evaluate specimen adequacy, and to examine cell culture flasks and plates. Two types of microscopes are necessary for virological procedures - an inverted microscope for examining cell culture flasks and plates, and an upright fluorescence microscope for detecting fluorescent antibody reactions. Although a phase contrast inverted microscope is ideal, a standard inverted light microscope can be used to examine cultures in microtiter plates, flasks, and shell vials without tipping the containers and disturbing the cultures.

The fluorescence microscope is essential in the virology laboratory because fluorescein-labeled antibody reagents are used extensively for culture confirmation and direct fluorescent antibody testing. Fluorescent microscopes come in transmitted light and epifluorescence (incident-light) models. For general usage, epifluorescence microscopes are preferred because (a) a darkfield condenser is not needed; (b) objectives with the highest numerical apertures can be used, thereby producing a brighter image; and (c) specimen thickness does not interfere with the fluorescence intensity (1). For fluorescence microscopy, three types of light sources are available, high pressure mercury vapor lamps, xenon high pressure lamps, and low voltage tungsten halogen lamps. High pressure mercury vapor lamps are used most widely because they provide a powerful light source in the UV and blue regions required for excitation of fluorescein. The 100 W lamp provides the best incident-light illumination.

The high cost and hazardous nature of high pressure lamps has increased the popularity of low voltage tungsten halogen lamps. These lamps provide a brilliant source of visible light and they may be turned on and off without a warm-up or cool-off period. Tungsten halogen lamps provide an economical light source for blue light (fluorescein) illumination. However, high-pressure lamps provide a higher intensity light and increased ability to visualize antibody-labeled viral antigens.

Microscope objectives are divided into three types - acromats, fluorites, and apochromats. Acromat objectives are the least corrected lenses and the least expensive. Fluorites are corrected chromatically and spherically and they provide excellent resolution at a reasonable cost. Apochromats are the most corrected and most expensive objectives. Plan-apochromat objectives are used when a flat image is required to the edge of the field. For most fluorescence microscopy, apochromat objectives are not required because the fluorescent image is essentially monochromatic. For incident-light fluorescence, the objectives should have the highest numerical aperture and a minimum number of lens elements for improved light transmission.

Upright microscopes used for examining tube cultures should be equipped with long working distance objectives. These objectives allow the technician to examine the cell monolayers by looking down through the side of a tube. In this manner, the cells are continuously bathed in culture fluid during examination and there is minimal risk of damaging an objective. The 6X,3/0.16 and the 10X/0.22 objectives work well for this purpose.

Water Baths

Water baths are used for thawing cultures and inactivating fetal calf serum. Water baths should be of sufficient size to handle these tasks. In addition, water baths should be constructed of corrosion-resistant materials so that commonly used fungicides and algacides will not destroy the interior of the bath. Regular cleaning is essential to prevent cell culture contamination. Therefore, water baths must be built to allow easy access and thorough cleaning. At a minimum, water baths should provide a constant temperature throughout the 20-60°C range and they should maintain the temperature within 1°C of the setpoint.

Centrifuges

Centrifuges are necessary for preparing blood cultures, harvesting cells, and for centrifugation-enhanced (shell vial) cultures. Centrifuges should be refrigerated and capable of handling a variety of tubes and shell vials. Centrifuge rotors should have swinging buckets and safety covers must be available for all tube carriers. Centrifuges should be constructed to allow easy access for maintenance and changing of brushes. Because shell vial methods typically require 1 hour centrifugation times, most laboratories will require more than one centrifuge.

Incubators

Incubators are one of the most important items in the virology laboratory. Two types of incubators are commonly used - CO_2 and ambient air incubators. Carbon dioxide incubators are necessary when using "open" culture vessels containing bicarbonate-buffered media. Bicarbonate/carbonic acid buffering systems are difficult to control in ambient air and 5% CO_2 atmosphere is required over a 26 mM bicarbonate buffer system to maintain a pH of 7.2-7.4. Efficient CO_2 transfer requires a humidified atmosphere. Therefore, CO_2 incubators must be humidified to 90-95% saturation at 35-37°C. Because CO_2, humidity, and temperature are critical for pH control, safety limits, audio alarms, and visual alarms must be incorporated into these incubators. In addition, the carbon dioxide levels must be checked weekly (Fyrite) to assure that the alarms are working properly.

Closed culture vessels containing HEPES buffers do not require CO_2 atmospheres and patient isolations can be accomplished in large forced air incubators. Placing roller racks in a forced air incubator will minimize corrosion and prolong the life of the equipment. Most clinical laboratories will have two large forced air incubators and one small- to moderate-sized CO_2 incubator for plates and "open" vessels. Recovery time is very important, no matter what type of incubator is used. Recovery time indicates the time interval needed to restore the set conditions (temperature, CO_2 levels, and humidity) whenever the door is opened. Ideally, laboratories should purchase incubators with the shortest recovery time. However, the larger incubators require more recovery time than smaller incubators and fast recovery models use much more CO_2. Incubators should be corrosion resistant and easy to clean. If carbon dioxide is being used, an inner glass door will reduce the loss of gas into the room and improve humidification.

Refrigerators and Freezers

Laboratory refrigerators are required for the virology laboratory because media and reagents must be refrigerated. While household refrigerators can be used in the virology laboratory, they are not recommended because they rarely hold the temperature within the 2-8°C range. In addition, the temperature fluctuations resulting from defrosting cycles can reduce specimen viability and reagent quality. A good quality laboratory or commercial refrigerator is recommended for most laboratories.

Laboratory or commercial grade freezers (-20°C to -40°C) are recommended over household freezers because the temperature fluctuations tolerated in domestic freezers will cause reagent deterioration in the laboratory. Self-defrosting freezers should not be used under any circumstances. Ultralow ($\leq$ -70°C) freezers are required for storing specimens, some reagents, virus stocks, and some cell lines. Chest freezers are more efficient than upright freezers but they have a much larger footprint and many laboratories do not have sufficient space for chest freezers. Freezers should have some type of inventory system so that materials can be quickly retrieved. Rapid retrieval minimizes the time the door is open, decreases the loss of cold air, and decreases the amount of ice buildup in the freezer. Freezers should be connected to an alarm in order to minimize reagent losses during freezer malfunctions or when the power is interrupted. Ideally, freezers should be connected to the emergency power outlets.

Liquid nitrogen freezers should be purchased with canes and canisters that allow easy retrieval of the vials. Liquid nitrogen systems require a standing order for liquid nitrogen to replace the liquid nitrogen that boils off over time. Although nitrogen purchases present an added expense, liquid nitrogen tanks are not subject to thawing when the power is interrupted and they provide the best environment for cell line preservation.

Roller Drum

Roller drums must be designed to work reliably at 37°C and provide a constant rotation of 10-15 rph. Roller drums and motor housings should be made of stainless steel to prevent corrosion. When ordering roller drums, make sure there is power available in the incubator for the roller apparatus and that there is sufficient space in the incubator for the roller apparatus.

Autoclave

An autoclave is required for sterilizing reagents and glassware. In addition, autoclaves are used to decontaminate laboratory wastes before they are transported to a landfill. While many laboratories prefer to incinerate medical waste, it is a good policy to autoclave potentially hazardous materials before transporting them out of the laboratory. Autoclaves should be certified and inspected annually. In addition, autoclaves should be of sufficient size to accommodate the amount of reagents, glassware, and wastes generated by the laboratory.

Autoclaves must be capable of producing steam under pressure to produce a working temperature of 121°C. The mechanism of the autoclave should be automatic and allow for dry goods (fast exchange and dry), liquids (slow, gravity exchange to minimize liquid boiling), and manual operation. Autoclaves should also provide a record of temperatures during the run. In many boilers, the steam is treated to prevent corrosion of the pipes and to minimize scale. These boiler treatments can leave cytotoxic residues on glassware that inhibits cell growth. Whenever possible, boiler treatments should not be used for autoclave steam. Check with the plant engineer to determine the type of boiler treatments used in your facility.

Dry heat ovens can also be used to sterilize glassware and pipettes. Dry heat has an advantage over steam because dry heat will not produce chemical contamination of the glassware. However, the dry heat sterilization process is less precise than autoclaving and it is not suitable for liquids or infectious waste. To be effective, a temperature of at least 160°C must be maintained for one hour.

REFERENCES

1. Lyerla HC, Forrester FT. *Immunofluorescence methods in virology*. Atlanta GA: Centers for Disease Control 1979;1-16.

CHAPTER 2

Laboratory Safety

INTRODUCTION

Laboratory safety is an increasingly visible component of virology laboratory operations. Like most infectious disease laboratories, universal precautions (1) have been employed for some time. However, with the advent of AIDS and the discovery of a wide variety of bloodborne pathogens, came the realization that other laboratory workers, clerical, housekeeping, and even mailroom personnel could also be at risk for acquiring these bloodborne pathogens. On March 6, 1992, the final Occupational Safety and Health Administration (OSHA) rules for minimizing the risk of exposure became effective (2). These rules require a combination of engineering and work practice controls together with personal protective clothing and equipment, training, medical surveillance, Hepatitis B vaccination, signs, and labels to make workers aware of the risks and to employ measures to minimize these risks.

In addition to biological safety precautions, the laboratory must also have chemical hygiene and radiation safety programs designed to educate and protect employees from radiation and chemical hazards. Like all workplaces, the clinical laboratory must provide a safe working environment and a variety of OSHA regulations and laboratory practices have been used to increase employee awareness of these problems and to prevent accidents.

BIOLOGICAL SAFETY

Because nearly every clinical laboratory procedure has the potential to disperse infectious material into the workplace, laboratories must train workers to handle infectious agents in a safe and effective manner. In addition, laboratory personnel must handle all specimens as if they were infectious. This is especially important because most laboratory exposures are inapparent. In one study, more than 80% of laboratory exposures could not be ascribed to a particular event or an accident (3). However, the most common causes of laboratory exposures can be virtually eliminated if the laboratory worker follows some simple biosafety rules (4). These practical measures are inexpensive and highly effective in reducing the risk of accidental infection. Basic biosafety rules include:

1. **Do not mouth pipette.**
2. **Manipulate infectious agents carefully to avoid spills and the production of aerosols and droplets.**
3. **Restrict the use of needles and syringes to those procedures where there is no alternative; use needles, syringes, and other "sharps" carefully to avoid self-inoculation; and dispose of "sharps" in leak- and puncture-resistant containers.**
4. **Use protective laboratory coats and gloves.**
5. **Wash hands following all laboratory activities, following the removal of gloves, and immediately after contact with infectious materials.**
6. **Decontaminate work surfaces before and after use and immediately after spills.**
7. **Do not eat, drink, store food, apply makeup or contact lenses, or smoke in the laboratory.**

Control of Aerosols and Droplets

Although any liquid handling procedure can generate aerosols and droplets, operations such as pipetting, aspirating fluids, mixing, shaking, grinding, filtering, sonicating, flaming, and centrifuging have the highest potential for aerosol formation. Pipetting presents the greatest hazard because it is done more frequently than any other procedure. A biological safety cabinet should be used for all procedures that might generate aerosols or droplets and for all infected cell culture manipulations. The following procedures should be implemented to minimize aerosol formation:

1. Mouth pipetting should not be done under any circumstances and mechanical pipetting devices must be available in the laboratory.
2. Pipettes should be drained into receiving vessels by touching the tip to inner wall of the tube or vial and providing gentle pressure.
3. Infectious material should not be forcibly expelled from the tip of a pipette.
4. Air should not be bubbled through an open container containing a suspension of infectious agents.
5. Safety covers or sealed centrifuge heads should be used whenever centrifuging infectious agents or infected cultures.
6. For ultracentrifuges, HEPA filters should be installed between the chamber and the vacuum pump.
7. Care should be taken when flaming inoculating loops because improper techniques can cause infectious aerosols and droplets.
8. Special care should be taken when opening shell vial cultures because pressure buildup within the tubes can cause significant splattering of the culture materials.

Needle Usage

Needle and syringe usage presents the greatest risk of exposure through inoculation. Needles and syringes are used for drawing blood, transferring materials from diaphragm-stoppered containers, and for animal inoculation. The use of needles and syringes should be limited to procedures where there are no alternative methods. Needle-stick injuries occur most often when needles are returned to their protective sheaths after use. Whenever possible, needles should not be resheathed. If resheathing is required, the procedure should utilize a needle resheathing device to minimize injury and accidental inoculation. Needle cutting devices should not be used because they can produce infectious aerosols. After use, needles and syringes should be placed in leak- and puncture-resistant containers for decontamination and disposal.

Laboratory Coats and Gloves

Laboratory coats, gowns, or uniforms should be worn to protect street clothes from contamination and laboratory coats should not be worn outside the laboratory area. Water-impermeable laboratory coats and mucus membrane protection must be made available for persons who are at risk of blood spills and splashes.

Protective gloves should be worn by all personnel engaged in activities that may involve direct contact of skin with potentially infectious specimens, cultures, or tissues. Gloves should be carefully removed and changed when they are visibly contaminated. Personnel who have dermatitis or other lesions on the hands and who may have indirect contact with cultures or clinical specimens should also wear protective gloves. Hands should be washed with soap and water immediately after handling infectious agents and after work is completed - **EVEN WHEN GLOVES HAVE BEEN WORN** (5).

Spills of Infectious Agents

Laboratories should have a written plan for handling laboratory spills and damaged or leaking specimens. The plan should also explain how to handle cases of laboratory exposure to infectious agents, accidental cuts, needle stick injuries, accidental ingestion, and contact of potentially infectious agents with mucus membranes. These procedures should include decontamination procedures, whom to contact for emergency services, and location of emergency equipment (i.e., eye wash stations, first aid kit, etc.).

When a spill occurs, the type of decontamination will depend upon the volume, location, and the infectious agent(s) in the spill. If small spills occur in the biological safety cabinet, the spill should be blotted with absorbent paper and the spill area should be wiped with a 10% household bleach solution. Because bleach can corrode the stainless steel, the bleached area should be wiped with 70% alcohol. After disinfection, the germicidal light should be turned on and the hood should be left running undisturbed, for 10-15 minutes. The laboratory worker should carefully remove their gloves and wash their hands thoroughly. Larger spills should be covered with absorbent paper and the paper should be soaked with 5-10% household bleach solution. Allow the bleach to react with the spill for 15 minutes then wipe up the fluid. Care must be taken to avoid generating infectious aerosols during containment and cleanup. Gloves, mask, and eye protection are recommended during spill containment and cleanup. Following cleanup, the area should be washed with detergent and water to remove the bleach. The protective clothing worn by of the clean-up personnel should be autoclaved and goggles should be disinfected via chemical means. After any large spill, a description of the incident including the personnel present, infectious agent(s), and decontamination procedures should be filed. Incident reports should be reviewed regularly by the institutional safety committee.

Infectious Wastes

Handling and decontamination of infectious wastes is an important part of any safety program and an area of increasing regulatory concern. Any article that may have come into contact with infectious agents should be decontaminated, preferably by autoclaving, before removing those articles from the laboratory area. In many states, infectious wastes must be placed in autoclavable bags possessing a biohazard insignia. Following decontamination, bags must be labeled as decontaminated before they can be transported to a landfill. **Do Not Pour Infectious Wastes Down the Sink.**

Cell Cultures

In general, cell cultures present few hazards as evidenced by their extremely wide usage and the rare cases of transmitted infections to laboratory personnel. Primary cell cultures from infected humans or animals are recognized hazards. In addition, cell cultures from macaques and other Old World monkeys may be latently infected with *Herpesvirus simiae* (B-virus) and thus present a hazard for personnel handling these animals and their tissues. At least 24 documented cases of infections of laboratory workers handling primary cell cultures (e.g., primary rhesus monkey kidney cells) have occurred in the past 30 years (4,6).

Clearly, primary cell cultures prepared from humans infected with hazardous agents such as HIV present a danger of infection. However, continuous cell cultures do not present a significant risk to the laboratory worker unless the cultures are contaminated with an infectious agent.

Immunization

Employees handling clinical specimens or infectious agents should be immunized with, or have documented immunity (antibody) to, mumps virus, measles (rubeola) virus, rubella virus, poliovirus, diphtheria, pertussis, and tetanus. Annual influenza vaccinations are recommended for virology laboratory personnel. In addition, hepatitis B vaccine should be made available to anyone who may become exposed to blood, serum, or other body fluids.

PHYSICAL AND CHEMICAL SAFETY

Glassware and Sharp Items

Broken glass, scalpels, needles and other "sharps" are the most common sources of laboratory injury. Accidental cuts can occur when cleaning up broken glassware, forcing stoppers on glass vials, and from Pasteur pipettes. Sharps should be placed in leak- puncture-resistant containers for decontamination and disposal.

Chemicals

Few toxic chemicals are used in the virology laboratory. When these chemicals are introduced into the laboratory, a Material Safety Data Sheet (MSDS) must be made available to the laboratory worker and to the safety committee. These sheets are available from reagent and diagnostic kit manufacturers. The information on these sheets must include the chemicals found in the product, the health risks involved with exposure, telephone numbers of the manufacturer and a variety of information on the chemical nature of the product.

In 1987, OSHA required employers to inform employees about hazardous chemicals that they contact in the workplace. This regulation requires that containers be appropriately labelled, MSDS sheets be readily available and that employees be trained and familiar with the safety and handling of chemical agents. This training must be formally handled, documented, and reviewed periodically.

Laboratory workers should be trained in the proper use of chemical fume hoods. In addition, laboratory workers must realize that biological safety cabinets are not an approved substitute for chemical hoods because the air is circulated back into the room.

A few potentially hazardous chemicals are commonly used in the virology laboratory. DMSO (dimethylsulfoxide) is used as a cryoprotectant when freezing cell cultures. DMSO is a powerful solvent and can penetrate skin and latex gloves, carrying with it, a wide variety of substances. Fixatives such as ethanol, methanol, and acetone are also common in the laboratory. Bulk containers of these reagents should be stored in an appropriately labeled and constructed flammable storage area. Hypochlorite (household bleach) should be used carefully to prevent splashing bleach on clothing and to prevent skin and mucus membrane irritation. Bleach should never be added to detergent solutions because the chlorine gas that is generated is toxic. Bleach can also corrode stainless steel and it should not be used for routine disinfection of laminar flow hoods. Bleach spills on waxed floors should be cleaned up immediately because bleach will ruin the flooring.

Detergents are caustic and can cause irritation of the skin, eyes, and lungs. Gloves should be worn when concentrated detergents are used and procedures that generate detergent powder aerosols should be avoided.

Gas Cylinders

Carbon dioxide is the most routinely used gas in the virology laboratory. In the laboratory, CO_2 is used in such small quantities that the risks are minimal as long as the room is well ventilated. However, the pressurized CO_2 cylinders can be a significant hazard. Cylinders are heavy and falling cylinders can cause serious injuries. In addition, even heavy cylinders can become jet-propelled missiles if the valve is accidently broken. For these reasons, all gas cylinders must be properly secured in the laboratory. Protective valve caps must be securely fastened whenever cylinders are moved. Gas cylinders should be securely fastened to approved carts when moving them from one place in the laboratory to another.

Liquid Nitrogen

Liquid nitrogen is commonly used in the laboratory for storing cell cultures and virus stocks. Liquid nitrogen is extremely cold (-196°C) and paradoxically, direct contact with the skin can cause severe burns. Metal canisters and wands that have been submerged in liquid nitrogen can also cause severe burns if contact occurs with bare skin. Thick gloves, lab coat, and a face shield are recommended when removing vials from liquid nitrogen. During storage, some inadequately sealed ampules will contain liquid nitrogen and, once the vials begin to warm, they can explode. Therefore, thawing of glass ampules should be done in a covered, sterilizable container.

Nitrogen is a gas at room temperature and liquid nitrogen boils when exposed to room temperature air. This normal boiling process can increase the nitrogen content of the immediate atmosphere. Although normal ventilation can handle the excess nitrogen under most circumstances, it may be advisable to have extra ventilation when working with liquid nitrogen for long periods of time.

Fire

Many laboratory chemicals are flammable or explosive. Only working amounts of flammable liquids should be stored in the laboratory at any time and bulk quantities should be stored in designated flammable storage cabinets. Flammables should be used well away from open flames, inoculating loop incinerators, autoclaves, or ovens. Bunsen burners are very dangerous especially when used to "flame" alcohol-dipped instruments and slides. Alcohol used for this type of sterilization should be stored in small narrow-necked bottles that have a broad flat bottom to prevent accidental tipping. Electrical equipment can be a source of fire and shock hazards in the laboratory. Routine maintenance and inspections must be done to ensure that cords are in good condition and grounds functional. Humidified incubators have heating elements that are exposed to moist conditions and these elements must be checked frequently for corrosion. Corroded elements could present a fire hazard.

PHYSIOLOGICAL HAZARDS

Back Injury

Back injuries are the second leading type of occupational injury after cuts. In the laboratory, all lifting should employ sound body mechanics and the legs, rather than the back, should be used for lifting. Although heavy lifting is not a significant occupational hazard in the virology laboratory, moving cases of glass tubes, bottled water, and gas cylinders should be done carefully to prevent injury.

Slipping

Glassware preparation, autoclaves, humidified CO_2 incubators, and water baths could potentially leave small amounts of water on the floor and present slipping hazards. These water hazards should be cleaned immediately to prevent injury.

REFERENCES

1. Centers for Disease Control. *Biosafety in Microbiological and biomedical laboratories.* U.S. Department of Health and Human Services Publication Number (NIH) 88-8395. 1988;108-109.
2. Occupational exposure to bloodborne pathogens; Final Rule. *Federal Register* 29 CFR Part 1910.1030, December 6, 1991:64004-64192.
3. Pike RM. Laboratory-associated infections: Incidence, fatalities, causes, and prevention. *Ann Rev Microbiol* 1979;33:41-66.
4. Committee on Hazardous Biological Substances in the Laboratory. *Biosafety in the laboratory.* Washington DC: National Academy Press 1989;19-20.
5. Centers for Disease Control. 1988 Agent summary statement for human immunodeficiency virus and report on laboratory-acquired infection with human immunodeficiency virus. *MMWR* 1988;37(S-4):1-22.
6. Davidson WL, Hummler K. 1960. B-virus infection in man. *Ann NY Acad Sci, USA* 1960;9970-9979.

CHAPTER 3

General Methodologies

INTRODUCTION

The field of clinical virology is changing rapidly and a wide variety of testing methodologies are now available for virus detection. While most procedures still rely upon cell cultures, a number of virus detection methods have been developed to detect the presence of viral antigens or nucleic acids. The type of test used in a particular laboratory will depend upon the testing volume, the patient population served, and the technical expertise of the laboratory staff. In some community hospitals, EIA or latex methods are the norm while cell culture isolations and DFA procedures are often used in larger laboratories.

Cell culture isolation, the backbone of the virology laboratory, has also changed significantly in the past 10 years. The introduction of the centrifugation-enhanced (shell vial) technique has, for the first time, allowed laboratories to provide culture information within a clinically relevant timeframe for many slow-growing viruses. Some of these general methodologies are described below.

Traditional Cell Culture Isolation

Cell culture isolation and identification remain the "gold standards" for virus detection. These techniques depend upon the selection of appropriate cell lines, inoculation of an appropriate specimen, maintenance of the inoculated cells, testing of the inoculated cells for the presence of virus (e.g., hemagglutinin activity, presence of CPE, etc.), and culture confirmation using immunologic reagents.

Cell Line Selection

A variety of cell lines can be used for virus isolations (Table 1) and the choice of cell line depends upon the sensitivity of the cells to a particular virus, distinctiveness of the CPE produced in these cells, cell growth characteristics, and the relative cost to grow and maintain the cells in culture. When specimens are inoculated, cell cultures should be actively growing or freshly confluent and free of contamination. Cells that have been confluent for prolonged periods should not be used because these cells are less sensitive to virus infection than actively growing cells. Table 1 contains a basic summary of cell lines and their suitability for individual virus isolation. Table 2 contains the same information for centrifugation-enhanced (shell vial) methods.

Cell Line Inoculation

Upon arrival in the laboratory, specimens are processed according to the specimen type and the site of collection (Chapter 5). Generally, specimen processing includes a vortexing step to release virus from the swab or a filtration step for fecal specimens. Once the specimen is processed, culture tubes can be inoculated using the following procedure:

1. Appropriately label cell culture tubes.
2. Remove the maintenance medium from the cells and add 2 ml of medium.
3. Add 0.2-0.5 ml of the specimen to each tube.
4. Place the tubes in a roller rack (10-15 rph) and incubate at 33-37°C.
5. Cell cultures should be examined for a period of 7-28 days. During this time, the culture medium should be replaced if the pH changes drastically (i.e., red to yellow). Cell cultures that become toxic should be reinoculated onto new cell cultures.

Hemadsorption Testing

During replication, influenza, mumps, and parainfluenza viruses produce hemagglutinin

Table 1. Sensitivity of various cell lines using standard virus isolation methods.

Virus	PMK	HEK	RK	HEp-2	A549	Vero	BGM	RD	ML	CV-1	HL	MRC5	HFF	HeLa	HEL	293
Adenovirus	+	+		+	+	+					+	+	+	+	+	+
Coxsackie A	+						+									+
Coxsackie B	+	+		+	+											+
Cytomegalovirus												+	+		+	
Echovirus	+						+					+	+		+	
Herpes Simplex		+	+	+	+	+	+	+	+	+		+	+		+	
Influenza	+				+/-											
Measles	+	+		+	+	+				+						
Mumps	+	+		+	+					+		+	+			
Parainfluenza	+		+	+/-												
Poliovirus	+	+		+	+	+	+					+	+		+	+
RSV	+		+	+							+			+		
Rhinovirus												+	+	+		
Rotavirus	+			+/-	+					+						
Rubella	+a				+											
Varicella-Zoster										+		+	+		+	

aPrimary African green monkey kidney cells are best cells for rubella virus isolation.

molecules that are inserted into plasma membrane of the host cell. Because these hemagglutinins bind to red blood cells, virus-infected cells can be readily identified because red cells adsorb to them. Hemadsorption assays are rapid, convenient, and relatively inexpensive. In addition, the proportion of cells with attached erythrocytes at 4°C compared with the number at 22°C can provide some indication as to whether the isolate is parainfluenza, mumps, or influenza virus. In parainfluenza 1, 2, and 3 infections, RBCs often elute from the infected cells when the culture is warmed to 22°C. Influenza and mumps virus infections will generally have the same amount of hemadsorption at 4°C and at 22°C.

Hemadsorption assays for these viruses can be performed with guinea pig, fowl, and human O erythrocytes. However, most laboratories use guinea pig cells because they are smaller than the other RBCs and more of them can attach to an infected cell. Thus, guinea pig RBCs produce a more uniform HAd pattern and reactions are easier to interpret than fowl or human erythrocyte HAd tests. Negative cultures need not be discarded after HAd testing. Once the RBCs have been removed, the monolayers should be washed twice with sterile HBSS, and 2 ml of fresh culture medium should be added.

Procedure

1. Prepare a 0.5% suspension of guinea pig RBCs in sterile saline or phosphate buffered saline (Appendix C).
2. Add 0.4 ml of the RBC suspension to each culture tube.
3. Incubate the tubes on their side (cell monolayer downward) for 20-60 minutes at 4°C in a slanted (4°) test tube rack.
4. Examine the cell monolayer under the microscope (100-200X) for the presence of RBC adsorption and record the results (Table 3).
5. Incubate the tubes for one hour at room temperature. Examine the tubes for hemadsorption as before and record the results.

A positive hemadsorption test is characterized by chains, rosettes, or clumps of red cells that adhere to the surface of the cell sheet. A negative HAd

Table 2. Sensitivity of various cell lines using centrifugation-enhanced (shell vial) isolation methods.

Virus	PMK	HEp-2	A549	Vero	BGM	ML	CV-1	HL	MRC5	HeLa	McCoy
Adenovirus	+	+	+						+		
Chlamydia					+			+		+	+[a]
Cytomegalovirus						+			+		
Herpes Simplex		+	+	+		+	+		+		
Influenza	+		+						+		
Measles	+		+	+							
Mumps	+		+	+							
Parainfluenza	+		+								
Respiratory Syncytial	+	+	+					+	+		
Rotavirus	+			+			+				
Varicella-Zoster							+		+		

[a]*Chlamydia pneumoniae* does not grow well in McCoy cells.

test is characterized by cells that float freely in the medium and the absence of RBC chins, rosettes, or clumps adhering to the cell sheet. Positive HAd results should be interpreted carefully because monkey kidney cells can be infected with hemadsorbing simian viruses. Comparing the level of HAd to an uninoculated control culture is important to eliminate false positive results. Influenza and mumps viruses will generally have the same amount of hemadsorption after incubation at 4°C and after warming to 22°C. Positive tubes should be removed from the tube and stained with specific monoclonal antibody reagents.

Table 3. Recording HAd Reactions

Percent of Cells With HAd	Score
75-100	4+
50-75	3+
25-50	2+
< 25	1+
Questionable	+/−

Hemagglutination (HA) Procedure

Some laboratories prefer to use hemagglutination (HA) rather than HAd for screening respiratory cultures. HA testing does not disturb the cell sheet and, if necessary, a single culture tube could be tested every other day for HA activity. The HA procedure is especially convenient when handling a large number of cultures because the cultures are handled only once per test. In addition, cultures can be tested repeatedly without washing RBCs from the cultures and hemadsorbing simian viruses do not produce as much background.

1. Appropriately label a 96-well U-bottom microtiter plate with culture identification and control information.
2. Prepare a 0.5% suspension of guinea pig RBC suspension (Appendix C).
3. Place 100 µl of saline into each well.
4. Add 25 µl of cell culture supernatant into each well.
5. Place 50 µl of the 0.5% guinea pig RBC suspension into each well.
6. Cover and gently tap the plate to mix the contents. Incubate for one hour at 4°C.
7. Examine the plate using a magnifying mirror

and record the results.
8. Warm the plate to 22-25°C, examine the plate, and record the results as before.

The presence of hemagglutinating virus prevents RBCs from forming a "button" in the bottom of the microtiter well. Because this is only a screening procedure for hemagglutinin activity, positive cell cultures should be scraped and stained with type specific monoclonal antibodies for final identification. The presence of a red cell button on the bottom of the well indicates that hemagglutinin activity was not detected.

Observation of Cytopathic Effect (CPE)
Cytopathic effect is the morphological change seen in virus infected cells. The particular type of CPE observed in cells is dependent upon the virus and the cell line. CPE usually develops in localized areas or foci that spread to involve the entire monolayer. Viruses such as polio and HSV or viruses present in high concentrations may not produce local lesions. Instead, the entire monolayer may be involved or destroyed. Although an experienced virologist can accurately identify many viruses on the basis of CPE, culture confirmation must be accomplished by staining with specific immunological reagents.

Culture Confirmation Procedure
Staining every cell culture with every reagent is neither practical or cost efficient. Therefore, virology laboratories must rely upon experienced technicians who can correlate CPE, cell tropism, and specimen type to select a small number possible viruses for confirmatory testing. In general, confirmatory testing is done as follows:

1. Remove all but a few drops of the cell culture medium.
2. Scrape the cells from the tube surface.
3. Suspend the cells in the remaining fluid.
4. Using a pipette, place a small drop of the cell suspension in the wells of an acetone-cleaned slide. The remainder of the cell suspension should be retained for inoculation onto fresh cultures as needed.
5. The slide should be processed as directed by the manufacturer of the fluorescein-labeled antibody reagents or as described below.
6. Allow the slide to air dry, then fix it in acetone for 10-15 min. Fixed slides may be stored for up to a week at 2-8°C or for one year under desiccated conditions.
7. Place enough of the appropriate fluorescein-labeled monoclonal antibody on each well to cover the cell smear (15-30 µl).
8. Incubate the slide for 15-30 minutes at 35-37°C in a covered, humidified chamber. Do not allow the antibody to dry on the slide as this could cause nonspecific antibody binding.
9. Remove the excess antibody with a gentle stream of PBS. Do not direct the stream directly at the cell smear as this could dislodge the cells.
10. Soak the slide in PBS for 5 minutes, shake off the excess fluid, and allow the slide to air dry.
11. Carefully add a small drop of FA mounting fluid to the center of each well.
12. Place a number 1 coverslip on the mounting fluid and carefully remove all the air bubbles.
13. Immediately examine the slide using a fluorescence microscope. For optimum clarity, use 200-300X magnification for screening and 400X for confirmation of cell morphology.

The presence of virus is indicated by apple-green fluorescence in the nucleus and/or cytoplasm of infected cells. Only individuals experienced in reading FA reactions should examine the cell smears because cell debris and cell clumps may exhibit a dull fluorescence which can be misinterpreted as a specific, positive reaction. A negative test is indicated by the lack of specific fluorescence described above.

Shell Vial Centrifugation Cultures
Rapid virus detection is essential for the timely institution of appropriate infection control measures and for the initiation of antiviral chemotherapy. In order to address the need for rapid virus detection,

centrifugation-enhanced (shell vial) methodologies originally described for cytomegalovirus (1) have been adapted for the detection of influenza A, influenza B, HSV, adenovirus, parainfluenza virus, rotavirus, and varicella-zoster virus. Because shell vials are usually stained before CPE develops, a variety of cell lines, including normally nonpermissive cells, may be used. Table 3 lists some of the cell lines that can be used for shell vial isolations.

In general, shell vial methods can detect 50-99% of all isolates 24-96 hours after inoculation, usually well before CPE is visible. Although shell vial methods are more rapid than conventional tube cultures, these methods are inherently less sensitive. Therefore, each set of shell vials should be backed up by tube cultures. In addition, tube cultures can be used to detect dual infections that could be missed if only shell vial methods are used.

Procedure
1. Appropriately label 2-4 shell vials for each specimen.
2. Remove the medium from the shell vials.
3. Inoculate each vial with 0.2-0.5 ml of the specimen.
4. Centrifuge the vials at 700 x g for 1 hour at room temperature.
5. Add 1 ml of maintenance MEM to each vial.
6. Incubate the vials at 33-37°C.
7. One or two vials are fixed and stained after 24-48 hours. The second set of vials are stained at 48-96 hours postinoculation.

Staining of Coverslips

Virus identification in shell vial systems is accomplished by staining the coverslips with antibody reagents. Staining should be accomplished as directed by the manufacturer of the fluorescein-labeled antibodies or as described below.

1. Carefully remove the culture fluid from the vials and wash the coverslips twice by adding 1-2 ml of PBS and soaking for 5 minutes.
2. Remove the PBS and add 2-3 ml of acetone to each vial. Fix the coverslips for 10 minutes at room temperature.
3. Remove the acetone and allow the coverslips to air dry.
4. Rinse the coverslips briefly with PBS. (The moisture trapped between the coverslip and the vial will keep the stain from wicking under the coverslip.)
5. Add enough monoclonal antibody to the vial to cover the coverslip (100-150 µl) and incubate for 30 minutes at 35-37°C. The shell vials should be covered and incubated at 35-37°C in a humidified chamber to prevent the antibody from drying.
6. Wash the coverslips twice as previously described (step 1).
7. Place a small amount of mounting fluid on a slide. Remove the coverslips from the vial and lay them, cell side down, on the mounting fluid.
8. Immediately examine the coverslips at 100-300X using a fluorescence microscope.

Coverslips that have individual, or small groups of cells that exhibit brilliant apple-green cytoplasmic and/or nuclear fluorescence are considered positive. Coverslips without the specific fluorescence described above are considered negative. Some specimens may trap the monoclonal antibodies between the cell monolayer and inoculum cells. Careful examination of the cell morphologies can distinguish this type of reaction from specific reactions. In addition, nonspecific staining can occur if the antibody reagents are allowed to dry on the coverslip. In this case, the entire cell sheet may stain apple-green.

Fluorescent Antibody Staining of Direct Smears

Although virus isolation is the gold standard, virus lability and long isolation times limit the diagnostic utility of some virus isolation systems. Several studies have shown that some direct fluorescent antibody (DFA) methods are faster and more sensitive than traditional cultures. In contrast with 1-30 day isolation times, DFA staining of cell smears and tissue specimens can usually be accomplished in less than an hour. DFA methods that employ monoclonal antibodies are very specific and DFA can often detect viral antigens for a longer period of time than culture. Realizing this, some laboratories routinely centrifuge exfoliated cells from the viral transport medium, make smears, and stain the cells with DFA reagents. Although rapid, sensitive, and specific, the DFA methodology is not without problems. DFA reagents can only detect specific antigens. Therefore, DFA could miss specimens containing more than one virus unless multiple specimens are prepared and tested with a multiple antibodies. Exclusive use of DFA reagents could also miss new infectious agents that might appear in the community. For these reasons, most laboratories employ a combination of DFA and culture methods for virus detection.

The success of the DFA procedure depends upon the submission of a well made cell smear. Smears that are too thick or "lumpy" can cause nonspecific trapping of the DFA reagent. Smears that are grossly contaminated with red blood cells can also cause interpretation difficulties due to red cell autofluorescence. Finally, cell smears containing too few cells could cause the laboratory to report a false-negative result because no positive cells were observed. Specimen adequacy and its definition are significant sources of laboratory-to--laboratory variation and an area of increasing regulatory concern. Many laboratories require at least 50 cells per smear while other laboratories require 1-2 cells/high power (400X) field. Whatever the number, too many cells can cause nonspecific trapping of the conjugate and with too few cells, the laboratory may not be able to find an infected cell.

For optimum performance, slides should be fixed and stained within 1-2 hours of specimen collection. Alternatively, unfixed slides may be stored for up to 48 hours at 2-8°C. Fixed slides may be stored at 2-8°C for 1 week or frozen at -20°C for up to 1 year. Storing slides under desiccated conditions will decrease the background staining and minimize antigen degradation.

1. Upon arrival in the laboratory, examine the slide wells at 100-300X magnification. Ideally, a minimum of 50 cells should be visible on each well before the specimen is considered adequate for further processing.
2. The slide should be processed as directed by the manufacturer of the labeled antibody or as described below.
3. Fix the cells by immersing the slides in acetone (room temperature) for 10 minutes. Allow the slides to air dry.
4. Place enough of the fluorescein-labeled monoclonal antibody on the slide to cover the cell smear (15-30 μl).
5. Incubate the slide for 30 minutes at 35-37°C in a covered, moist chamber. Do not allow the antibody to dry as this could cause nonspecific antibody binding.
6. Remove the excess antibody with a gentle stream of PBS. Do not direct the stream directly at the cell smear as this could dislodge the cells.
7. Soak the slide in PBS for 5 minutes, shake off the excess fluid, and allow the slide to air dry.
8. Carefully add a small drop of FA mounting fluid to the center of each well.
9. Place a number 1 coverslip on the mounting fluid and carefully remove all the air bubbles.
10. Immediately examine the slide using a fluorescence microscope. For optimum clarity, use 200-300X magnification for screening and 400X for confirmation of cell morphology.

A positive test is indicated by intense apple-green fluorescence in the cytoplasm and/or nucleus of the exfoliated cells. Only intact cells should be examined because cell debris and clumps of normal cells may exhibit a dull fluorescence which could be misinterpreted as a specific, positive reaction. A negative test is indicated by the absence of specific staining in the cell smear. Specimens with fewer than 50 cells on each well (and no positive cells) cannot be adequately evaluated and an additional specimen should be requested.

Latex Agglutination Tests

Latex agglutination tests for herpes simplex, *Clostridium difficile*, and rotavirus are commercially available. Although, generally less sensitive than EIA testing, latex tests are widely used because they are rapid, easy to perform, and relatively inexpensive. In the latex test procedure, the clinical specimen is mixed with the antibody-coated latex particles and the mixture is rocked for several minutes. The presence of the specific antigen will cause visible agglutination of the latex particles as the antigens become linked to two or more latex particles. Care must be taken when interpreting latex agglutination results because nonspecific binding can produce false-positive results. Some manufacturer's address this problem by combining the clinical specimen with the negative control to determine if nonspecific reactions are occurring.

Enzyme Immunoassay Tests

As a group, EIA tests are highly specific, sensitive and relatively rapid (15 min to 4.5 hours). However, the sensitivity and specificity of EIA tests vary widely from manufacturer to manufacturer. In general, unlabeled antibody is bound to a solid phase such as a bead, tube, or microwell. The clinical specimen is incubated with the antibody-coated solid phase and antigen, if present, will bind to the solid support. The excess specimen is washed away and enzyme-conjugated antibodies to the antigen are added and allowed to react. The unbound conjugate is removed with another wash step and a chromogen/substrate is added. If any conjugate is bound to the well, the enzyme will change the fluid from clear to a colored solution. In theory, the intensity of the color is proportional to the amount of antigen present in the clinical specimen and the intensity of the color can be determined visually or spectrophotometrically. EIA tests have been used in a wide range of formats from totally automated machines requiring significant bench space and little technical support, to microwell formats requiring significantly more technical time and skill, to hand-held, self-contained modules for testing small numbers of specimens in a physician office environment.

REFERENCES

1. Gleaves CA, Smith TF, Shuster EA, Pearson GR. Rapid detection of cytomegalovirus using low-speed centrifugation and monoclonal antibody to an early antigen. *J Clin Microbiol* 1984;19:917-919.

CHAPTER 4

Virus Testing Protocols

INTRODUCTION

The virology laboratory processes and tests specimens based upon the type of specimen, the provisional diagnosis or symptomatology (if listed on the laboratory request), and the specimen source. This chapter attempts to link the information on the laboratory request with cell culture inoculations and potential antigen testing methodologies. Few if any laboratories run all these tests on every specimen. Therefore, the combination of tube cultures, shell vial cultures, and antigen tests will vary from laboratory to laboratory. Specific virus detection information can be found in the individual virus chapters.

RESPIRATORY SPECIMENS

Specimens
Nasopharyngeal swabs, washes and aspirates; throat swabs and washes; broncheoalveolar lavages; lung biopsy.

Possible Symptoms
Bronchitis/bronchiolitis, cold, croup, influenza syndrome, pneumonia/pneumonitis, sinusitis, otitis media, sore throat,

Potential Agent
Adenovirus, *Chlamydia pneumoniae*, *Chlamydia psittaci*, cytomegalovirus (CMV), enteroviruses, herpes simplex virus (HSV), influenza, measles, parainfluenza, respiratory syncytial virus (RSV), rhinovirus, varicella-zoster virus (VZV).

Turnaround Times
Conventional Culture: 3-28 days.
Shell Vial Culture: 24-96 hours.
Direct Smear: 30-90 minutes.

Inoculation
Tube Cultures
RMK (1), MRC-5 or HFF (1), Hep-2 (1), Vero or CV-1 (1), and A549 (1) tubes.

Shell Vials
A549 (2) for adenovirus, influenza, parainfluenza, and RSV.
RMK (2) for parainfluenza or influenza.
A549, HEp-2, or HL (2) for RSV.
MRC-5, HFF, or ML (2-3) for CMV.
MRC-5, HFF, or CV-1 (4-6) for HSV and VZV.
HL or HeLa-229 (2) for *Chlamydia pneumoniae*.
McCoy, BGM, or HeLa (2) for *Chlamydia trachomatis* or *C. psittaci*.

Antigen Detection
Fluorescent Antibody Methods
Adenovirus, chlamydia, CMV, HSV, influenza, parainfluenza viruses, RSV, and VZV.

Enzyme Immunoassay
Adenovirus, HSV, influenza A, RSV.

FECAL SPECIMENS OR RECTAL SWABS

Specimens
Stool or rectal swab.

Possible Symptoms
Fever, headache, malaise, meningitis, diarrhea, vomiting, cold, Bornholm disease.

Potential Agent
Coxsackie A, coxsackie B, enterovirus, adenovirus, rotavirus, poliovirus

Note: The presence of *Clostridium difficile* toxins in fecal specimens or rectal swabs may mimic viral CPE in many cell lines.

Turnaround Times
Conventional Culture: 1-10 days.
Shell Vial Culture: 48-96 hours.
Enzyme Immunoassay: 1.5-3 hours (Adenovirus, adenovirus 40/41, and rotavirus).

Culture
Tube Cultures
RMK (1), HFF or MRC-5 (1), A549 (1), Vero or CV-1 (1), and RD (1) cells. Graham 293 (1 or 2) tubes may be inoculated for enteric adenoviruses.

Shell Vials
A549, (2) for adenovirus.
Vero or CV-1 (2) for rotavirus.

Antigen Detection
Enzyme Immunoassay
Rotavirus, adenovirus, adenovirus 40/41

Latex Agglutination
Rotavirus

CSF OR BRAIN BIOPSIES

Possible Symptoms
Meningitis, encephalitis, fever, confusion, drowsiness, convulsions.

Potential Virus
Adenovirus, cytomegalovirus (CMV), coxsackie viruses, echoviruses, human immunodeficiency virus (HIV), human T-lymphotropic viruses (HTLV), herpes simplex virus (HSV), mumps, measles, rubella, varicella-zoster virus (VZV).

Turnaround Time
Conventional Culture: 1-28 days.
Shell Vial Culture: 48-96 hours.
Enzyme Immunoassay: 3-24 hours.
Fluorescent Antibody: 1 hour (Brain biopsy).

Culture
Tube Cultures
RMK (1), HFF or MRC-5 (1), RD (1), A549 (1), and Vero or CV-1 (1).
PHA stimulated lymphocytes (HIV/HTLV)

Shell vials
MRC-5, or HFF (2) for CMV.
MRC-5, CV-1, or HFF (4-6) for HSV and VZV.
RMK or A549 (2) for measles.
RMK (2) for mumps.

Antigen Detection
Fluorescent Antibody Methods
HSV (Brain biopsy)

Enzyme Immunoassay
HIV

DERMAL LESIONS

Specimens
Vesicle fluid, swab of lesion, basal cell scraping.

Possible Symptoms
Vesicular or maculopapular rash, lesions, fever, chickenpox, shingles.

Potential Virus
Adenovirus, enteroviruses, herpes simplex virus (HSV), varicella-zoster virus (VZV).

Turnaround Time
Conventional Culture: 1-10 days.
Shell Vial Culture: 48-96 hours.
Enzyme Immunoassay: 3-24 hours.
Fluorescent Antibody Methods: 1 hour (Brain biopsy).

Culture
Tube Cultures
MRC-5 or HFF (1), A549 (1), Vero or CV-1 (1), RD (1), and RMK (1) tubes.

Shell vials
MRC-5, HFF, or CV-1 (4-6) for HSV and VZV.
A549 (2) for adenovirus.

Antigen Detection
Fluorescent Antibody Methods
HSV, VZV

Enzyme Immunoassay
HSV

OCULAR SPECIMENS

Specimens
Conjunctival swabs or scrapings, aqueous or vitreous fluids, vitreous washings, corneal biopsy.

Possible Symptoms
Pain and swelling of eyelids, ocular discharge, conjunctivitis, acute hemorrhagic conjunctivitis, keratitis, uveitis.

Potential Agents
Adenovirus, chlamydia, cytomegalovirus (CMV), coxsackie viruses, echoviruses, herpes simplex virus (HSV), varicella-zoster virus (VZV).

Turnaround Time
Conventional Culture: 1-28 days.
Shell Vial Culture: 24-96 hours.
Enzyme Immunoassay: 1-5 hours.
Fluorescent Antibody Methods: 1 hour.

Culture
Tube Cultures
RMK (1), Vero or CV-1 (1), A549 (1), RD (1), and HFF or MRC-5 (1) tubes.

Shell vials
MRC-5, CV-1, or HFF (4-6) for HSV and VZV.
MRC-5, HFF, or ML (2-3) for CMV.
A549 or HEp-2 (2) for adenovirus.
McCoy, BGM, or HeLa-229 (2) for *Chlamydia trachomatis*.

Antigen Detection
Fluorescent Antibody Methods
Adenovirus, chlamydia, CMV, HSV, VZV.

Enzyme Immunoassay
Adenovirus, chlamydia, HSV.

UROGENITAL SPECIMENS

Specimens
Urethral swab, cervical swab or brush, lesion scraping.

Possible Symptoms
Urethritis, cervicitis, lesions, vaginal or urethral discharge.

Potential Agent
Adenovirus, chlamydia, cytomegalovirus (CMV), herpes simplex virus (HSV), mumps.

Turnaround Time
Conventional Culture: 1-28 days.
Shell Vial Culture: 24-96 hours.
Enzyme Immunoassay: 1-5 hours.
Fluorescent Antibody Methods: 1 hour.

Culture
Tube Cultures
RMK (1), Vero or CV-1 (1), A549 (1), and HFF or MRC-5 (1) tubes.

Shell vials
MRC-5, CV-1, or HFF (4) for HSV.
MRC-5, HFF, or ML (2-3) for CMV
A549, or HEp-2 (2) for adenovirus.
McCoy, BGM, HeLa-229 (2) for *Chlamydia trachomatis*.
RMK (2) for mumps.

Antigen Detection
Fluorescent Antibody Methods
Adenovirus, chlamydia, HSV.

Enzyme Immunoassay
Adenovirus, chlamydia, HSV

BLOOD, BONE MARROW, BIOPSY SPECIMENS, AMNIOTIC FLUIDS, PRODUCTS OF CONCEPTION.

Specimens
Blood or bone marrow in heparin (green top) or EDTA (purple top) tubes. Tissues in sterile saline. Amniotic fluids.

Possible Symptoms
Fever of unknown origin, AIDS, ARC.

Potential Virus
Cytomegalovirus (CMV), Epstein-Barr virus (EBV), enterovirus, human immunodeficiency virus (HIV), human T-lymphotropic virus (HTLV), herpes simplex virus (HSV), measles, mumps, varicella-zoster virus (VZV).

Turnaround Time
Conventional Culture: 1-30 days.
Shell Vial Culture: 24-96 hours.
Enzyme Immunoassay: 1-5 hours.
Fluorescent Antibody Methods: 1 hour.

Culture
Tube Cultures
RMK (1), RD (1), Vero or CV-1 (1), and HFF or MRC-5 (1) tubes.
PHA stimulated lymphocytes for HIV/HTLV.
Cord blood lymphocytes for EBV.

Shell vials
MRC-5, CV-1, or HFF (4-6) for HSV and VZV.
MRC-5, HFF, or ML (2-3) for CMV.
RMK (2) for mumps.
RMK or A549 (2) for measles.

Antigen Detection
Fluorescent Antibody Methods
CMV, HSV, VZV.

Enzyme Immunoassay
HIV, HSV

URINE

Specimens
First void urine is preferred. Random urine or catherized urines.

Possible Symptoms
Cystitis, urethritis, viral exanthem, dysuria, urethral discharge.

Potential Agent
Adenovirus, chlamydia, cytomegalovirus (CMV), measles, mumps, polyomaviruses.

Turnaround Time
Conventional Culture: 1-28 days.
Shell Vial Culture: 24-96 hours.
Enzyme Immunoassay: 1-5 hours.
Fluorescent Antibody Methods: 1 hour.

Culture
Tube Cultures
RMK (1), Vero or CV-1 (1), A549 (1) and HFF or MRC-5 (1) tubes. HEK(1) for polyomaviruses.

Shell vials
A549 (1) for adenovirus.
MRC-5, HFF, or ML (2-3) for CMV.
RMK (2) for mumps.
RMK or A549 (2) for measles.
HEK (2) for polyomaviruses.

Antigen Detection
Fluorescent Antibody Methods
Adenovirus, chlamydia, CMV, measles.

Enzyme Immunoassay
Chlamydia, HSV

CHAPTER 5

Specimen Collection and Processing

INTRODUCTION

Collecting specimens from an appropriate anatomic site and at the proper time after infection are extremely important because viruses are often shed for only a short period of time. Viral specimens should be collected as early in the course of disease as possible. Specimens should be collected with sterile implements and quickly transported to the laboratory. The site of specimen collection should correlate with the clinical presentation and local epidemiologic patterns. A general listing of syndromes and appropriate specimens is shown in Table 1. Individual requirements that differ from these general recommendations are listed in the virus isolation chapters. Table 2 shows a number of viruses, appropriate specimens associated with each virus, and the appropriate time for specimen collection.

Special Reagents and Supplies
Swabs
Swab material should be made of sterile cotton or Dacron. **DO NOT USE CALCIUM ALGINATE SWABS FOR VIRUS OR CHLAMYDIA ISOLATIONS.** Calcium alginate swabs are toxic for chlamydia and many enveloped viruses. In addition, specimens collected with calcium alginate swabs cannot be used for direct fluorescent antibody testing. Wooden-shafted swabs are not recommended because wooden swabs can contain toxins and formaldehydes which inhibit the recovery of viruses and chlamydia. In addition, wooden swabs absorb transport media thereby reducing the amount of fluid for inoculation onto cell cultures.

Microscope Slides
Slides should be soaked in acetone (or ethanol) for 5 minutes before use. An entire box of slides can be cleaned at once and stored in a dust-free environment for later use. The purpose of this procedure is to remove any oils or solvents that may have been acquired during slide manufacturing and processing. While uncleaned slides may work for most applications, repeated use of uncleaned slides can result in the unexpected loss of a specimen when the cells fail to adhere to the slide.

SPECIMEN COLLECTION

Specimen collection is the foundation upon which all other procedures are built. Collecting the correct specimen by an appropriate method at the proper time during infection significantly enhances virus isolation frequency and improves the clinical relevance of the laboratory results. Because specimen collection is so critical, laboratories must assume responsibility for training clinicians, nursing personnel, and laboratory technicians in the proper collection and transport of specimens. Without good specimen collection, even the best virus isolation procedure is worthless. General guidelines for specimen collection are listed below. Individual specimen collection procedures that differ from those listed below are listed in the virus isolation chapters.

Body Fluids
1. Collect CSF, pleural fluids, pericardial fluids and other body fluids in a sterile container.
2. Most body fluids contain enough protein to stabilize viruses. Therefore, viral transport medium need not be added to the specimen.
3. Send the specimens to the laboratory on wet ice. DO NOT FREEZE AT -20°C. Specimens may be stored for up to 24 hours at 2-

Table 1. Summary of appropriate specimen collection sites associated with various clinical presentations. This information is provided as a general collection guide. Agent-specific information is presented in the virus isolation chapters.

Clinical Presentation	Specimen of Choice	Other Specimens
Bronchitis/Bronchiolitis	Nasopharyngeal swabs, washes, and aspirates	Broncheoalveolar lavage
Colds, Upper Respiratory Tract Infections	Nasopharyngeal swabs, washes, and aspirates	Throat swabs
Croup	Nasopharyngeal swabs, washes or aspirates	Throat swabs
Exanthems	Vesicle swab and/or fluid	
Gastroenteritis	Stool	Rectal Swab
Influenza syndrome	Nasopharyngeal swabs, washes, and aspirates; sputum	Throat swabs or throat washes
Meningitis	Cerebrospinal fluid (CSF)	Throat swab, stool, or rectal swab
Pharyngitis	Nasopharyngeal swab, wash or aspirates; throat wash or swab	
Pneumonia or Lower Respiratory Tract Infections	Nasopharyngeal or tracheal aspirates or washes, broncheoalveolar lavages	Nasopharyngeal or throat swabs

8°C. If longer delays are anticipated, specimens should be frozen at -70°C. Refrigeration is critical for body fluids because the viruses isolated from body fluids are often extremely labile.

Cervical Specimens
1. Use a vaginal speculum to facilitate visualization of the cervix and to prevent the swabs from touching the sides of the vagina.
2. Using a sterile Dacron, rayon, or cotton swab, carefully remove the mucus from the endocervix. This swab should be discarded because cervical exudates and mucus are not appropriate specimens.
3. Using a second sterile Dacron, rayon, or cotton swab carefully swab the transitional zone of the cervix, rolling the swab to assure that the swab contains cervical epithelial cells.
4. Carefully remove the swab to prevent contamination with vaginal flora.
5. Place the swab into a vial of chlamydial transport medium and send the specimen to the laboratory on wet ice. Specimens may be stored at 2-8°C for up to 72 hours. Longer storage should be done at -70°C.

NOTE: Cervical or vaginal lesions should be swabbed as described above and the swabs should be placed in viral transport medium.

Chlamydia Slide Preparation from Swab or Cytobrush Specimens
1. Use a vaginal speculum to facilitate visualization of the cervix and to prevent the swabs from touching the sides of the vagina.
2. Using a sterile Dacron, rayon, or cotton swab, carefully remove the mucus from the

Table 2. Summary of appropriate specimen collection sites and timing. The information below should be used as a general guideline. Agent-specific information is presented in the virus isolation chapters.

Agent	Specimen of Choice	Time of Collection
Adenovirus	Throat swab/wash, rectal swab/stool, urine	During symptomatic disease
Chlamydia	Cervical/Urethral swab	During symptomatic disease
C. difficile Toxin	Stool	During symptomatic disease
Cytomegalovirus	Urine, throat swab/wash, buffy coat	During symptomatic disease
Enterovirus	Throat swab, CSF, stool/rectal swab	First week of symptoms
Herpes simplex	Vesicle fluid/swab, throat/mouth swab, vaginal swab	First 3 days of lesion
Hepatitis A	Serum, stool, liver, kidney	First 8 days of symptoms
Influenza	Throat/NP wash or swab, BAL	First 3 days of symptoms
Measles	Throat swab, urine, blood	First 2 days of symptoms
Mumps	Throat swab, urine, blood	First 7 days of symptoms
Parainfluenza	Throat/NP wash or swab	First 3 days of symptoms
RSV	NP wash/aspirate/swab, throat swab	First 3 days of symptoms
Rhinovirus	NP wash/swab	First 2 days of symptoms
Rotavirus	Stool	First 4 days of symptoms
Rubella	Throat swab, stool, urine	First 4 days of symptoms
Varicella-Zoster	Vesicle fluid/swab, lesion swab	First 2 days of symptoms

endocervix. This swab should be discarded because cervical exudates and mucus are not appropriate specimens.

3. Using a cytobrush or a second swab, carefully rub the transitional zone of the cervix, rolling the swab or brush to assure that they contain cervical epithelial cells. When using a cytobrush, care must be taken to avoid bleeding because red cell autofluorescence can make interpretation difficult.
4. Carefully remove the swab or brush to prevent contaminating them with vaginal flora.
5. Swabs and cytobrushes should not be allowed to dry and smears should be prepared immediately after specimen collection.
6. Rub the cytobrush or swab on the well of an acetone-cleaned slide making sure the specimen is evenly dispersed over the entire well.
7. Allow the slide to air dry.
8. Lay the slide flat and add 0.5 ml of methanol to the well. Let the methanol evaporate at room temperature.
9. Label the slide and send it to the laboratory in a petri dish or a slide mailer.
10. Slides may be stored at room (20-30°C) or refrigerator (2-8°C) temperature for up to 7 days before staining. If longer delays are anticipated, slides should be stored at or below -20°C.

Eye (Conjunctiva)
1. Moisten a fine Dacron, rayon, or cotton swab with sterile physiologic saline.
2. Gently pull the lower eyelid downward and carefully swab the lower conjunctiva to collect both cells and fluids.
3. If both eyes are to be cultured, a separate sterile swab should be used for the other eye.

4. Place both swabs in the same viral (or chlamydial) transport vial and send the vial to the laboratory on wet ice. Specimens may be stored at 2-8°C for up to 48 hours. If longer delays are anticipated, specimens should be frozen at -70°C.

NOTE: Scleral, corneal, or conjunctival scrapings should be taken only by a physician.

Nasopharyngeal Aspirate
1. Attach a sterile soft polyethylene #8 French feeding tube suction to a disposable aspiration trap.
2. Using the tube, measure the distance from the patient's nostril to their ear. Mark the distance on the tube using your thumb and forefinger.
3. Gently insert the tube into the nostril until the thumb and forefinger touch the patient's nose.
4. While applying intermittent suction, slowly remove the tube from the nasopharynx.
5. Place the end of the tube in a vial containing 2-3 ml of viral transport medium and aspirate the contents into the trap.
6. Remove the feeding tube and carefully cap all the orifices on the aspiration trap.
7. Send the aspiration trap to the laboratory on wet ice. Specimens may be stored at 2-8°C for up to 24 hours. If longer delays are anticipated, specimens should be frozen at -70°C.

Nasopharyngeal Smear
1. A dry flexible type 1 aluminum-shafted, cotton-tipped wire swab should be used for this procedure. Two swabs should be used if specimens are to be collected from both nostrils. Both swabs should be placed into the same viral transport tube.
2. Place a small drop of sterile saline on each well of an acetone-cleaned 2-well microscope slide.
3. Using the swab, measure the distance from the patient's nostril to their ear. Mark the distance on the swab using your thumb and forefinger.
4. Gently insert the swab into the nostril until the thumb and forefinger touch the patient's nose.
5. Hold the swab in place for 15-30 seconds then gently rotate the swab three times.
6. Remove the swab from the nasopharynx.
7. Roll the swab onto each of the slide wells.
8. Allow the slide to air dry at room temperature.
9. Send the slide to the virology laboratory in a petri dish or in a slide mailer. Slides can be held at 2-8°C for up to 48 hours.

Nasopharyngeal Swab
1. A dry flexible type 1 aluminum-shafted, cotton-tipped wire swab should be used for this procedure. Two swabs should be used if specimens are to be collected from both nostrils. Both swabs should be placed into the same viral transport tube.
2. Using the swab, measure the distance from the patient's nostril to their ear. Mark the distance on the swab using your thumb and forefinger.
3. Gently insert the swab into the nostril until the thumb and forefinger touch the patient's nose.
4. Hold the swab in place for 15-30 seconds then gently rotate the swab three times.
5. Place the swab in viral transport medium and cut the shaft so that the swab fits into the tube.
6. Send the specimen to the laboratory on wet ice. DO NOT FREEZE AT -20°C. Specimens may be stored for up to 48 hours at 2-8°C. If longer delays are anticipated, specimens should be frozen at -70°C.

Nasopharyngeal Washes
1. Position the patient so that he/she is sitting with their head tipped backward.
2. Add up to 5 ml of saline into one nostril and gently close the nasal passage.
3. Ask the patient to expectorate the wash into a collection device or cup.
4. Add the wash to the viral transport medium.
5. Send the specimen to the virology laboratory on wet ice. DO NOT FREEZE AT -20°C.

Specimens may be stored for up to 72 hours at 2-8°C. If longer delays are anticipated, specimens should be frozen at -70°C.

Rectal Swabs
1. Position the patient so that the anus is readily accessible.
2. Insert a dry cotton, rayon, or Dacron swab at least 5 cm into the rectum.
3. Rotate the swab and carefully withdraw it from the rectum. The swab should show signs of fecal material.
4. Place the swab into a vial of viral transport medium and break the shaft of the swab so that it fits into the tube.
5. Send the specimen to the laboratory on wet ice. DO NOT FREEZE AT -20°C. Specimens may be stored for up to 72 hours at 2-8°C. If longer delays are anticipated, specimens should be frozen at -70°C

Saliva
1. Rub a sterile dry rayon, Dacron, or cotton swab over the buccal mucosa opposite the upper molars in the vicinity of the Stensen's ducts and then over the floor of the mouth anterior to the tongue.
2. Place the swab into a vial of viral transport medium and break the shaft of the swab so that it fits into the tube.
3. Send the specimen to the laboratory on wet ice. DO NOT FREEZE AT -20°C. Specimens may be stored for up to 24 hours at 2-8°C. If longer delays are anticipated, specimens should be frozen at -70°C.

NOTE: Saliva may be collected by aspiration or expectoration into a sterile container. Specimen handling is the same for these specimens as for swab specimens (step 3 above).

Stool
1. Collect 10-30 grams of stool (approximately the size of a walnut) in any clean vessel possessing a tight-fitting lid. Do not use preservatives.
2. Stool specimens should be refrigerated to retard bacterial growth and sent to the laboratory as soon as possible after collection.

Stool, Pediatric
1. Scrape the stool from the diaper immediately after the child's bowel movement.
2. Place the specimen in any clean vessel possessing a tight-fitting lid. Do not use preservatives.
3. Stool specimens should be refrigerated to retard bacterial growth and sent to the laboratory as soon as possible after collection.

NOTE: Highly absorbent paper diapers will quickly soak up liquid stools and cause viruses to bind irreversibly to the paper fibers. This problem can be alleviated by placing a piece of plastic (e.g., a plastic bag) in the diaper. Stool collected on the plastic can be placed into the container as described above. Alternatively, paper diapers can be turned inside out so that the outer plastic cover is next to the child's skin. This procedure can reduce stool leakage to outer garments. Stool specimens collected using this method should be placed in a leak-proof container as described above.

Throat Swab
1. Moisten a cotton or Dacron swab with physiologic saline. A moistened swab will have more cells adhere to it than a dry swab.
2. Vigorously rub the swab across the tonsils and posterior pharynx.
3. Place the swab in viral transport medium and break the shaft of the swab so that it fits in the vial.
4. Send the specimen to the laboratory on wet ice. DO NOT FREEZE AT -20°C. Specimens may be stored for up to 72 hours at 2-8°C. If longer delays are anticipated, specimens should be frozen at -70°C.

Throat Smear
1. Moisten a cotton or Dacron swab with physiologic saline.

2. Vigorously rub the swab across the tonsils and posterior pharynx.
3. Rub the swab on the wells of an acetone-cleaned slide making sure that the specimen is evenly dispersed over the entire wells. Be sure to roll the swab so that entire surface of the swab comes into contact with the slide.
4. Allow the slide to air dry.
5. Send the slide to the laboratory in a petri dish or in a slide mailer. Slides may be stored at room temperature or at 2-8°C for up to 48 hours before fixation.

Throat Washes
1. Ask the patient to clear all mucus and post-nasal secretions from their throat and mouth.
2. Give patient 2-3 ml of sterile physiological saline and ask them to gargle for 30-60 seconds and expectorate the wash into a clean collection device or cup.
3. Add the wash to the 3 ml of viral transport medium.
4. Send the specimen to the laboratory on wet ice. DO NOT FREEZE AT -20°C. Specimens may be stored for up to 72 hours at 2-8°C. If longer delays are anticipated, specimens should be frozen at -70°C.

Tissues
1. Collect tissues aseptically taking care to prevent cross-contamination when specimens are taken from multiple sites.
2. Autopsy specimens should be collected within 24 hours of the time of death.
3. Small tissue specimens may be placed into viral transport medium.
4. Send all specimens to the laboratory on wet ice or with a cold pack. DO NOT FREEZE AT -20°C. Specimens may be stored for up to 24 hours at 2-8°C. If longer delays are anticipated, specimens should be frozen at -70°C.

Urethral Swabs
1. Position the patient so that the urethra is readily accessible.
2. Insert a fine, aluminum-shafted cotton or Dacron swab 2 to 4 cm into the urethra.
3. Carefully rotate the swab three times and remove the swab from the urethra.
4. Place the swab in viral transport medium and cut the shaft of the swab so that it fits into the tube.
5. Send the specimen to the laboratory on wet ice. DO NOT FREEZE AT -20°C. Specimens may be stored for up to 72 hours at 2-8°C. If longer delays are anticipated, specimens should be frozen at -70°C.

NOTE: Urethral discharges are not acceptable specimens.

Urethral Smears
1. Position the patient so that the urethra is readily accessible.
2. Insert a fine, aluminum-shafted cotton or Dacron swab 2 to 4 cm into the urethra.
3. Carefully rotate the swab three times and remove the swab from the urethra.
4. Do not allow the swab to dry.
5. Rub the swab on the well of an acetone-cleaned slide making sure the specimen is evenly dispersed over the entire well.
6. Allow the slide to air dry.
7. Lay the slide flat and add 0.5 ml of methanol to the well. Let the methanol evaporate at room temperature.
8. Label the slide and send it to the laboratory in a petri dish or a slide mailer.
9. Slides may be stored at room (20-30°C) or refrigerator (2-8°C) temperature for up to 7 days before staining. If longer delays are anticipated, slides should be stored at or below -20°C.

Urine
1. Collect freshly voided early morning urine in a sterile container with a tight-fitting lid.
2. Send the specimen to the laboratory under ambient conditions. **DO NOT FREEZE AT -20°C.** Specimens may be stored for up to 72 hours at 2-8°C. If longer delays are anticipated, specimens should be frozen at -70°C.

Vesicular Cell Smears
1. Place a small drop of sterile saline on each well of an acetone-cleaned 2-well microscope slide.
2. Aseptically unroof the lesion and blot the excess fluid with a swab.
3. Scrape the base of the lesion with a sterile scalpel blade to obtain virus-infected epithelial cells. Gross bleeding should be avoided because red blood cells can interfere with the interpretation of the assay.
4. Transfer the cell scrapings from the scalpel blade to the drops of saline.
5. Spread the exfoliated cells thinly over the wells and allow the slide to air dry at room temperature.
6. Send the slide to the laboratory in a petri dish or in a slide mailer.
7. Slides may be stored for up to 48 hours at 2-8°C.

Vesicle Fluid by Aspiration
1. Select a vesicle containing clear fluids. The efficiency of virus recovery decreases significantly when pustular fluids are collected.
2. Thoroughly clean the outside of the vesicle to be sampled.
3. Hold a tuberculin needle with a 26-27 ga needle parallel with the skin and gently insert the end of the needle into the vesicle.
4. Gently aspirate the vesicle fluid and transfer it to a vial of viral transport medium. Rinse the syringe and needle once with viral transport medium.
5. Send the specimen to the laboratory on wet ice. DO NOT FREEZE AT -20°C. Specimens may be stored for up to 72 hours at 2-8°C. If longer delays are anticipated, specimens should be frozen at -70°C.

Vesicle Fluid (Swab)
1. Select a vesicle containing clear fluids. The efficiency of virus recovery decreases significantly when vesicles pustulate.
2. Aseptically unroof the lesion and blot the excess fluid with a swab.
3. Gently rub the base of the lesion with the swab to collect infected epithelial cells.
4. Place the swab in viral transport medium and break the shaft of the swab so that it will fit into the vial.
5. Send the specimen to the laboratory on wet ice. DO NOT FREEZE AT -20°C. Specimens may be stored for up to 72 hours at 2-8°C. If longer delays are anticipated, specimens should be frozen at -70°C.

SPECIMEN TRANSPORT

Once the specimen has been collected, specimen handling and transportation can have a significant impact upon the ability of the laboratory to isolate viruses or chlamydiae. Specimens should be transported to the laboratory as quickly as possible to maximize virus recovery. Viral specimens should be transported on cold packs or wet ice whenever possible. Specimens that cannot be sent to the laboratory right away may be stored at 2-8°C or on wet ice for 24-72 hours. If longer delays are anticipated, specimens should be frozen at -70°C. Do not freeze specimens at -20°C.

Shipping specimens to another laboratory via a commercial carrier is regulated by the federal government (1) and shipping regulations can be obtained from the Post Office or from the Federal Register. Generally speaking, specimens must be wrapped in enough absorbent material to absorb the entire contents of the package should the vial break or leak during shipment. In addition, the specimen and the absorbent material must be placed in a watertight container which in turn, is enclosed in another shipping container. Shipping containers for virus specimens should be insulated to keep the specimen cold during shipment and to prevent specimens from freezing during winter shipments.

SPECIMEN PROCESSING

Once specimens have been received in the laboratory, they must be processed before they can be used for fluorescent antibody testing, EIA testing, or inoculation onto cell cultures. Several general processing procedures and procedural comments are listed below. Special protocols for individual viruses or chlamydia are listed in the appropriate virus procedures.

Fixatives

Acetone, methanol, ethanol are the most commonly used fixatives for fluorescent antibody methods. These fixatives act by dehydrating the specimen and dissolving some lipids. Acetone, methanol, and ethanol should be stored in a sealed container to prevent them from accumulating water from the air. Water-contaminated fixatives can interfere with fluorescent antibody staining by causing an uneven, hazy background.

Blood (Buffy Coats)

White cells are separated from red cells by centrifugation through a ficoll-hypaque density gradient. A number of ready-made gradient materials can be purchased commercially (Ficol-Paque from Pharmacia Fine Chemicals, Histopaque from Sigma Chemical Company or Lymphocyte Separating Medium from Organon-Teknika) for this purpose or gradient materials can be prepared in the laboratory (Appendix C). Separation should be done as directed by the manufacturer or as described below.

1. Aseptically collect 10 ml of blood in green top vacutainer tubes containing preservative-free heparin.
2. Place 3 ml of ficoll-hypaque in a clear, sterile 15 ml centrifuge tube.
3. Dilute the heparinized blood with an equal volume of sterile saline.
4. Carefully overlay 5 ml of heparinized blood onto the ficoll-hypaque.
5. Centrifuge the tubes at 400 x g in a swinging bucket rotor for 30 minutes at ambient temperature.
6. Lymphocytes will band at the plasma-ficoll interface.
7. Carefully remove and discard the clear upper (plasma) layer.
8. Remove the lymphocyte band and the ficoll layer down to the red blood cells. Pool the buffy coat preparations from a single patient in a sterile 50 ml centrifuge tube.
9. Add PBS to the 30 ml mark.
10. Centrifuge the cells at 200 x g for 15 minutes and remove the supernatant fluids.
11. Suspend the cell pellet in 1 ml of viral transport medium.

Semen

1. Centrifuge the semen specimen at 1000-3000 x g for 15 minutes at 2-8°C to pellet the cellular fraction.
2. Suspend the pellet in 5.0 ml of viral transport medium or HBSS.
3. Centrifuge the specimen for 15 minutes as before.
4. Suspend the cellular fraction in 1 ml of viral transport medium.
5. Use this specimen to inoculate cell cultures.

NOTE: Semen specimens are extremely toxic to cell cultures. However, a significant portion of the toxicity can be eliminated by washing away the seminal fluids and inoculating the cellular fraction onto cell cultures.

Sputum or Throat Washings

1. Dilute sputa and throat washings 1:5 in viral or chlamydial transport medium containing twice the normal concentration of antibiotics.
2. Place the diluted specimen in a tightly stoppered sterile container and vortex vigorously with several glass beads.
3. Centrifuge the specimen for 15 minutes at 600 x g.
4. Transfer the supernatant fluids to another sterile container and incubate for 1 hour at room temperature to reduce the bacterial load.

Stool
1. Combine approximately 0.5 gram of stool in 5-10 ml of viral transport medium.
2. Add 2-3 glass beads to the tube and vortex the specimen vigorously to extract as much virus as possible.
3. Centrifuge the suspension at 1-3000 x g for 10-15 minutes.
4. Filter the supernatant fluids through a 0.45 μm syringe filter.

Swabs

Swabs for virus isolation are occasionally sent to the laboratory in bacterial transport systems. These swabs should be placed in viral transport medium as soon as they arrive in the laboratory. After the swabs have soaked for at least 15 minutes, they can be processed as described below.

1. Vortex the specimen with several glass beads for 20-30 seconds to release any bound cells or virus.
2. Remove the swabs from the transport medium and firmly roll them against the inside of the tube to remove as much fluid as possible.

NOTE: Rectal swabs should be handled as described above except that the viral transport medium should be filtered through a 0.45 μm syringe filter prior to inoculation onto cell cultures.

Tissues
Method A
1. Carefully weigh tissue specimens and mince them with sterile scissors.
2. Homogenize using a disposable tissue homogenizer.
3. Add enough viral transport medium to the homogenizer to produce a 10-20% (w/v) suspension.
4. Remove the suspension from the homogenizer and vortex it vigorously to extract as much antigen as possible.
5. Centrifuge the suspension at 300-600 x g high speed for 10 minutes at 4°C but do not remove the supernatant fluids from the tissue.
6. Carefully remove the required amount of specimen for inoculation onto cell cultures.

Method B
1. Carefully weigh tissue specimens and mince them with sterile scissors.
2. Place the tissue in a sterile mortar containing 1-2 ml of viral or chlamydial transport medium. A small amount of sterile sand may be used if necessary.
3. Grind the tissues with a pestle until the tissue is a paste of uniform consistency.
4. Add 5-10 ml of viral transport medium to the mortar and suspend the tissue in this fluid.
5. Vortex the suspension vigorously to extract as much antigen as possible.
6. Centrifuge the suspension at 300-600 x g high speed for 10 minutes at 4°C but do not remove the supernatant fluids from the tissue.
7. Carefully remove the required amount of supernatant fluids for inoculation onto cell cultures.

Urine Sediments
1. Vortex the specimen to suspend any sediments and remove 10-20 ml.
2. Centrifuge the specimen at 1000-3000 x g for 15 minutes at 2-8°C to pellet the cellular fraction.
3. Suspend the pellet in 1.0 ml of viral transport medium.
4. Use this specimen to inoculate cell cultures.

Wash and Aspirate Specimens
1. Vortex the specimen with several glass beads for 20-30 seconds.
2. Centrifuge the cell suspension at 200 x g for 5 minutes to remove any mucus.
3. Carefully remove the mucus and use the remainder of the specimen for inoculation of cell cultures.

Preparation of Cell Smears

The success of the direct fluorescent antibody (DFA) procedure depends upon the preparation of a well made cell smear. Smears that are too thick or "lumpy" can cause nonspecific trapping of the antibody reagents. Smears that are grossly contaminated with red blood cells can cause interpretation difficulties due to red cell autofluorescence. Excess mucus can cause false positive and false negative reactions. False positive reactions result from nonspecific trapping of the antibody reagents while false negative reactions can occur due to cell clumping and subsequent masking of virus-specific antigens.

The principal reasons why direct smears fail to detect specific viruses are (a) inadequate specimen collection and (b) the presence of too few cells on the slide. Specimen adequacy and its definition are significant sources of laboratory-to-laboratory variation and an area of increasing regulatory concern. Many laboratories require at least 50 cells per smear while others require 1-2 cells/high power field. Whatever the number, too many cells can cause nonspecific trapping of antibody and with too few cells the laboratory may not be able to find an infected cell.

Ideally, smears should be made at the collection site. However, adequate smears can be made from specimens in viral transport medium. Preparation of smears from frozen specimens is not recommended because the resulting slides are very difficult to interpret due to cell lysis.

Smears from Nasal Aspirates, Nasal Washes, and Throat Washes
1. Centrifuge the specimen at 1000 to 1500 x g to pellet the cells.
2. Suspend the cells in 3-5 ml of viral transport medium and centrifuge as before.
3. Remove all but 100-200 µl of the fluid.
4. Suspend the cell pellet in the remaining fluid.
5. Spread one drop of the cell suspension onto the wells of an acetone cleaned two-well slide.
6. Allow the smears to air dry at room temperature. Slides may be stored for up to 48 hours at 2-8°C.

Smears from Broncheoalveolar Lavage (BAL) Specimens
1. Centrifuge the BAL specimen at 600 x g for 10 minutes at 4°C to pellet the cells.
2. Suspend the cells in 10 ml of PBS and mix gently.
3. Centrifuge the cells as before and remove all but a few drops of fluid.
4. Suspend the cells in the remaining fluid.
5. Using a pipette, place a drop of the cell suspension onto the wells of an acetone-cleaned slide.
6. Allow the smear to air dry at room temperature.
7. Slides may be stored for up to 48 hours at 2-8°C.

Smears from Nasal Aspirates, Nasal Washes, and Throat Washes
1. Vortex the specimen gently to break up the mucus.
2. Place one drop of the specimen on the wells of an acetone-cleaned slide and spread the specimen evenly over the well.
3. Allow the slide to air dry at room temperature.
4. Slides may be stored for up to 48 hours at 2-8°C.

Smears from Swabs in Viral Transport Medium
1. Vortex the specimen vigorously to release as many cells from the swab as possible.
2. Remove the swab from the viral transport medium and firmly roll it against the inside of the tube to remove as much fluid as possible.
3. Centrifuge the specimen at 1000 to 1500 x g to pellet the cells.
4. Remove all but 100-200 µl of the viral transport medium.
5. Suspend the cell pellet in the remaining fluid.
6. Spread one drop of the cell suspension onto the wells of an acetone cleaned two-well slide.
7. Allow the smears to air dry at room temperature. Slides may be stored for up to 48 hours at 2-8°C.

Virus Smears from Swabs in Chlamydia (Sucrose-Containing) Transport Medium
1. Virus specimens transported to the laboratory in sucrose-containing transport medium are not suitable for preparing direct smears because the cells will not adhere to the slide.

The sucrose must be removed before the specimen can be processed.
2. Vortex the specimen vigorously to release as many cells from the swab as possible.
3. Remove the swab from the viral transport medium and firmly roll it against the inside of the tube to remove as much fluid as possible.
4. Centrifuge the specimen at 1000 to 1500 x g to pellet the cells.
5. Suspend the cells in 3-5 ml of viral transport medium and centrifuge as before.
6. Repeat step 4.
7. Remove all but 100-200 µl of the viral transport medium.
8. Suspend the cell pellet in the remaining fluid.
9. Spread one drop of the cell suspension onto the wells of an acetone cleaned two-well slide.
10. Allow the smears to air dry at room temperature. Slides may be stored for up to 48 hours at 2-8°C.

Chlamydia Smears from Swabs in Chlamydia Transport Medium
1. Vortex the specimen vigorously to release as many cells from the swab as possible.
2. Remove the swab from the viral transport medium and firmly roll it against the inside of the tube to remove as much fluid as possible.
3. Centrifuge the specimen at 10,000-15,000 x g to pellet the EBs.
4. Suspend the pellet in 0.2 ml of PBS.
5. Spread a small amount of the suspension onto the wells of an acetone cleaned slide.
6. Allow the smears to air dry at room temperature. Slides may be stored for up to 48 hours at 2-8°C.

Chlamydia Smears from Urine
1. First void morning urine is recommended.
2. Vortex the specimen to suspend any sediments and remove 1-2 ml.
3. Centrifuge the specimen at 10,000-15,000 x g to pellet the EBs.
4. Suspend the pellet in 0.1 ml of PBS.
5. Spread a small amount of the suspension onto the wells of an acetone cleaned slide.
6. Allow the smears to air dry at room temperature. Slides may be stored for up to 48 hours at 2-8°C.

REFERENCES

1. Committee on Hazardous Biological Substances in the Laboratory. *Biosafety in the laboratory*. Washington DC: National Academy Press 1989;20-22.

CHAPTER 6

Mammalian Cell Culture Procedures

INTRODUCTION

Viruses are intracellular parasites and require living cells in order to replicate. In the clinical laboratory, living cells can be provided in the form of suckling mice, embryonated chicken eggs, and cell cultures. Today, most clinical laboratories prefer to use cell cultures over mouse and egg inoculation. The availability of a wide variety of cell lines provides the clinical laboratory with the ability to isolate most of the clinically relevant viruses.

Cell culture is the foundation upon which the virology laboratory is built. Although many laboratories continue to make their own cell culture tubes and vials, the commercial availability of cell cultures has made viral diagnostic procedures available to laboratories that cannot or prefer not to grow their own cells. Whether the laboratory makes their own cells or purchases ready to use cell cultures, the procedures for handling cells are the same. Many of the basic procedures are listed below.

Cell Culture Media

In the body, cells and tissues are continuously bathed by circulating body fluids that provide nutrients and remove waste products. These fluids provide all the components necessary for the survival, differentiation, and growth of cells *in vivo*. Likewise, when living cells are grown *in vitro*, they are completely dependent upon the culture fluids for their nutritional requirements. The first cell culture fluids consisted of blood serum, plasma clots, aqueous humors, and lymph. In the 1930s, many of the nutritional requirements for cell growth were established and a more synthetic approach to cell culture fluids was used. In 1950, Morgan and coworkers (1) formulated Medium 199 which supported the growth of primary chicken embryo cells. This medium contained 61 synthetic nutrients and, although it supported the growth of chicken embryo cells, Medium 199 required 5-10% serum to support the growth of any other cell lines.

In 1955, Eagle identified 28 synthetic substances that were essential for the growth of HeLa cells and mouse L cells (2). Eagle's basal medium (BME) included six salts that provided metabolic cofactors, pH control, and regulation of the electrolyte balance; glucose as an energy source; 13 amino acids as protein precursors; and 8 vitamins for necessary cofactors. Even in this completely defined medium, many cells required 1-5% serum to support growth. Eagle subsequently developed another medium, Eagle's minimum essential medium (EMEM) that contained higher concentrations of essential amino acids and vitamins (3). The principal advantage of EMEM was that cultures could be kept for longer periods of time without media changes. EMEM is the most commonly used culture medium in the clinical laboratory.

Serum Supplements

Most laboratories supplement the defined medium with serum to support the growth of a variety of cells. Serum is a heterogenous, largely undefined media additive that provides a variety of nutrients. Serum also provides a number of growth factors including hormones, trace elements, minerals, and lipids. In the clinical laboratory, the most widely used supplements include calf serum (CS) and fetal calf serum (FCS). FCS contains much higher concentrations of growth factors than calf serum and it is often used as a supplement for culturing fastidious cell lines. One of the principal disadvantages to serum supplements is the cost. It is not unusual for serum costs to make up 90-98% of the cost of a liter of medium.

Although serum is necessary for cultivation of many cell lines, serum may also contain substances that are toxic. Some of the toxicity can be reduced by heating the serum to 56°C for 30 minutes. This procedure inactivates complement and reduces the cytopathic effects of immunoglobulins without damaging polypeptide growth factors. In addition to growth promotion and growth inhibition, serum can also protect cell cultures from proteases and toxic agents found in some specimens. Because serum may simultaneously have toxic, protective, and growth promoting effects, the type and amount of serum used in culture media is often determined empirically and each lot of serum must be tested for toxicity before placing it into general use. In the clinical laboratory, 5-10% fetal calf serum and 1-2% fetal calf serum are used most often for cell growth and cell maintenance, respectively.

pH Control

Control of pH is an important function of any cell culture medium. While cell cultures continuously metabolize substrates to obtain energy, complete oxidation of carbohydrates does not often occur. Instead, glucose is converted to pyruvic acid and then to lactic acid. Lactic acid is excreted into the culture medium where it accumulates. Most cell lines grow well at pH 7.2-7.4 and lactic acid must be neutralized or cell damage could occur. Control of pH is so critical that most culture media contain a pH indicator, phenol red, so that pH changes can be monitored visually. Phenol red is yellow below pH 6.8, orange at pH 7.0, red at pH 7.4, and purple above pH 7.8 (see Color plate 1A following Chapter 7). Because color assessment is highly subjective, it is often useful to make a set of pH standards using sterile phenol red-containing MEM or HBSS.

A number of buffering systems have been used for controlling pH in cell culture media. In most media, Na^+, K^+, HCO_3^-, and HPO_4^- are the principal components of pH control. One to ten millimolar phosphate buffers and 26 mM carbonic acid/bicarbonate buffer system similar to that found in blood, are usually incorporated into the culture medium for pH control. Most culture media use physiologic concentrations of these chemicals because they are inexpensive, they have low toxicity, and they provide nutritional benefits to the cells.

Despite its low buffering capacity at physiological pH, bicarbonate buffers are used more frequently than any other type of buffer. Bicarbonate buffering systems are also difficult to control accurately because they depend upon the partial pressure of CO_2 in the atmosphere over the culture. A 5% CO_2 atmosphere is required over a 26 mM bicarbonate system to maintain a pH of 7.2-7.4.

In 1966, Good, et al. (4) described a number of zwitterionic buffers that have become popular replacements for bicarbonate. The principal advantage of these buffers is that they eliminate the need for expensive CO_2 incubators. HEPES (N-2-hydroxyethlypiperazine-N'-2-ethanesulfonic acid) is widely used in cell culture media. Compared with a pK_a of 6.2 at 37°C for bicarbonate, HEPES has a pK_a of 7.3 and provides more buffering capacity throughout the physiological pH range than bicarbonate. HEPES (10-25 mM) is usually not toxic to cell cultures when used in serum-supplemented media. However, storage of HEPES-containing media in areas with fluorescent lighting can cause the accumulation of toxic substances in the media. The addition of sodium pyruvate to the medium counteracts this effect and prevents the formation of toxic substances.

Antibiotics

The use of antibiotics in cell cultures has dramatically changed cell culture technology from the tedious, labor intensive procedures developed by Alexis Carrel in the early 1900s to the relatively straightforward methods used in most modern laboratories. Although the use of antibiotics is widespread, antibiotics cannot replace good aseptic technique.

A wide variety of antibiotics including penicillin, streptomycin, gentamicin, kanamycin, tetracycline, and neomycin have been used in cell cultures. Penicillin, streptomycin, and gentamicin are commonly used in the clinical laboratory.

Penicillin G, when used in concentration of 100-250 Units/ml, inhibits the growth of most gram positive bacteria. However, penicillin G is very labile and in generally inactivated after 48 hours at 35-37°C. Streptomycin sulfate is more stable than penicillin and remains effective for approximately 4 days at 35-37°C. Streptomycin effectively inhibits the growth of many gram negative bacteria. Many laboratories use a combination of penicillin (100-250 U/ml) and streptomycin (100 μg/ml) for routine cell cultures.

Gentamicin is active against a broad spectrum of gram negative and gram positive bacteria. This antibiotic also inhibits the growth of many mycoplasmas that commonly contaminate cell cultures. Gentamicin is widely used at concentration of 5-10 μg/ml and is stable for about 2 weeks at 35-37°C. Antifungal agents like amphotericin B (Fungizone) are also used under special circumstances. Amphotericin B suppresses the growth of many yeast and filamentous fungi. When used at working concentration of 1-4 μg/ml, amphotericin B has a half-life of 4 days at 35-37°C. It should be noted that amphotericin B does not kill fungi, but rather, it suppresses the growth of these organisms. Long-term usage of amphotericin B can mask low-level, often inapparent, fungal contaminations. This level of contamination can cause changes in culture phenotypes as the cells adapt to these new culture conditions. Therefore, amphotericin B should not be used for routine cell propagation. If the use of amphotericin B is required, this agent should be withdrawn from the cultures at frequent intervals to check for fungal contamination. Routine use of amphotericin B in medium in isolation tubes and vials is appropriate because these cultures are not routinely passaged.

CELL CULTURE PROCEDURES

Cell Dissociation

Enzymes such as trypsin and collagenase are commonly used for dissociating tissues and making primary cultures. In addition, 0.25% trypsin is commonly used to make single cell suspensions from cell culture monolayers. Most commercial trypsin preparations are prepared from pigs. Therefore, all trypsin preparations should be tested for the presence of porcine parvovirus (PPV). PPV is an extremely hardy single-stranded DNA virus that causes persistent infections in a wide range of cells. PPV rarely produces CPE and persistent PPV may not be readily apparent. Persistent PPV infections can cause changes in the cells including altered susceptibility to virus infection and changes in the cell growth rate.

Trypsin and ethylenediaminetetraacetic acid (EDTA) are commonly used to produce single cell suspensions from cell monolayers. The trypsin cleaves the surface proteins that link cells together while EDTA chelates the divalent cations necessary for attachment of the cells to the surface of the flask. Together, these reagents can quickly and gently remove anchorage dependent cells from glass or plastic vessels and produce single cell suspensions that can be efficiently counted or diluted. Trypsin is quickly inactivated by the trypsin inhibitors found in mammalian serum and does not present any residual activity once the cells are passaged.

Water for Cell Culture

Cell culture reagents should be prepared in the highest quality water available. The effects of using poor quality water can be acute (e.g., cell death, rounding, or CPE-like lesions) or insidious. Routine use of water containing endotoxins, organic chemicals, or bacteria can cause unwanted changes in the cell phenotype and lead to decreased sensitivity to virus infection. A number of companies produce equipment that can produce cell culture-quality water. If cell culture water is made in the laboratory, aliquots should be tested for the presence of endotoxins at least monthly. Alternatively, sterile, endotoxin-free water can be purchased from a hospital supply company.

Inactivation of (Fetal) Calf Serum

The use of heat inactivated serum is a standard procedure in cell culture laboratories. Heating serum to 56°C destroys the biological activity of complement. Inactivating complement is important because some cellular byproducts can nonspecifically trigger the complement cascade and destroy the cells in the monolayer. Heating the serum to 56°C will also inactivate some infectious agents that may be present in the serum.

When inactivating serum, care must be taken to assure that the medium is isothermal and is maintained at 56°C for the required length of time. Longer periods of heating can destroy the nutritive value of the serum while shorter periods may not fully inactivate the complement. Small bottles of serum (e.g., less than 100 ml) do not present a significant heating problem. However, a 500 ml bottle can contain a significant temperature gradient when placed in a 56°C water bath. Therefore, a second (water) bottle should be prepared and equilibrated to 2-8°C. Both bottles are placed in the water bath and a thermometer is inserted in the water-containing bottle, giving an indication as to the temperature in the FCS bottle. Swirling the bottles every 5-10 minutes also helps to provide even heat distribution.

1. Thaw a 500 ml bottle of fetal calf serum in the refrigerator. Do not place a frozen bottle of serum in a water bath because the bottle may crack or break.
2. Place 500 ml of water in an identical bottle, cap, and place the bottle in the refrigerator.
3. Once the serum is thawed, place the bottle of serum in a 56°C water bath.
4. Remove the cap of the water bottle (step 2) and place the bottle in the water bath next to the serum. The water bath should have enough fluid so that the water level reaches at least half way up the bottles.
5. Place a thermometer in the water-containing bottle.
6. Swirl the contents of both bottles every 5-10 minutes to assure even temperature distribution.
7. When the thermometer reaches 56°C, start a timer and incubate the bottles for another 30 minutes, swirling the contents of both bottles every 5-10 minutes.
8. Remove the bottles from the water bath and store the serum at 2-8°C.

Propagation of Cells in Culture

A wide variety of cell cultures can be obtained from commercial sources as "starts" and propagated in the laboratory. Although the following procedure is applicable to most anchorage-dependent cell cultures, some cells require special media, trypsinization protocols, and/or culture conditions. For the best results, a monolayer of actively growing or newly confluent cells should be used for subcultivation. Cells can be passaged from and into a wide variety of vessels including bottles, flasks, tubes, plates, petri dishes, and shell vials.

1. Remove the culture medium from the cells.
2. Add enough warm (35-37°C) ATV Trypsin to the vessel to cover the monolayer.
3. Rock the vessel to evenly coat the monolayer.
4. Carefully decant the trypsin from the monolayer.
5. Incubate the vessel at 37-37°C for 1-5 minutes. During this procedure the cell monolayer will become translucent and lifts off the surface of the vessel when the vessel is tapped against the heel of the hand.
6. Add enough warm growth medium to the vessel to suspend the cells and pipette them up and down gently to dissociate any cell clumps.
7. The cells can be transferred into other culture vessels.

Depending upon the number of viable cells seeded into each vessel, a cell monolayer can take up to a week to form. Observation of the vessel during this period demonstrates different stages of the monolayer formation. Within the first 4-8 hours after inoculation, the cells are usually round and refractile. The cells slowly settle to the bottom of the vessel where they attach and spindle-shaped fibroblast cells begin to appear. The culture will contain both round and spindle-shaped cells until the cells grow adjacent to one another. The influence of "contact inhibition" causes the cells to stop actively growing and to form an even monolayer.

Cells that do not exhibit contact inhibition often spread randomly over the surface of the vessel. Once these cells are adjacent to each other, they begin growing in mounds. Cell growth continues until all the nutrients are depleted, or the environment becomes too polluted with metabolic wastes.

Performing Cell Counts

Cell counts are a basic cell culture procedure. Although automated cell counters are available, most cell culture laboratories use a hemocytometer to determine cell concentrations. Viable cell counts can be performed using trypan blue or erythrosine B. These dyes are excluded from cells that have intact cell membranes. Cells that take up the stain are considered to be dead.

The Neubauer brightline hemocytometer consists of two chambers that are covered by a special coverslip. Each large square is 1 x 1 mm and the distance between the square and the coverslip is 0.1 mm. This means that each square contains 10^{-4} ml. Counting the number of cells in this standard volume can be used to produce a fairly accurate estimate of the cell concentration in the original tube.

Counting cells with a hemocytometer is not without problems. However, most cell count errors are due to improper calculations, inadequate dispersion of cells (cell clumps or failure to mix the cell suspension thoroughly), inaccurate dilution, improper filling (over- and under-filling) of the hemocytometer, and improper cleaning of the chambers and the coverslip (skin oils, dirt, and dust can keep the chamber from filling properly) (5). When used correctly, this procedure is accurate, reliable, and inexpensive. Sample calculations for determining the percentage of viable cells, cell concentrations, and for preparing standard cell concentrations are shown in Fig. 1.

1. Thoroughly clean a Neubauer brightline hemocytometer and coverslip with mild detergent, rinse with water, then rinse it with 70% alcohol. Dry the hemocytometer and coverslip with a lint-free cloth or tissue.
2. Assemble the hemacytometer.
3. Mix the cell suspension to assure an even distribution of cells.
4. Transfer 0.1 ml of the cell suspension to a clean 10 x 75 mm test tube.
5. Add 0.1 ml of Trypan Blue solution to the tube.
6. Mix gently.
7. Insert a pasteur pipette into the cell suspension and allow some of the fluid to flow into the pipette via capillary action.
8. Touch the end of the pipette to the loading notch on the hemocytometer. Allow the cell suspension to flow into the chamber until the chamber is just filled. **DO NOT OVER-FILL HEMOCYTOMETER CHAMBERS.** Over-filling and under-filling the chamber will produce erroneous counts. If a chamber is overfilled, the hemocytometer must be disassembled, washed, and reloaded.
9. Repeat the loading sequence with the other chamber on the hemocytometer.
10. Using a microscope (10X objective), count the viable (clear or unstained) and nonviable (blue) cells in the 4 large, corner squares of the hemocytometer grid. Do not count the cells lying on the top and right sides of each square. Individual cells in cell clumps should be counted. However, suspensions with a large number of clumps should be pipetted up and down several times to dissociate the clumps and counted again.
11. Repeat the counts on the other hemocytometer well.
12. If the cell suspension was well dispersed and if the hemocytometer was loaded without overfilling, the counts from both sides of the hemocytometer should agree within 10%. If the error is greater than 10%, the cells should be mixed well and counted again.

Tube Cultures

Tube cultures are the backbone of the virology laboratory. A wide variety of cells can be established in 16 x 125 mm borosilicate glass culture tubes. These cell cultures can be purchased commercially or they can be prepared as described below. The average cell yields and seeding densities are shown in Table 1.

Figure 1. Cell Number Calculations.

Percentage of Viable Cells
The percentage of viable cells can be determined as follows:

$$\frac{\text{Number of Viable (Unstained) Cells}}{\text{Number of Viable + Dead (Stained) Cells}} \times 100\%$$

Determining Cell Concentration
The hemocytometer is calibrated so that the amount of fluid in one square is 10^{-4} ml, the number of cells per milliliter in the original suspension can be determined.

$$\frac{\text{Number of Cells Counted}}{\text{Number of Squares}} \times \frac{\text{Square}}{10^{-4} \text{ ml}} \times \text{Dilution Factor} = \text{Number of cells/ml}$$

Example: If 320 cells were counted in 8 squares the initial cell concentration would be 880/8 (110), times 10^4, times 2 (a 1:2 dilution was done with the trypan blue), or 2.2×10^6 cells/ml.

Preparing a Standard Cell Concentration

A. Determine the number of cells needed.

$$\text{Desired Cell Concentration (cells/ml)} \times \text{Desired Volume of Media (ml)} = \text{Cell Number}$$

B. Determine the volume of the cell suspension needed.

$$\frac{\text{Total Number of Cells Needed (above)}}{\text{Existing Cell Concentration (cells/ml)}} = \text{ml of Cell Suspension Required}$$

C. Prepare the Cell Suspension.

$$\text{Desired Volume (ml)} - \text{ml of Cell Suspension Required} = \text{Volume of Media Needed}$$

Example. If 50 ml of a cell suspension is required at 3×10^5 cells/ml, a total of 1.5×10^7 cells are needed. If the cell suspension contains 2.2×10^6 cells/ml, the final suspension consists of 6.8 ml of the cell suspension plus 43.2 ml of medium.

1. Wash the culture tubes and caps to remove any oils or dirt that may have accumulated during manufacturing, shipping, or storage. Tubes should be thoroughly rinsed to remove any detergents.
2. Dry the tubes and place the cap loosely on the tube.
3. Autoclave the tubes at 121°C for 15 minutes using the dry goods cycle.
4. Allow the tubes to cool to room temperature.
5. Add 1 ml of a cell suspension (Table 1) to each tube.
6. Incubate the tubes on their side in a stationary rack (4° angle) at 35-37°C to allow the cells to attach to the glass.

Shell Vial Cultures

Centrifugation enhanced (shell vial) cultures are used increasingly for the isolation and identification of viruses. Shell vials are short, flat-bottomed tubes of sufficient diameter to accommodate a 12 mm coverslip. Various types of closures are available for shell vial cultures with silicone stoppers and commercially available plastic caps being used most often.

1. Wash the vials and caps to remove any oils or dirt that may have accumulated during manufacturing, shipping, or storage. Tubes should be thoroughly rinsed to remove any detergents.
2. Soak the 12 mm round coverslips for at least 5 minutes in absolute ethanol or acetone to remove any oils that may have been accumulated during manufacturing.
3. Dry the vials, coverslips, and caps. Place one coverslip in each vial making sure the coverslip lies flat in the vial.
4. Loosely cover the shell vials with aluminum foil.
5. Place the silicone stoppers in a beaker and cover the beaker with foil.
6. Autoclave the vials and stoppers at 121°C for 15 minutes on the fast exchange (dry goods) and dry cycle.
7. Allow the vials and stoppers to cool to room temperature.
8. Add 1 ml of a cell suspension (Table 1) to each vial.
9. Place a stopper on the vial.
10. Incubate at 35-37°C to allow the cells to attach to the coverslip.

Freezing Mammalian Cells

In the laboratory, mammalian cells are routinely frozen to minimize the number of cell lines that must be handled each day and to guard against contamination and technical errors that could cause the accidental loss of the line. Frozen cell stocks are also used to minimize genetic drift, senescence, and unwanted phenotypic changes (i.e., changes in virus susceptibility) that can occur when continuous and finite cell lines are cultured for long periods of time.

When frozen, osmotic shock and intracellular ice crystals invariably destroy unprotected cell cultures. Early experiments in cell culture preservation demonstrated that glycerol (6) and dimethylsulfoxide (7) could prevent cell death during freezing. In addition to the cryoprotectant, cells must be cooled slowly (approximately 1°C per minute) until the cells reach a temperature of -25°C (8). Once the cell suspensions are cooled below this critical temperature, they can be cooled rapidly to -70°C or -196°C (the temperature of liquid nitrogen) without further loss of viability.

When the cryoprotectant is omitted or when the cells are frozen too quickly, the aqueous fluids within the cells freeze and the resulting intracellular ice crystals can rupture cell membranes. During slow cooling, the external medium becomes supercooled and ice crystal nuclei form in the extracellular fluid. As a result, the extracellular milieu contains artificially elevated salt concentrations which in turn, produce an osmotic gradient. This gradient causes water to diffuse out of the cells and the nonelectrolytic cryoprotective agents to diffuse into the cells. This "dehydration" process tends to minimize osmotic shock and the formation of intracellular ice crystals (9-11).

A number of methods can be used to produce controlled cooling and several companies sell equipment for this purpose. However, this equipment is expensive and is not available in most

Table 1. Average cell yields and seeding densities for commonly used cell lines. Because some laboratories do not routinely count cells, the split ratios are given for passage of a 75 cm² flask into a new 75 cm² flasks.

Cell Name	Average Yield/75cm² Flask	75 cm² Flask		Tube Culture		Shell Vial	
		Cells/Flask (Split)	Days to Confluency	Cells/Tube (Split)[a]	Days to Confluency	Cells/Vial (Split)[a]	Days to Confluency
A549	2×10^7	3×10^6 (1:10)	2-3	3×10^5 (70 tubes)	2-3	3×10^5 (70 Vials)	2-3
BGM	2×10^7	3×10^6 (1:10)	2-3	3×10^5 (70 Tubes)	2-3	3×10^5 (70 Vials)	2-3
CV-1	6×10^6	3×10^6 (1:2)	2-3	2×10^5 (30 Tubes)	2-3	2×10^5 (30 Vials)	2-3
Graham 293	2×10^7	3×10^6 (1:10)	2-3	3×10^5 (70 Tubes)	2-3		
HeLa-229	2×10^7	3×10^6 (1:10)	2-3	3×10^5 (70 Tubes)	2-3	2×10^5 (100 Vials)	2-3
HEp-2	2.2×10^7	3×10^6 (1:10)	2-3	2×10^5 (100 Tubes)	2-3	2×10^5 (100 Vials)	2-3
HL	1.6×10^7	3×10^6 (1:5)	2-3	3×10^5 (55 Tubes)	2-3	2×10^5 (85 Vials)	2-3
LLC-MK2	1×10^7	3×10^6 (1:3)	2-3	3×10^5 (30 Tubes)	2-3	3×10^5 (30 Vials)	2-3
McCoy	4×10^7	3×10^6 (1:10)	2-3			2×10^5 (200 Vials) 3×10^5	2 1
Mink Lung	2×10^7	3×10^6 (1:10)	2-3	2×10^5 (100 Tubes)	2-3	3×10^5 (70 Vials)	2-3
RD	2×10^7	4×10^6 (1:5)	2-3	4.5×10^5 (45 Tubes)	2-3		
Vero	2×10^7	3×10^6 (1:10)	2-3	3×10^5 (70 Tubes)	2-3		
HFF	6×10^6	3×10^6 (1:2)	2-3	1×10^5 (60 Tubes)	2-3	1×10^5 (60 Vials)	2-3
MRC-5	1.5×10^7	3×10^6 (1:5)	2-3	3×10^5 (50 Tubes)	2-3	3×10^5 (50 Vials)	2-3
RMK				100 Tubes	7-10	100 Vials	7-10

[a] Number of 16 x 125 mm tubes or vials that can be prepared from a single 75 cm² cell culture flask.

laboratories. Controlled cooling can be accomplished by immersing the vials in a relatively large volume of 95% ethanol and placing the vials in a -70°C freezer overnight. The ethanol will not freeze at this temperature and acts as a thermal buffer. Alternatively, the vials can be placed in the center of a solid styrofoam test tube rack and another rack is placed over the vials. The two pieces of styrofoam are sealed together with tape and the styrofoam block is placed in a -70°C freezer overnight.

Once the cells are frozen, liquid nitrogen storage is the most reliable method for long-term preservation of cells. While some laboratories routinely store cells at -70°C, we have found that cells stored at -70°C gradually lose their viability and must be thawed, passaged and frozen every two or three years.

1. Select rapidly growing cells for freezing. Ideally, the culture medium should be removed and replaced with fresh medium 24 hours before they are frozen.
2. Remove the culture medium and trypsinize the cells as described above.
3. Centrifuge the cell suspension at 150 x g for 5-10 minutes.
4. Suspend the cells in cold MEM or HBSS.
5. Perform a cell count and adjust the cell concentration to 2×10^6 cells/ml.
6. Slowly add 2X Freezing Medium to the cell suspension. **CAUTION: DMSO SOLUTIONS MUST BE HANDLED WITH CARE.** DMSO solutions must be added very slowly because the latent heat that is released during mixing can cause cell injury and death.
7. Dispense 1 ml into each freezing vial and tighten the caps securely.
8. Place the vials in a commercial cell freezing apparatus, an alcohol bath, or in the center of a solid styrofoam test tube rack and freeze to -70°C.
9. Transfer the cells to liquid nitrogen for long-term storage.

Recovery of Frozen Cells

Recovery of frozen cells is enhanced if the original culture was healthy, free of contamination, and in late log phase before freezing. When recovering frozen cell stocks, cell suspensions must be thawed rapidly. Rapid thawing prevents ice crystal formation as the temperature of the suspension moves through the critical -50°C to 0°C range. Slow thawing will cause cell damage and loss of viability.

1. Remove the vial from the freezer or from liquid nitrogen.
2. Place the vial in a 35-37°C water bath until thawed.
3. Wipe the exterior of the vial with 70% alcohol.
4. Open the vial in a laminar flow hood and transfer the contents to a 75 cm^2 flask.
5. Add 12 ml of warm (35-37°C) growth medium. Rock the flask to distribute the cells evenly over the bottom of the flask.
6. Incubate the flask at 35-37°C overnight to allow the cells to attach.
7. Carefully remove the culture fluid and replace it with 12 ml of fresh growth medium. Changing the medium is important because the cryoprotectants can be toxic if left they are not removed.

Establishing Primary Cultures

The following procedure can be used for a variety of mammalian tissues including rabbit kidney cells and human foreskin fibroblast cells. All supplies and equipment should be sterile and tissue dissociation should be done in a clean environment, preferably a laminar flow hood.

Tissue Collection and Preparation
1. Aseptically remove the tissue from the animal or patient and transport it to the laboratory in cold (2-8°C) serum-free MEM with antibiotics.
2. Remove the tissue from the MEM and place it in a sterile petri dish containing fresh cold MEM with antibiotics.
3. Rinse the tissue with cold MEM or HBSS to remove any blood.

4. Add fresh MEM to the petri dish.
5. Aseptically excise any capsular or connective tissue from the tissue.
6. Mince the tissue with scissors to produce 2 mm thick sections.
7. Place the minced tissue into a sterile 250 ml beaker and wash the tissue twice with cold MEM or HBSS.

Tissue Dissociation
1. Remove the MEM or HBSS and add a sterile stir bar and 20-30 ml of prewarmed trypsin to the beaker.
2. Stir the minced tissue for 5 minutes at 25-37°C.
3. Remove and discard the trypsin.
4. Add 10-20 ml of fresh, warm trypsin and stir for 15 minutes.
5. Filter suspension through sterile gauze into a 50 ml centrifuge tube containing 10 ml of growth medium. The tube should be placed in an ice bath until the dissociation is complete.
6. Repeat steps 6 and 7 until tissue is almost totally disintegrated. A second centrifugation tube may be used to collect filtrate volumes in excess of 50 ml.
7. Centrifuge the cold cell suspension at 500-1000 rpm for 15 minutes at 2-8°C.
8. Discard the supernatant fluids and quickly suspend the cells in 15 ml of warm growth medium.
9. Draw the suspension up in a sterile 10 ml pipette. Place the tip of the pipette very close to the bottom of the tube and quickly force the fluid from the pipette. Repeat this process several times to dissociate any cell aggregates.
10. Centrifuge and suspend the cells in 25-50 ml of growth medium.
11. Perform a cell count.

Establishing Primary Tube Cultures
1. Suspend the cells at 1×10^5 cells/ml in growth medium.
2. Add 1 ml of the cell suspension to sterile 16 x 125 mm screw-capped culture tubes.
3. Incubate the cells at 35-37°C for 7-10 days in a stationary rack. The growth medium should be replaced on the third or fourth day of culture or when the medium becomes acidic.

Frozen Storage of Primary Cells
1. Suspend the cells at $2-3 \times 10^7$ cells/ml in MEM without serum.
2. Slowly add an equal volume of 2X Freezing Medium.
3. Dispense 1 ml of the suspension into each freezing vial.
4. Slowly freeze the cells (-1°C/minute) as described in the cell freezing section of this chapter.
5. Primary cells can be stored for up to 7 years in liquid nitrogen or up to 3 years at -70°C.

Preparation of Human Foreskin Fibroblast Cultures

Human foreskin fibroblast (HFF) cells can be produced via the dissociation methods described above or by the tissue explant method below. Because HFF cells can be passaged several times without a significant loss of virus susceptibility, most laboratories do not use HFF cells as primary (passage 1) cultures. Instead, these cells are passaged two or three times to expand their number and then frozen for future use. Although HFF cells will continue to grow for about 50 passages, the virus susceptibility often wains after passage 30. When used judiciously, a single human foreskin can produce enough cells for several years.

1. Collect and prepare the tissue as described above.
2. Place the foreskin into a sterile petri dish containing 6-10 ml of serum-free MEM or HBSS.
3. Mince the tissue with scissors to produce 1 x 1 mm pieces. Mincing too much (i.e., making pieces that are less than 1 x 1 mm can cause a preponderance of epithelioid cells after 3-4 weeks in culture.
4. Wet the bottom (monolayer side) of three 75 cm^2 flasks with growth medium.

5. Remove the pieces of foreskin with sterile forceps or a large bore pipette and plant them in the bottom of the flasks.
6. Let the tissue stand for 30-40 minutes to allow the explants to adhere to the plastic.
7. Carefully add 2-4 ml of fresh DMEM containing high glucose, 10% FCS, 10 mM HEPES, and L-glutamine (growth medium) to each flask. Take care not to dislodge the explants.
8. Incubate the flasks at 35-37°C for 48 hours. Do not disturb the flasks.
9. Add 2 ml of fresh medium on day 3, 5, 7, and 9.
10. On day 11 (or earlier if the explants appear to be firmly attached to the flask), remove the culture medium and add 10 ml of fresh growth medium. Incubate as before.
11. Remove the spent medium every 3-4 days and add 10 ml of fresh growth medium.
12. Cell monolayers will be established within 3-4 weeks.
13. Once the monolayers are established, trypsinize the monolayers, pool the cells, and inoculate 10 new flasks (or five 150 cm^2 flasks).
14. When the monolayers are established in these flasks, split two 75 cm^2 flasks to produce 8 new flasks (or use 1 150 cm^2 flask to produce 4 new 150 cm^2 flasks). Freeze the remaining cells (passage 2) as seed stocks.
15. Split the 8 flasks to produce 16 150 cm^2 flasks of passage 3 cells.
16. When the monolayers are confluent, trypsinize the cells and freeze most of the cells as passage 4 working stock. The remainder of the cells can be passaged and used for virus isolations after checking for mycoplasma contamination.

TROUBLESHOOTING GUIDE FOR CELL CULTURES

A number of problems can occur during cell culture growth and maintenance. Some of the problems that can be resolved with early detection are listed below.

Rounding of Cells

Cell rounding can occur for a number of reasons including excessively high or low temperatures, inadequate volume of culture medium to cover the cells, contamination, toxic chemicals, or the introduction of cold medium or trypsin.

Normal-Shaped Cells Sloughing from the Vessel

Excessive heat or cold can cause the cells to slough from the vessel. This type of problem can occur during transportation of commercially purchased cells. Toxic material in clinical specimens and chemical residues from latex gloves can also cause the cells to slough from the vessel.

Contamination

Contamination can occur at two levels; in all cultures that were passed at same time or in single cultures inoculated with contaminated specimens. Contaminated patient isolation cultures should be centrifuged to remove cell debris, filtered with a 0.45 μm filter, and reinoculated into new cultures. Cultures contaminated during subcultivation should be discarded and the reagents, media, tubes, vials, and media supplements should be checked for contamination. Trying to save cultures by antibiotic/antimycotic treatments are generally unsuccessful and are not recommended. In addition, treated cultures may not behave as expected after treatment.

Piling and Rounding of Cells

This is a normal phenomenon in cells that do not possess contact inhibition. However, piling and rounding of cells that normally do not grow in mounds is cause for concern. This may indicate that (a) the culture is changing, (b) the culture is contaminated with another cell line, or (c) toxic chemicals are in the vessel or in the culture medium.

Granular Cytoplasm, Rounded Cells Above the Surface of the Monolayer

This type of reaction often results from alkaline or acidic pH or pollution of the vessel with metabolic waste products. If the cells are not too old, this can be reversed with replacement of medium. As cells grow and reproduce, a certain number of cells will "die" leaving them rounded and above the surface of the cell monolayer. They eventually float from the surface and pose no immediate harm to the cell monolayer.

CPE in Uninoculated Cultures

Primary cell cultures are often contaminated with persistent viruses that can cause CPE when the culture is stressed by low oxygen tension, low pH, increased metabolic waste products, and low nutrient levels. CPE in continuous cell lines can result from contamination of the culture medium, toxic chemicals, and detergent residue in the culture vessel.

REFERENCES

1. Morgan JF, Morton HJ, Parker RC. Nutrition of animal cells in culture. I. Initial studies on a synthetic medium. *Proc Soc Exp Biol Med* 1950;73:1-8.
2. Eagle H. Nutritional needs of mammalian cell cultures. *Science* 1955;122;501-504.
3. Eagle H. Amino acid metabolism in mammalian cell cultures. *Science* 1959;130:432-437.
4. Good NE, Winget GD, Winter W, Connolly TN, Izawa S, Singh RMM. Hydrogen ion buffers for biological research. *Biochemistry* 1966;5:467-477.
5. Burleson FG, Chamber TM, Wiedbrauk DL. *Virology: A laboratory manual.* New York: Academic Press 1992.
6. Scherer WF, Hogasian AC. Preservation at subzero temperatures of mouse fibroblasts (strain L) and human epithelial cells (HeLa). *Proc Soc Exp Biol Med* 1954;87:480-487.
7. Lovelock JE, Bishop MWH. Prevention of freezing damage to living cells by dimethylsulfoxide. *Nature* 1959;183:1394-1395.
8. Stulberg CS, Soule HD, Berman L. Preservation of human epithelial-like and fibroblast-like strains at low temperature. *Proc Soc Exp Biol Med* 1958;98:428-431.
9. Farrant J. The preservation of living cells, tissues and organs at low temperatures: Some underlying principles. *Lab Pract* 1966;15:402-404.
10. Robinson DM. Low-temperature preservation of cells in culture. *Lab Pract* 1966;15:410-412.
11. Kuchler RJ. Biochemical methods in cell culture and virology. Stroudsburg PA: Dowden, Hutchingson, & Ross, Inc. 1977;34-36.

CHAPTER 7

Quality Assurance

INTRODUCTION

In the clinical laboratory, numerous titles such as quality assurance, quality control, quality improvement, continuous quality improvement, and quality engineering have been applied to methods and programs designed to answer two relatively simple questions; "How do I know my tests are working the way they were intended to work?" and "How do I know my results are correct?" A good quality assurance program will answer these questions each time a test is performed. Testing control reagents with each test run is the heart of any quality assurance program. However, if control reagents are the heart of a quality control program, then record keeping is certainly the backbone. Running controls and keeping accurate records of the results are central themes in every quality assurance program. Other important parts of any quality assurance program include measures to assure that tests are always performed the same way (procedure manuals), that personnel who perform the procedures are trained to perform these tests in a safe and appropriate manner, that laboratory equipment is working properly, and that proficiency testing is performed regularly. Taken together, these parts of a quality assurance program serve to confirm that the tests and the results are valid. In addition, a good quality assurance program provides for a safe and efficient workplace.

A quality control program must be designed for and implemented in every laboratory regardless of size. The College of American Pathologists (CAP) recommends that laboratories establish written quality control procedures and that the procedures and quality control records should be reviewed periodically by a designated supervisor (1). These reviews and corrective actions (if any) must be in written form.

A quality control manual must include the written quality control protocols for media and reagents, equipment, testing procedures, and personnel proficiency testing. The manual should be designed to cover all aspects of laboratory activity from the time a specimen is collected until the time a final report is made.

Written Procedures

Ideally, there should be at least two identical sets of written procedures available in the clinical laboratory. The first set is a laboratory copy and is available to the people who are actually performing the test. Because this copy is used at the bench, it is advisable to keep a second copy in the laboratory to be used if the laboratory copy is destroyed or contaminated. A third set of laboratory procedures should be available to the medical staff so that they can better understand how the tests are done and how the tests should be ordered.

All procedure manuals should state the purpose of the procedure, limitations of the test, turnaround time, the schedule of when the test is performed, the type and quantity of specimen required, and instructions for transport and holding of specimens. In addition, the procedure manual should be written in sufficient detail that an inexperienced technologist or student could perform the procedure without additional information.

NCCLS (2) recommends inclusion of the following items in each technical procedure:

1. A summary of the principal of the test.
2. Specimen requirements including patient preparation, if necessary.
3. Materials needed to perform the test.
4. Instrumentation necessary including any

calibration needed.
5. Step-by-step instructions.
6. Calculations.
7. Frequency and tolerance of controls and corrective actions to be taken when the controls are exceeded.
8. Interpretation of results, with notation regarding the expected,"normal" outcome.
9. Limitations of the test.
10. References.
11. Effective date and schedule for review.
12. Author.

Procedure manuals should be designed for ease of use. Each procedure should start on a new page so that changes and updates can be easily made. Procedures should be reviewed annually or whenever a methodology, instrumentation or reagent change is made. Obsolete procedures should be retired but saved for future reference.

National regulatory agencies such as JCAH, CAP and CLIA/Medicare require similar information in a departmental procedure manual. Information required by these agencies include:

A. General requirements:
1. The manual must be written. Copies of textbooks or manufacturer's package inserts are not acceptable.
2. The manual should include only those tests currently being performed in a particular section of the laboratory.
3. Reviews must be made and documented annually with changes dated and approved.
4. The manual must be readily available at the bench.

B. Specimen requirements:
1. Patient preparation.
2. Specimen collection.
3. Specimen storage, handling and transport.
4. Criteria for unacceptable specimens.
5. Notation of specimen source, patient name and physician name.

C. Quality control requirements:
1. The preparation of controls.
2. The expected outcome of the test including the limits of acceptability.
3. The lot number and date of preparation of reagents and controls.
4. Remedial action to be taken with unacceptable results.

D. Procedure requirements:
1. The preparation and storage of reagents, standards and controls.
2. A complete, current methodology.
3. Calibration and linearity if applicable.
4. Alternate procedure for automated tests (not specifically required by CAP).
5. Normal ranges of results.
6. Criteria for handling abnormal/alert values.
7. References.

Specimen Collection and Transport

The ability to isolate viruses from a clinical specimen is directly related to the quality of the specimen provided to the laboratory. Timely collection of an appropriate specimen during the prodromal and acute phase of illness provides the best isolation rates. In general, specimen collection sites should be wiped clean of excessive blood or purulent matter before collecting the specimen. However, specimens should not be collected from sites that have been wiped with disinfectants because disinfectants and lubricants can inactivate a wide range of viruses.

Even with appropriate collection, the viability of the virus can be lost during transportation to the laboratory. Therefore, specimens should be placed in viral transport medium as soon as possible after collection in order to stabilize the virus. If specimens are held for less than one day, they should be stored at 2-8°C. Longer storage periods should take place at or below -70°C. Specimens should not be placed in -20°C freezers or in refrigerators with automatic defrost cycles.

Smears submitted for fluorescent antibody staining should be evaluated for quality and adequacy before staining. Criteria for acceptance of smears for direct staining should include:

1. Smears should be made from an appropriate site for the virus being stained.
2. Smears should only be made on cleaned slides by rolling the swab across a designated area. the designated area should be of reasonable size, 5-15 mm, to avoid wasteful use of reagents and excessive examination time.
3. Blood, debris and purulent matter should be minimal as it can cause evaluation and interpretation problems.
4. The smear should contain an adequate number of cells, approximately 50-100 total cells or 5 cells per low power field.

Rejection of a smear, or other specimen, should be based on written procedures and the reason for rejection should be communicated to the clinician. If events warrant testing of an inappropriate specimen or smear, this information should be included in the written report. For example, fluorescent staining of nasopharyngeal smears may produce positive cells when fewer than 50 cells are present on the slide. A positive result may be generated on this specimen. However, "negative" specimens containing fewer than 50 cells does not indicate the absence of infection.

Reagents

Stains and chemical reagents must have the date received and the date opened noted on the container. All reagents should be tested for effectiveness when they are prepared or purchased. Stains that are diluted into working strength and used on a daily basis should be checked weekly for correct performance. Reagents which are known to produce variable results should be tested more frequently. Stains and reagents should be stored as indicated by the manufacturer with expiry dating noted on each component of the kit. Expired reagents, especially those that are expensive, may be used past the stated expiration date as long as the performance is documented and performance checks are made with each use. A written protocol must be established for the testing and usage of any outdated reagents. Reagents and stains should be tested against control materials to assure that they function correctly.

A listing of reagent expiry dates and storage conditions (Table 1) should be kept in a quality assurance manual. Expiry dating should not exceed the manufacturer's dating unless a written protocol is established. Incoming reagents must be inspected upon receipt in the laboratory for evidence of appropriate shipping and handling. For example, boxes containing frozen cell lines that were shipped frozen on dry ice should still contain dry ice when the container arrives in the laboratory. Thawed vials should not be accepted under these circumstances and the condition of the product on arrival, quantity, source, lot number, expiration and receipt dates should be documented.

Cell Cultures

Many cell cultures can be propagated within a clinical virology laboratory. However, strict quality control measures should be followed because changes in the culture conditions, reagents, and increasing passage levels can cause variations in the sensitivity to virus infection. Information available for each batch of cells used should include the source of the cells, the individual cell line, the origin of the cell line, the morphology and the current passage number. The following documentation should be available for all cells grown in the laboratory:

1. Current passage number.
2. Cell yields.
3. Time to confluence.
4. Comments of the cell condition.
5. Results of sterility, toxicity and mycoplasma checks.

A quality assurance program for cell cultures must include assays for contamination, cell growth, and virus sensitivity. All cells lines and culture media should be tested regularly for bacterial and fungal contamination. Contaminated cultures should be discarded. Whenever possible, cell culture stocks should be propagated in antibiotic-free media to prevent passage of low-level

Table 1. Reagent storage and outdate guide.

Reagent	Storage Conditions	Expiration Dating
Alsever Solution	2-8°C	Unopened - 2 years Opened - 3 months
Amphotericin B	-20°C or lower	1 year but not to exceed the expiration dating on the original vial.
Antibiotic Solution (for Chlamydia Transport Medium)	-20°C or lower	1 year but not to exceed the expiration dating on the original vials.
Antibiotic Solution (for Viral Transport Medium)	-20°C or lower	1 year but not to exceed the expiration dating on the original vials.
ATV Trypsin	-20°C or lower	1 year.
Cell Freezing Medium	2-8°C	3 months.
Cell Growth Medium	2-8°C	2 months.
Cell Maintenance/Refeed Medium (RMK Cells)	2-8°C	2 months.
Cell Maintenance/Refeed Medium (Virus Isolations)	2-8°C	2 months.
Chlamydia Overlay Medium	2-8°C	2 months.
Chlamydia Transport Medium	2-8°C	1 year.
Cycloheximide Stock Solution	-20°C or lower	1 year but not to exceed the expiration dating on the original container.
Fetal Calf Serum	-20°C or lower 2-8°C	Unopened - 3 years Opened - 3 months.
Ficoll-Hypaque	2-8°C	6 months.
Fluorescent Antibody Reagents	2-8°C	1 year but not to exceed original expiry dating.

Table 1. (Continued) Reagent storage and outdate guide.

Reagent	Storage Conditions	Expiration Dating
Gentamicin Solution	2-8°C	1 year but not to exceed the expiration date on the original vial.
GLB - Gelatin Lactalbumin Broth	Room Temperature 2-8°C	Unopened - 2 years Opened - 1 month.
8.8% Glucose	Room Temperature	2 years.
Guinea Pig RBCs	2-8°C	7 days.
1 M HEPES Buffer	2-8°C	1 year.
1 N HCl (Sterile)	Room Temperature	1 year.
Mounting Fluid, FA	Room Temperature	3 months.
Penicillin/Streptomycin Solution	-20°C or lower	1 year but not to exceed the expiration dating on the original vials.
pH 3 Medium	2-8°C	6 months.
Phosphate Buffered Saline (PBS)	Room Temperature	Unopened - indefinitely Opened - 2 months.
Phytohemagglutinin P Stock Solution	-20°C or lower	2 weeks.
PHA Medium	2-8°C	2 weeks
Phenol Red Stock Solution, 1%	Room Temperature	1 year.
Polybrene, 0.1% Stock Solution	2-8°C	6 months
Propagation Medium (Retrovirus Assay)	2-8°C	2 weeks
Sodium Bicarbonate, 7.5%	Room Temperature 2-8°C	Unopened - 1 year Opened - 1 month.

Table 1. (Continued) Reagent storage and outdate guide.

Reagent	Storage Conditions	Expiration Dating
1 N NaOH (Sterile)	Room Temperature	6 months.
Sorbitol, 70%	2-8°C	2 years.
Trypan Blue, 0.4%	Room Temperature	Unopened - 1 year Opened - 2 months.
Viral Transport Medium (without antibiotics)	2-8°C	1 year.
Viral Transport Medium (with antibiotics)	2-8°C	2 months.
Water, Endotoxin-Free (Sterile)	Room Temperature	Unopened - Use vendor dating Opened - 2 weeks
Water, Endotoxin-Free	Room Temperature	1 week.

contaminants. Mycoplasma testing of all cultures should be done regularly either by Hoechst staining or by culture. Cell line sensitivity to virus infection should be checked periodically or whenever cell yields vary significantly. Sensitivity testing is usually done via $TCID_{50}$ experiments using standard virus inocula. It is critical to use relatively low passage virus isolates as challenges. ATCC cultures are not recommended because these viruses are culture adapted and may not accurately represent the viruses isolated in individual laboratories.

When ready-to-use cell cultures are purchased for patient isolations, a vendor certificate that the cells are free of mycoplasma, fungal, and bacterial contamination can be accepted in lieu of laboratory testing. However, this certificate does not guarantee that the cells that arrive in your laboratory are free of contamination. Upon arrival in the laboratory, cultures should be visually inspected for gross contamination and one culture should be held at 35-37°C (contamination control culture) for as long as that lot of cells are in the laboratory. Primary monkey cells should be checked for the presence of foamy agent or simian virus. Each lot of purchased cells should be challenged with laboratory isolates (not ATCC cultures) virus to confirm that the cells are sensitive to infection. Because all cell vendors have occasional contamination, quality, and/or shipping problems, it is prudent to have a second commercial source of cell cultures. In addition, some laboratories split cell shipments between two suppliers. While this is a prudent and often necessary practice, multiple vendors for a single cell type will multiply the QA documentation and workload.

To prevent or minimize the potential for contamination of cell stocks, propagation and contact should be made by a minimal number of personnel. Separate laboratory apparel, reagents and glassware should be used for cell culturing and work should take place in a clean environment preferably in a laminar flow hood. All media used should be checked for sterility prior to use. In addition, new lots of media and serum should be checked for their ability to support cell growth. Because of the inherent lot to lot variability of serum, it is advisable to run checks on a specific serum lot and once approved, the laboratory should purchase a large serum stock. This will keep quality assurance checks to a minimum as well as ensuring a stock of quality serum.

Instrumentation and Facilities

A good quality assurance program will help to provide a safe work environment where biohazards, chemical hygiene, electrical safety, and fire control are well controlled. The laboratory must have sufficient space so that the quality of work, safety of personnel and the appropriate quality control measures are not compromised. Ideally, the virology laboratory should be separated from other laboratory areas and set up so that dust accumulation is minimized and routine disinfection and cleaning can be easily performed. Personal protection devices such as pipetting aids, vinyl gloves, masks and lab coats should be provided and should not leave the laboratory. Bench tops, telephones and routinely used equipment should be cleaned and disinfected daily.

Instruments should be operated and maintained as specified by the manufacturer. All instruments should be checked regularly to assure that they perform as expected. Preventive maintenance measures outlined by the manufacturer should be performed and documented when completed. Service manuals should be available to the equipment user. Equipment maintenance logs should be kept for each instrument and should include the following data:

1. Instrument name, serial number, date of installation and original supplier.
2. Performance checks or maintenance protocols.
3. Performance ranges.
4. Date work is done and name of technologist performing it.
5. Remedial action taken.

Routine quality control checks should be outlined in the procedure manual. The following is a guide to routine instrument monitoring and maintenance. In the event of gross contamination or instrument failure, immediate action should be taken to correct the situation. Discrepancies in temperature levels should be corrected and more frequent monitoring should be done until temperatures become within the appropriate range. Inspections for electrical hazards, such as frayed cords, should be made with each use.

Incubators

Temperatures, CO_2, and humidity should be checked daily before the incubator is opened. Alarm systems should be tested periodically to assure that they are still working. Decontamination of the interior with disinfectant should be performed weekly. Weekly check of CO_2 levels should be performed using a Fyrite indicator to confirm instrument readouts. Daily confirmation of instrument readouts can be made through observation of phenol indicators within cell culture medium.

Laminar Airflow Hoods

Working surfaces and protective front glass should be cleaned and disinfected before and after each use. An ultraviolet light should be used to decontaminate the work area each day. Magnehelic gauges should be checked daily and the numbers recorded. Hoods should be inspected for air velocity and filter integrity by a certified inspector at least annually and whenever the hood is moved.

Microscopes

Objectives and stages should be cleaned daily and microscopes should be checked and cleaned every 6-12 months by a qualified microscopy company. Mercury vapor lamps should be monitored for lamp usage and bulbs should be changed within the manufacturer's guidelines.

Refrigerators and Freezers

Temperatures should be checked and recorded daily. Fluid levels should be checked annually. Freezers should be defrosted periodically.

Water Baths
Temperature and water levels should be checked daily. Water baths should be cleaned and disinfected on a monthly basis.

Centrifuges
Cleaning should be performed on a monthly basis and speed calibrations done at least annually.

Autoclave
Temperature, and pressure checks should be made daily and spore strip testing is performed weekly.

Countertops and Telephones
Countertops and telephones should be cleaned and disinfected daily.

Tissue Culture Rotators
RPM check should be done each month or whenever the rotators are moved.

Spectrophotometer
Absorbance and linearity checks should be done with each run.

pH Meters
Single reference buffer check should be done before each use and multiple point checks should be done at least once each month.

Mechanical Pipetting Devices
Mechanical pipetters should be checked for accuracy at least once each year.

Thermometers
Thermometers should be calibrated against a standard NBS thermometer before being placed into use. Thermometers used in refrigerators, freezers, water baths, and incubators should be immersed in glycerol or in a large volume of water.

Proficiency Testing
Proficiency testing takes place at two levels. Within a laboratory the routine use of controls serves as a form of proficiency testing when controls are treated as patient specimens. Some hospitals will introduce "blind" specimen controls in order to document the performance and competency of a laboratory or an individual technologist. Individual personnel competency and performance can also be improved by participation in continuing education programs.

Several organizations provide proficiency testing reagents which can be used for laboratory certification and to comply with local regulations. The College of American Pathologists (CAP), the Centers for Disease Control, and various state, county and city health departments are involved in providing "blind" samples to laboratories and then compiling results. These external tests not only serve to document a laboratory performance, but they serve as an indicator of commercial kit or reagent performance. CAP surveys provide information about the performance of various test kits allowing the laboratory to determine if a specific product produces erroneous results.

Statistical documentation of isolation rates can not only show seasonal trends but they can serve as an indicator of testing proficiency and quality. Normal seasonal fluctuations are expected in the isolation of different viruses. However, certain indicator viruses (i.e., sexually transmitted agents) should be consistently isolated without major fluctuations in prevalence. Changes in prevalence may indicate that culture and detection processes may have changed.

All laboratory personnel who perform a specific procedure should participate in the proficiency testing for that procedure. Proficiency testing should include both full-time and part-time personnel as well as personnel on all shifts.

Experienced and dedicated personnel are the key to quality testing. Personnel training, a well-written procedure manual, proficiency testing, and continuing education of all laboratory personnel help to improve the quality of work and patient care.

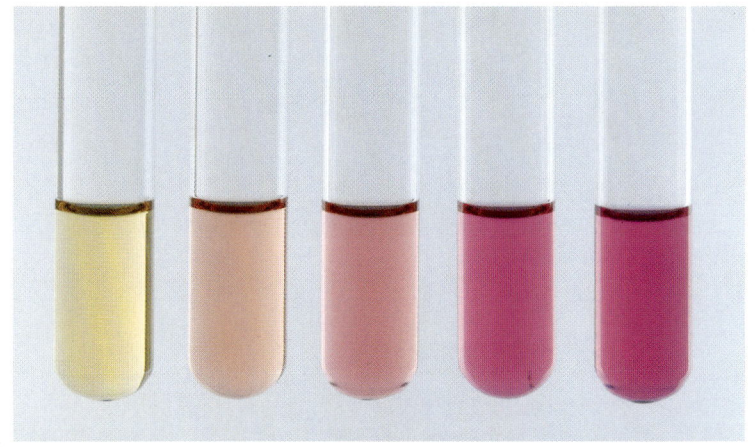

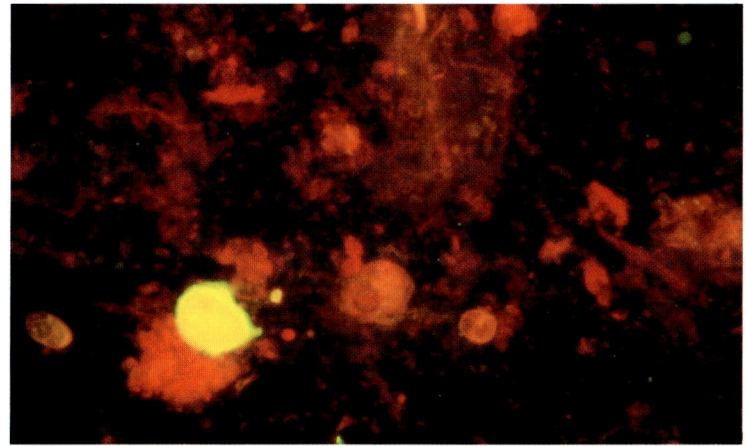

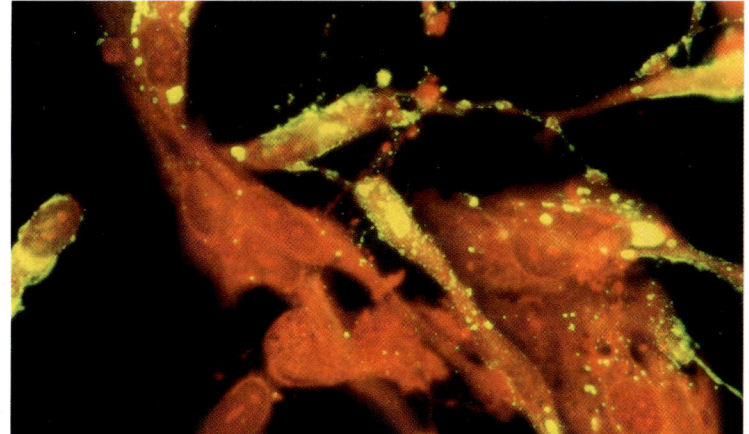

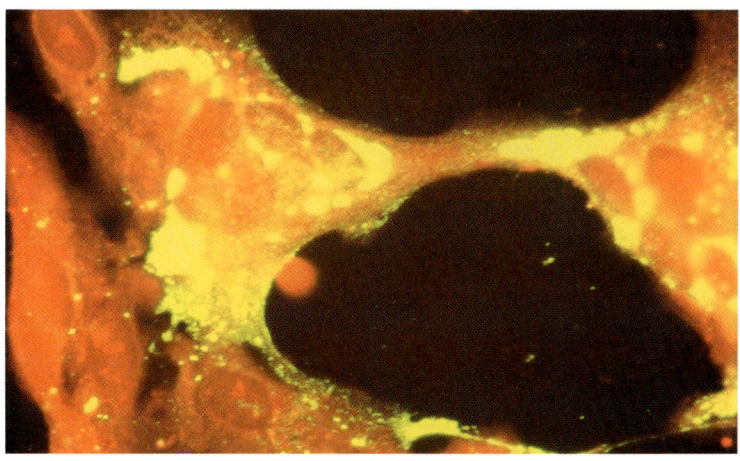

Color Plate 1. A: Medium coloration at various pH levels. From left to right, pH 6.0, 6.5, 7.0, 7.5 and 8.0. **B:** Direct fluorescent antibody staining of a broncheoalveolar lavage specimen with bivalent antibody to herpes simplex virus. **C:** Fluorescent antibody staining of a measles virus-infected A549 shell vial culture. **D:** Magnified view of measles-virus infected A549 shell vial culture stained with monoclonal antibody. Notice the multinucleated giant cells and the fluorescent cytoplasmic inclusions.

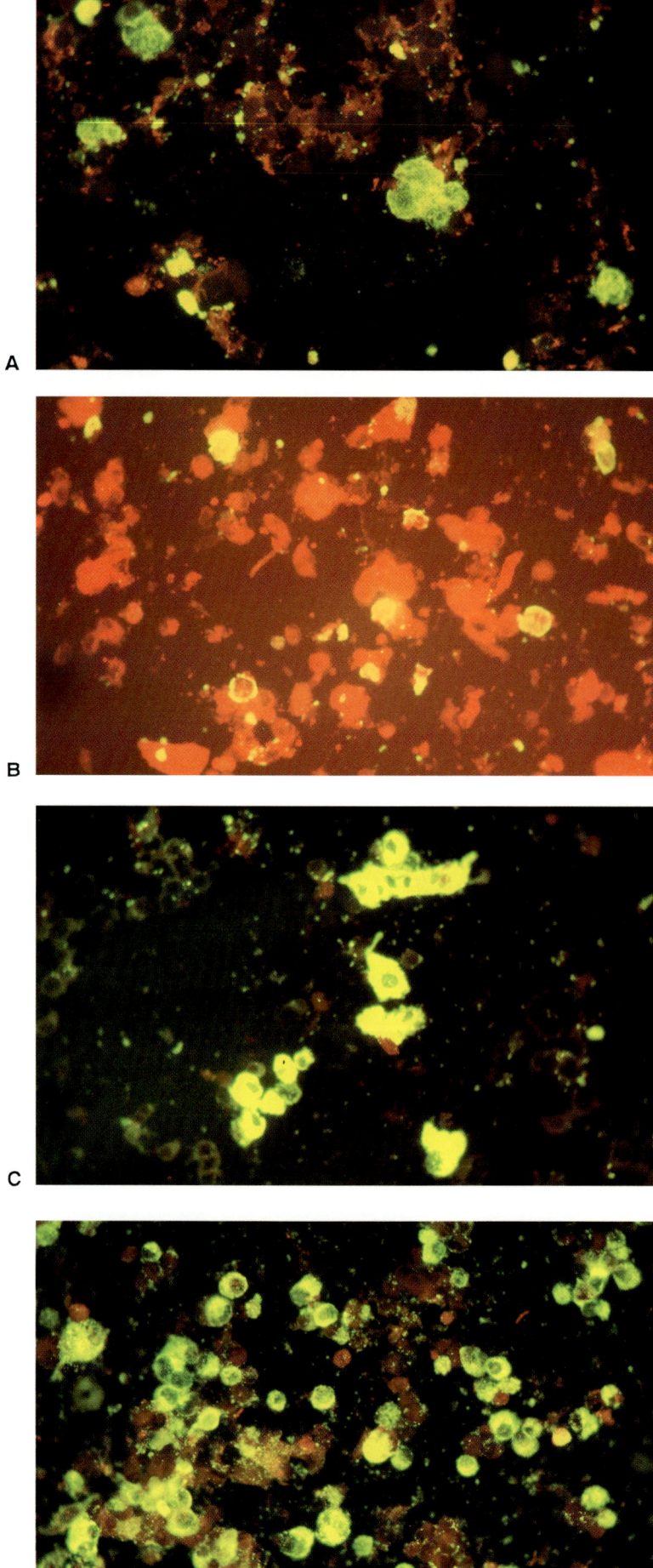

Color Plate 2. A: Fluorescent antibody staining of HSV-1 infected CV-1 cells. Cells were scraped from their tube and stained with bivalent antibody. **B:** Direct fluorescent antibody staining of varicella-zoster virus-infected HFF cells. Cells were scraped from their tube and stained with monoclonal antibody. **C:** Fluorescent antibody staining of rotavirus-infected CV-1 cells. Cells were scraped from their tube and stained with polyclonal antibody. Note the cytoplasmic fluorescence and nuclear sparing. **D:** Direct fluorescent antibody staining of respiratory syncytial virus-infected HEp-2 cells. Cells were scraped from their tube and stained with monoclonal antibody.

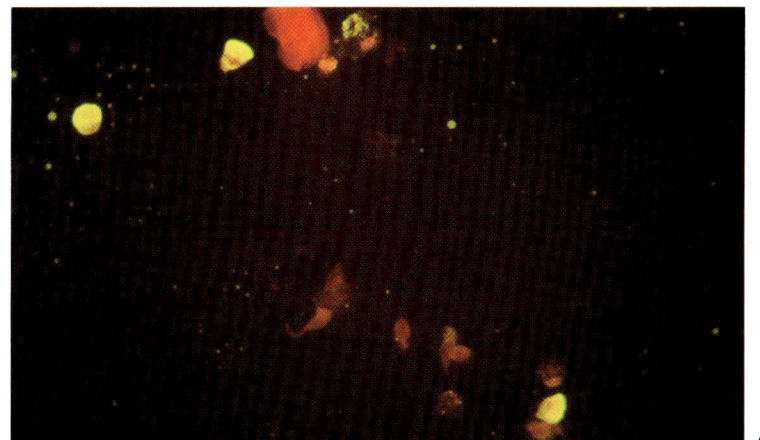

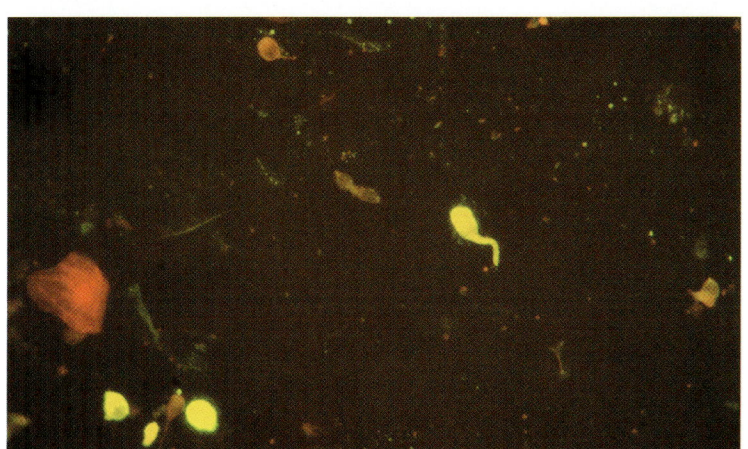

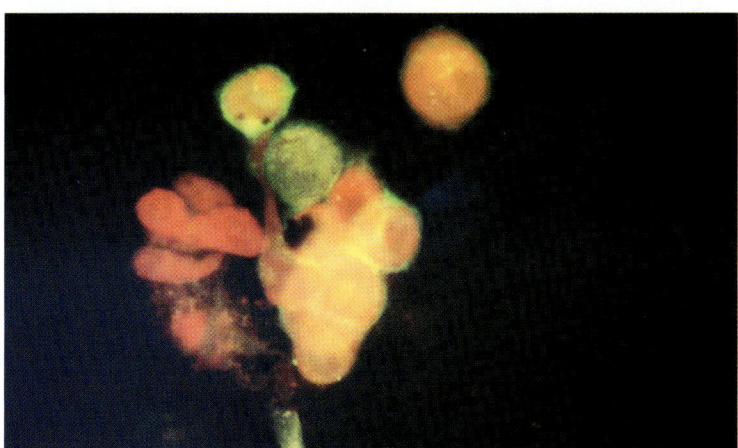

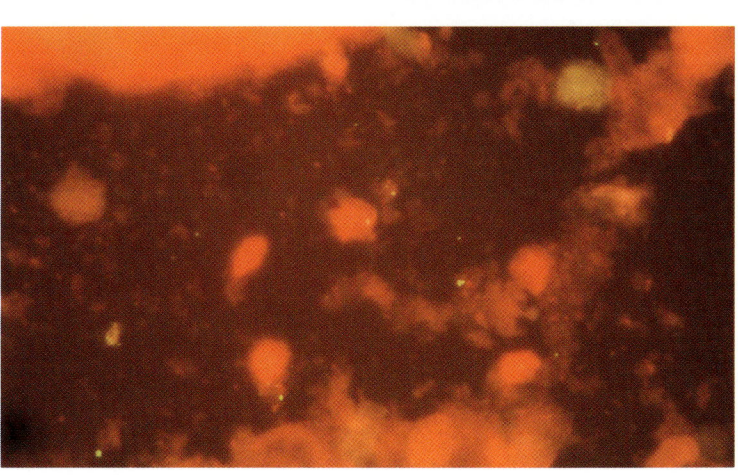

Color Plate 3. A: Direct fluorescent antibody staining of a nasopharyngeal cell smear with monoclonal antibodies to RSV. Note the fluorescing respiratory epithelial cells. **B:** Direct fluorescent antibody staining of a nasopharyngeal cell smear with monoclonal antibodies to RSV. Note the fluorescing goblet cells. **C:** Direct fluorescent antibody staining of a vesicular cell smear with monoclonal antibodies to varicella-zoster virus. **D:** Direct fluorescent antibody staining of a cervical cell smear with antibodies to *Chlamydia trachomatis*. Note the tiny fluorescent elementary bodies against the red cellular background.

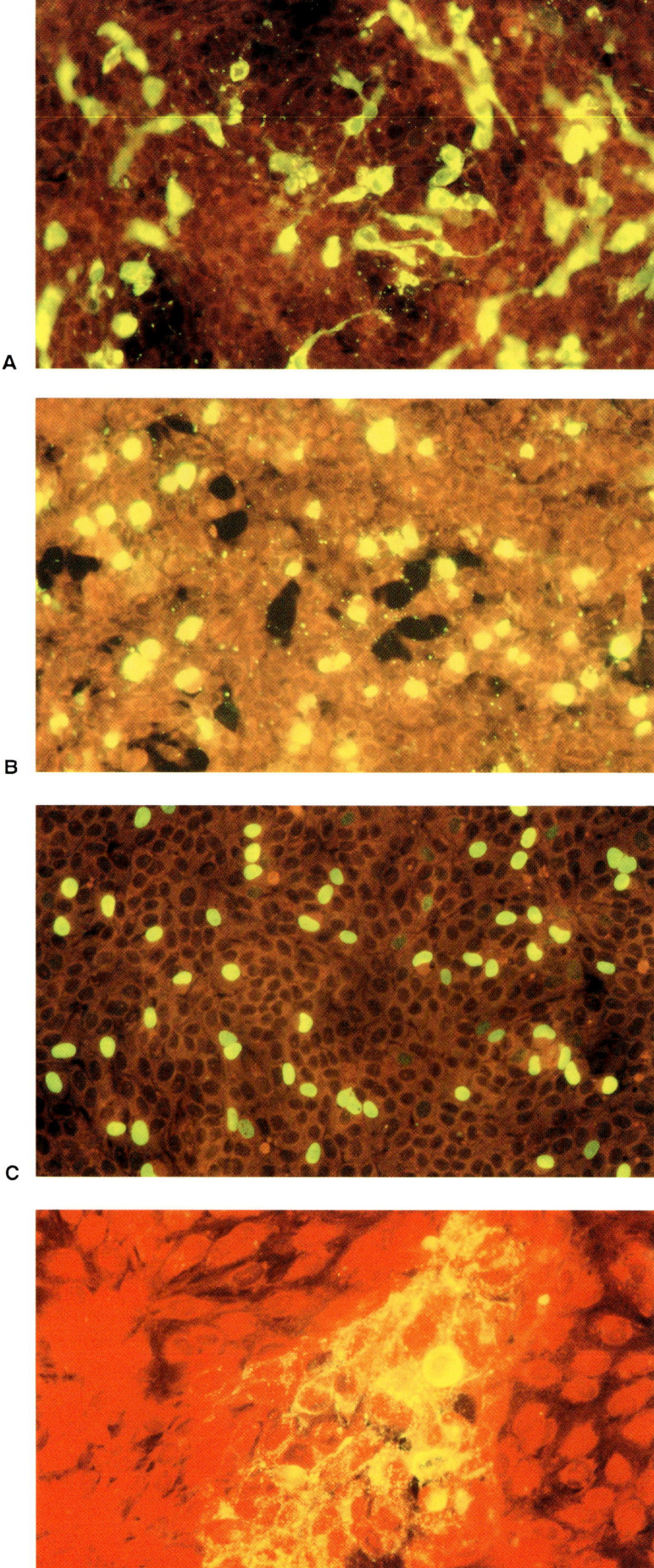

Color Plate 4. A: Fluorescent antibody staining of a rotavirus-infected Vero shell vial culture 16 hours after inoculation. **B:** Direct fluorescent antibody staining of an adenovirus-infected A549 shell vial culture 16 hours after inoculation. **C:** Fluorescent antibody staining of a cytomegalovirus-infected mink lung shell vial culture 24 hours after infection. The nuclei of infected cells are stained by the CMV early antigen reagents. **D:** Direct fluorescent antibody staining of a varicella-zoster virus-infected CV-1 shell vial culture 48 hours after inoculation.

REFERENCES

1. Inspection Checklist - Microbiology (Bacteriology, Mycobacteriology, Mycology, Parasitology, and Virology). Section IV. Northfield IL: College of American Pathologists 1991;53-59.
2. National Committee for Clinical Laboratory Standards. Clinical Laboratory Procedure Manuals 1984;4(2):27-49.
3. Warford, Ann. Troubleshooting in the Clinical Virology Laboratory. *Clin Microbiol Newsletter* 1990;12:41-44.
4. Weissfeld AS, August MJ, Hindler JA, Huber TW, Sewell DL. Quality control and quality assurance practices in clinical microbiology. In: *CUMITECH 3A*, American Society for Microbiology, Washington, D.C. 1990.

CHAPTER 8

Adenoviruses

INTRODUCTION

Human adenoviruses are nonenveloped viruses belonging to the family *Adenoviridae*. Adenoviruses are 70-90 nm in diameter and have a distinctive icosahedral morphology with a fiber structure extending from each of the vertices. The length of the fiber varies with the adenovirus serotype and the distal end or knob of the fiber contains the receptor-binding determinants of the virus. The adenovirus genome consists of a 35.9 kilobase pair linear DNA molecule that codes for 11-15 polypeptides. Ten of these polypeptides are incorporated into the adenovirus virion. DNA replication, transcription, and maturation of adenovirus virions occurs in the nucleus of infected cells. Adenoviruses are relatively stable and are resistant to detergents, organic solvents, low pH, and many proteolytic enzymes.

Adenoviruses were discovered in the 1950's and since that time, 47 human adenovirus serotypes have been identified (1). Adenovirus infections are found throughout the world and primary infection usually takes place in the first few years of life. Because infections produce type-specific immunity, multiple adenovirus infections are common and can occur throughout the lifetime of the individual. Adenovirus infections cause a variety of symptoms ranging from keratoconjunctivitis to respiratory disease, to meningitis. Although upper respiratory and pharyngoconjunctival syndromes are most commonly associated with adenovirus infections, gastrointestinal and urinary tract diseases have been extensively documented (2-5).

Upper respiratory disease caused by adenovirus types 1-7 includes acute febrile pharyngitis, pharyngoconjunctival fever, acute respiratory disease, pneumonia, common colds and tonsillitis. These mild infections are generally seen in infants and young children. The virus can be isolated from the oropharynx during the first few days of the illness and subsequently from the stool for many months following recovery.

Lower respiratory illness caused by types 3, 4, 7, and 21 can cause bronchitis, bronchiolitis, and pneumonia. Type 5 infections are associated with a coughing syndrome which is similar to pertussis. Types 4 and 7 have been associated with outbreaks in military recruits causing fever, pharyngitis, myalgia, headache and coughing (2,4). Transmission is thought to be primarily through infected aerosol inhalation with an incubation period of 5-6 days (4).

Ocular adenovirus infections include syndromes such as epidemic keratoconjunctivitis and waterborne "swimming pool" conjunctivitis. These are typically caused by types 8, 19, or 37. The incubation period is 8-10 days resulting in 1-2 weeks of follicular conjunctivitis, headache, malaise and mild upper respiratory symptoms. Subepithelial corneal keratitis following an infection can interfere with the patient's vision for many months.

Herpes-like genital lesions have been associated with adenovirus type 2, 19, and 37. While these lesions can be very similar in nature to the ones of herpes simplex virus, adenovirus lesions often are accompanied by orchitis, cervicitis or urethritis.

Gastroenteritis caused by types 40 and 41 is a common syndrome of infants and young children and is thought to cause 5-20% of pediatric hospitalizations due to diarrhea (6). Type 40 infections are generally seen in patients less than 12 months of age and type 41 infections in patients older than 12 months. Diarrhea is the major symptom and infections are not typically associated with any respiratory symptoms. Symptoms last for 2-28 days with fever and vomiting of a shorter duration. The incubation period is 8-10 days (7).

AT A GLANCE...

ADENOVIRUS

Virus Detection Methods
Tube cultures of human epithelial, A549, Graham 293, HeLa, Hep-2, KB, or HEK cells.
Centrifugation-enhanced (shell vial) cultures using A549 cells.
Direct fluorescent antibody staining of nasopharyngeal cell smears.
EIA for adenovirus types 40 and 41 is commercially available.

Specimen Source
Specimen of Choice - Nasopharyngeal swabs, washes or aspirates; conjunctival swabs.
Cerebrospinal Fluid - Adenovirus can be isolated from CSF in cases of meningitis.
Urine - Virus will be present in the urine especially when there is renal involvement.
Conjunctiva - Virus can be isolated from conjunctival swabs when the patient is symptomatic.
Rectal Swabs or Stool - Virus may persist for up to 18 months following infection.

Time to Result
Standard Culture: Positive Culture - 2 to 7 days. Negative Culture - 3-4 weeks.
Shell Vial Method: Positive Culture - 2-5 days. Negative Culture - 5 days.
Direct Fluorescent Antibody Stain - 1-2 hours.

Epidemiology
Adenovirus is endemic throughout the year and causes local outbreaks of respiratory disease in spring and winter. Peak incidence occurs during the school year. Closely housed groups of people are at increased risk of acquiring adenovirus infections. "Swimming-pool" conjunctivitis usually occurs in the summer.

Transmission/Incubation Period
Adenovirus is spread directly by oral contact and inhalation of infectious aerosols. The virus can be spread indirectly by hands, soiled handkerchiefs and tissues, and soiled eating utensils. Virus shed in the feces may also be involved in disease transmission. The average incubation period is 5-6 days and patients are infectious from shortly before the onset of symptoms and through the entire course of the infection. Some patients will shed infectious virus for up to 18 months.

Inactivators and Disinfectants
Adenoviruses are relatively stable and they are resistant to detergents, organic solvents, low pH conditions, and many proteolytic enzymes. Effective disinfectants include phenols, formalin solutions, and 10% household bleach solutions.

Unlike the other adenoviruses, types 40 and 41 are fastidious in nature making culturing impractical for most laboratories.

Adenoviruses are highly transmissible through contact with aerosols and contaminated fomites or through the fecal-oral route. The best documented cases of transmission are within close populations such as families, military recruit populations and resident institutions and schools. Patients are infectious a few days before the onset of symptoms and continue to shed virus in the stool for weeks following recovery.

Diagnosis of adenoviral infection is often based upon clinical presentation alone. However, a definitive diagnosis requires either visualization of virus by electron microscopy, isolation of virus in cell culture, demonstration of adenovirus infected cells by direct fluorescent antibody methods, or demonstration of seroconversion or a four-fold rise in adenovirus antibody levels during the course of disease. A high complement fixation titer in a single serum specimen is suggestive of recent infection but definitive diagnosis requires demonstration of seroconversion. Complement fixation tests have been used most frequently for demonstrating seroconversion and significant rises in antibody titers. However, enzyme immunoassays have proven to be superior to complement fixation for demonstrating antibody responses in children (8).

Genus-specific diagnosis can be obtained using enzyme immunoassay or complement fixation tests. However, serotype-specific diagnoses require the use of hemagglutination inhibition or preferably, virus neutralization tests. Regardless of the test used, false negative results can occur due to persistent infections or poor serologic responses in infants and children. False positive results can occur due to heterotypic amnestic responses (9).

VIRUS ISOLATION

Introduction

Most adenoviruses can be readily isolated in a number of cell lines including HeLa, KB, HEp-2, A549, HEK, and Graham 293 cells. Some adenoviruses will grow in HFF cultures however, HFF cultures are not recommended for adenovirus isolation (10). Characteristic adenovirus CPE is generally visible within 2-7 days of inoculation.

Adenovirus types 40 and 41 are very fastidious and isolation of these viruses is uncertain and difficult. Of the cell lines mentioned above, the Graham 293 cells provide the best, albeit inconsistent, isolation frequencies. These isolation difficulties have caused manufacturers to introduce EIA systems for adenovirus type 40 and 41 detection.

Culture confirmation is usually accomplished by fluorescent antibody staining (11,12). Two types of adenovirus monoclonal antibodies are available. Antibodies to the hexon subunit are genus specific and react with all human adenoviruses including types 40 and 41. While the hexon-specific antibodies are suitable for screening cultures and identifying the virus species, they cannot distinguish the individual adenovirus types. Type-specific and neutralizing antibodies and restriction endonuclease digestion are used to establish adenovirus serotypes and genotypes, respectively.

Adenoviruses can produce chronic and latent infections and asymptomatic virus shedding is well documented. Therefore, adenovirus isolations from stool or oropharyngeal specimens may not provide definitive proof that adenovirus is responsible for the clinical symptoms. Isolations from CSF, skin lesions, and conjunctival scrapings provide more definitive evidence of disease.

Specimen Collection and Storage

Adenovirus can be isolated from nasopharyngeal swabs, washes or aspirates, eye or conjunctival swabs, urine, urethral or cervical swabs and tissue and specimens should be collected as soon as possible after the onset of symptoms. Rectal swabs or stool specimens are appropriate during later phases of systemic disease. However adenoviruses are excreted in the stool for up to 18 months after systemic infection. Therefore, adenovirus isolation from stool does not provide definitive proof that adenovirus is the cause of the symptoms.

Specimens should be placed into viral transport medium immediately after collection and held on wet ice (2-8°C) until they are transported to the laboratory. If inoculation into cell culture is delayed by more than 72 hours, specimens should be rapidly frozen on dry ice and stored at -70°C. **DO NOT FREEZE SPECIMENS AT -20°C.** Storage at -20°C and repeated freeze/thaw cycles will destroy viral infectivity.

Specimen Preparation
Swabs
1. Vortex the specimen with several glass beads for 20-30 seconds to release any bound cells or virus.
2. Remove the swabs from the transport medium and firmly roll them against the inside of the tube to remove as much fluid as possible.

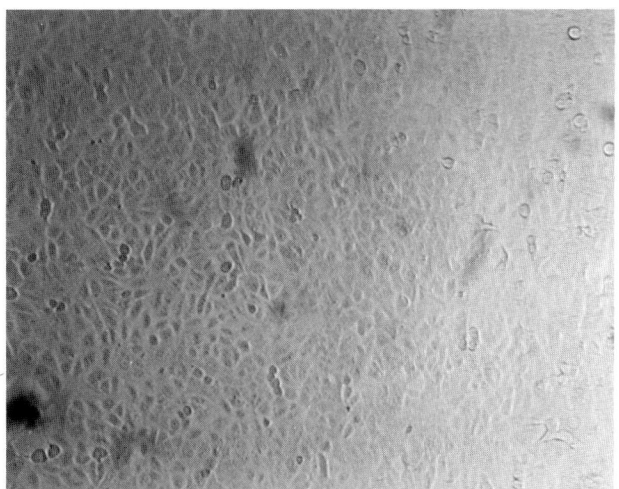 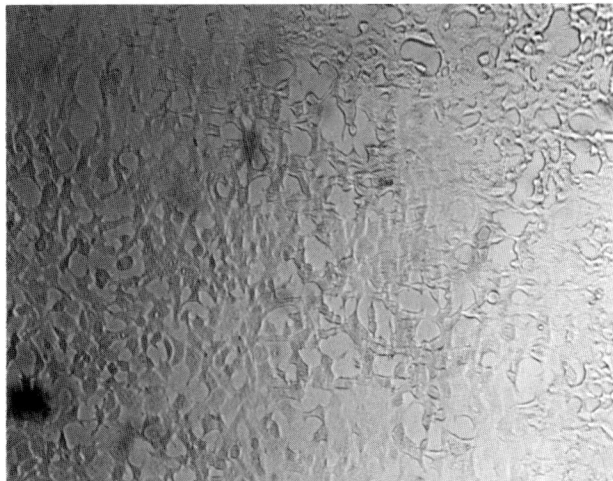

FIG. 1. Uninfected A549 cells (left) and early adenovirus CPE in A549 cells (right).

Wash and Aspirate Specimens
1. Vortex the specimen with several glass beads for 20-30 seconds.
2. Centrifuge the cell suspension at 200 x g for 5 minutes to remove any mucus.
3. Carefully remove the mucus and use the remainder of the specimen for inoculation of cell cultures.

Stool
1. Combine approximately 0.5 gram of stool in 5-10 ml of viral transport medium.
2. Add 2-3 glass beads to the tube and vortex the specimen vigorously to extract as much virus as possible.
3. Centrifuge the suspension at 1000-3000 x g for 10-15 minutes.
4. Filter the supernatant fluids through a 0.45 μm syringe filter.

Standard Tube Culture
Procedure
1. Appropriately label 2 tubes containing freshly confluent A549 cells.
2. Remove the medium from the cells.
3. Add 2 ml of maintenance medium containing 2% FCS and antibiotics to each tube.
4. Add 0.2 to 0.5 ml of specimen to each tube.
5. Place the tubes in a roller drum and incubate at 35-37°C and 10-15 rph.
6. Examine the cells every other day for CPE (Fig. 1) for 14 days. During this time the culture medium should be replaced every third day or whenever it becomes acidic (yellow). Adenovirus CPE is typically present within 2-7 days of inoculation. Some laboratories will perform blind passages after 14 days to detect enteric adenoviruses or any adenovirus present in low concentrations. Cell cultures that become toxic should be reinoculated into fresh cultures.

Interpretation of Results

Adenovirus CPE is characterized by the presence of large, rounded pycnotic cells (Fig. 1). Infected cells will eventually aggregate into rounded cell sheets which eventually detach from the culture tube. Extensive CPE is often accompanied by increased acidity in the culture. No report should be made based upon CPE alone.

Culture Confirmation
1. Remove all but a few drops of medium from the cell culture tube.
2. Scrape the cells from the tube surface.
3. Resuspend the cells in the remaining cell culture medium.
4. Using a pipet, place a small drop of the cell

suspension onto an acetone-cleaned slide.
5. The slide should be processed as directed by the manufacturer of the fluorescein-labelled antibody to adenovirus or as described below.
6. Allow the slide to air dry, then fix it in cold acetone for 10-15 min. Fixed slides may be stored for up to a week at 2-8°C or for one year at -20°C under desiccated conditions.
7. Place enough of the appropriate fluorescein-labelled monoclonal antibody on the slide to cover the cell smear (15-30 µl).
9. Incubate the slide for 15-30 minutes at 35-37°C in a covered, humidified chamber. Do not allow the antibody to dry on the slide as this could cause nonspecific staining.
10. Remove the excess antibody with a gentle stream of PBS. Do not direct the stream directly at the cell smear as this could dislodge the cells.
11. Soak the slide in PBS for 5 minutes, shake off the excess fluid, and allow the slide to air dry.
12. Carefully add a small drop of FA mounting fluid to the center of each well.
13. Place a number 1 coverslip on the mounting fluid and carefully remove all the air bubbles.
14. Immediately examine the slide using a fluorescence microscope. For optimum clarity, use 200-300X magnification for screening and 400X for confirmation of cell morphology.

Interpretation of Results
Positive Test. The presence of adenovirus is indicated by the characteristic CPE and the presence of apple-green fluorescence in the nucleus and cytoplasm of the infected cells on the slide. Only individuals experienced in reading FA reactions should examine the cell smears because cell debris and cell clumps may exhibit a dull fluorescence which can be misinterpreted as a specific, positive reaction.
REPORT: Adenovirus isolated.

Negative Test. The absence of adenovirus virus is indicated by the lack of specific fluorescence in the infected cells. Cultures that fail to produce CPE after 14 days of culture are also considered to be negative for adenovirus.
REPORT: Adenovirus not isolated.

QC Procedures.
Subpassages of recent adenovirus isolates or ATCC cultures should be inoculated into cell culture tubes with each batch of adenovirus isolations. Uninfected cell cultures serve as negative controls. Infected and uninfected cell monolayers should be scraped and stained as described above. Positive controls must exhibit typical CPE and fluorescence patterns. Negative controls must not exhibit CPE or specific fluorescent staining.

Preparation of Positive Control Inocula
1. Add 20 µl of a recent adenovirus isolate (or an ATCC culture) to 2 ml of sterile GLB. Hold the diluted virus on ice until used to inoculate the monolayer.
2. Remove the growth medium from one 75 cm^2 flask of 80-90% confluent A549 cells.
3. Rinse the monolayer twice with 10 ml of serum-free MEM.
4. Add 2 ml of the diluted virus to the monolayer and allow the virus to adsorb for 1-2 hours at 35-37°C. Note: The flask should be rocked every 15 minutes to assure even virus distribution and to prevent monolayer desiccation.
5. Add 20 ml of MEM containing 10% FCS.
6. Incubate the flask at 35-37°C until CPE involves 80-100% of the monolayer.
7. Scrape the cells from the flask into the cell culture medium.
8. Transfer the medium to a 50 ml polypropylene centrifuge tube and freeze at -70°C.
9. Thaw the medium quickly in a 37°C water bath.
10. Centrifuge the tube at 600-800 x g for 15 minutes to pellet the cell debris.
11. Transfer the supernatant fluids to a sterile vessel containing 80 ml of sterile GLB.

12. Dispense 1.0 ml of the diluted supernatant fluids into each of 200 freezing vials and freeze the vials at or below -70°C. Store the vials in liquid nitrogen for up to 5 years or at -70°C for up to 2 years.

Centrifugation Culture (Shell Vial) Method

Shell vial cultures can be used to identify human adenoviruses in clinical specimens as early as 48 hours after infection. Although adenoviruses can be isolated in RMK, Vero, HEp-2, and HFF shell vial cultures, A549 cells are usually preferred. Although rapid and specific, shell vial cultures are not as sensitive as conventional tube cultures (11,12). Compared with standard tube culture, shell vial cultures possess 77% sensitivity after 2 days of culture and 100% sensitivity after 5 days of culture (11,12). Therefore, an A549 tube culture should be inoculated for each set of shell vials.

Procedure
1. Appropriately label two A549 shell vial cultures and one A549 tube culture for each specimen.
2. Remove the medium from the shell vials.
3. Inoculate each vial with 0.2-0.5 ml of the specimen.
4. Centrifuge the vials at 700 x g for 1 hour at room temperature.
5. Add 1 ml of growth medium containing 10% serum to each vial.
6. Incubate the vials at 35-37°C.
7. Fix and stain one vial after 48 hours and the second vial after 5 days.

Staining of Coverslips

Identification of adenovirus in shell vial systems is accomplished by staining the coverslips with monoclonal antibodies to the adenovirus hexon. Most specimens will produce 1-40 small foci after 48 hours of culture. Staining should be accomplished as directed by the manufacturer of the monoclonal antibodies or as described below.

1. Carefully remove the culture fluid from the vials and wash the coverslips twice by adding 1-2 ml of PBS and soaking for 5 minutes.
2. Remove the PBS and add 2-3 ml of acetone to each vial. Fix the coverslips for 10 minutes at room temperature.
3. Remove the acetone and allow the coverslips to air dry.
4. Rinse the coverslips briefly with PBS. (The moisture trapped between the coverslip and the vial will keep the stain from wicking under the coverslip.)
5. Add enough monoclonal antibody to the vial to cover the coverslip (100-150 μl) and incubate for 30 minutes at 35-37°C in a humidified chamber to prevent the antibody from drying on the coverslip.
6. Wash the coverslips twice as previously described (step 1).
7. Place a small amount of mounting fluid on a slide. Remove the coverslips from the vial and lay them, cell side down, on the mounting fluid.
8. Immediately examine the coverslips at 100-300X using a fluorescence microscope.

Interpretation of Results

Coverslips that have individual or small groups of cells that exhibit brilliant apple-green nuclear or cytoplasmic fluorescence (Fig. 2; see Color Plate 4B following Chapter 7) are positive for adenovirus. Coverslips without the specific fluorescence described above are considered negative. Some specimens may trap the monoclonal antibodies between the cell monolayer and inoculum cells. Careful examination of the cell morphologies can distinguish this type of reaction from adenovirus-specific reactions. In addition, nonspecific staining can occur if the monoclonal reagents are allowed to dry on the coverslip. In this case, the entire cell sheet may stain apple-green.

Positive Test. The presence of adenovirus is indicated by brilliant, apple-green nuclear and/or nuclear fluorescence (Fig. 2).
REPORT: Adenovirus isolated.

Negative Test. The absence of adenovirus is

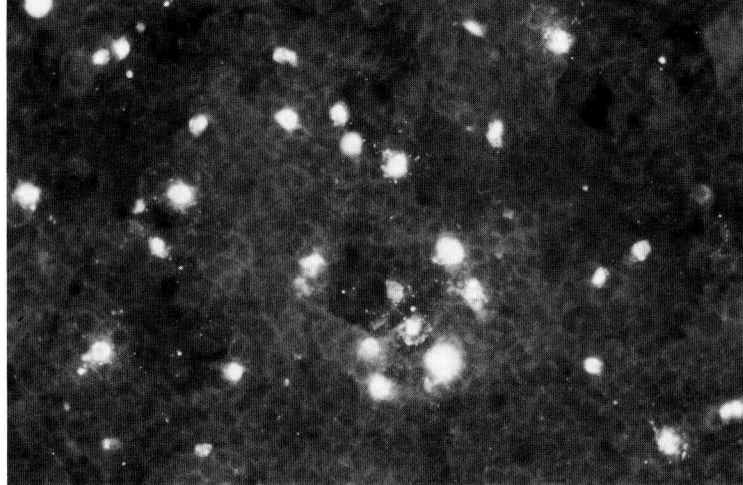

FIG. 2. Fluorescent antibody staining of adenovirus-infected A549 shell vial cultures.

indicated by the lack of specific fluorescence described above. The entire cell sheet should be stained red by the counterstain.
REPORT: Adenovirus not isolated.

QC Procedures.

A 1:10 dilution of a recent adenovirus isolate or ATCC cultures (see Preparation of Positive Control Inocula, above) should be inoculated with each batch of shell vial cultures. Uninfected shell vial cultures serve as negative controls. Infected and uninfected shell vial cultures should be processed and stained as described for the patient cultures. Positive controls must exhibit the typical fluorescence patterns. Negative controls must not exhibit specific fluorescent staining.

DIRECT SMEARS

Although rapid and highly specific, direct smears are less sensitive than either conventional tube cultures or shell vial methods (13). Therefore, direct smears should be backed up by tube cultures to detect adenoviruses that may be present in low concentrations or other viruses that may be in the specimen.

Direct smears require the presence of intact, adenovirus-infected respiratory epithelial cells before a positive result can be reported. Inadequate specimen collection and poor smear preparation are the principle reasons why this method sometimes fails to detect adenoviruses. In addition, excess mucus can interfere with antibody binding. Mucus can cause false positive reactions by nonspecifically trapping the antibody reagents and false negative reactions by clumping the cells thereby masking the adenovirus antigens. However, skilled technicians can usually distinguish these reactions from specific adenovirus reactions.

Despite these drawbacks, evaluation of direct smears for adenovirus is an effective method for the rapid detection of adenovirus infections. Direct smears may be especially valuable in remote laboratories where cold-chain specimen transport is not readily available.

Specimen Collection

The success of the DFA procedure depends upon the submission of a well made cell smear. Smears that are grossly contaminated with red blood cells can cause interpretation difficulties due to red cell autofluorescence. The principal reasons for the failure of direct smears to detect adenovirus in respiratory specimens is inadequate specimen collection and too few cells on the slide. Many laboratories require at least 50 cells per smear while other laboratories require 1-2 cells/high power (400X) field. Whatever the number, too many cells can cause nonspecific trapping of the conjugate and with too few cells, the laboratory may not be able to find an infected cell.

Specimen Processing
Nasal Aspirates, Nasal Washes, and Throat Washes
1. Centrifuge the specimen at 1000-1500 x g for 5 minutes to pellet the cells.
2. Remove all but 100-200 µl of the fluid.
3. Suspend the cell pellet in the remaining fluid.
4. Spread one drop of the cell suspension onto an acetone cleaned slide.
5. Allow the smears to air dry at room temperature. Slides may be stored for up to 48 hours at 2-8°C.

Swabs in Viral Transport Medium
1. Vortex the specimen vigorously to release as many cells from the swabs as possible.
2. Remove the swabs from the viral transport medium and firmly roll them against the inside of the tube to remove as much fluid as possible.
3. Centrifuge the specimen at 1000-1500 x g for 5 minutes to pellet the cells.
4. Remove all but 100-200 µl of the viral transport medium.
5. Suspend the cell pellet in the remaining fluid.
6. Spread one drop of the cell suspension onto an acetone cleaned slide.
7. Allow the smears to air dry at room temperature. Slides may be stored for up to 48 hours at 2-8°C.

Test Procedure
For optimum performance, slides should be fixed and stained within 1-2 hours of specimen collection. Alternatively, unfixed slides may be stored for up to 48 hours at 2-8°C. Fixed slides may be stored at 2-8°C for 1 week or frozen at -20°C or below for up to 1 year. Storing slides under desiccated conditions will decrease the background staining and minimize antigen degradation.

1. Upon arrival in the laboratory, examine the slide wells at 100-300X magnification. Ideally, a minimum of 50 cells should be visible on each well before the specimen is considered adequate for further processing.
2. The slide should be processed as directed by the manufacturer of the labeled antibody or as described below.
3. Fix the cells by immersing the slides in cold acetone for 10 minutes. Allow the slides to air dry.
4. Place enough of the monoclonal antibody on the well to cover the cell smear (15-30 µl).
5. Incubate the slide for 30 minutes at 35-37°C in a covered, humidified chamber. Do not allow the antibody to dry. Drying could cause nonspecific antibody binding.
6. Remove the excess antibody with a gentle stream of PBS. Do not direct the stream directly at the cell smear as this could dislodge the cells.
7. Soak the slide in fresh PBS twice for 5 minutes, shake off the excess fluid and allow the slide to air dry.
8. Add enough FITC-labeled conjugate to cover the cell smear (15-30 µl).
9. Incubate the slide for 30 minutes at 35-37°C in a covered, humidified chamber.
10. Remove the excess conjugate with a gentle stream of PBS. Do not direct the stream directly at the cell smear as this could dislodge the cells.
11. Soak the slide in PBS for 5 minutes, shake off the excess fluid, and allow the slide to air dry.
12. Carefully add a small drop of FA mounting fluid to the center of each well.
13. Place a number 1 coverslip on the mounting fluid and carefully remove all the air bubbles.
14. Immediately examine the slide using a fluorescence microscope. For optimum clarity, use 200-300X magnification for screening and 400X for confirmation of cell morphology.

Interpretation of Results
Positive Test. The presence of adenovirus is

indicated by intense apple-green fluorescence in the nucleus (and cytoplasm) of the infected cells. Only intact cells should be examined because cell debris and clumps of normal cells may exhibit a dull fluorescence which could be misinterpreted as a specific, positive reaction.
REPORT: Adenovirus detected.

Negative Test. The absence of adenovirus is indicated by the lack of specific fluorescence described above.
REPORT: Adenovirus not detected.

Inconclusive Test. Specimens with fewer than 50 cells on each well (and no positive cells) may give erroneous results.
REPORT: Unacceptable Specimen - Too Few Cells.

QC Procedures.

Positive and negative controls should be stained at least once each day to assure that the antibody reagents are performing properly. Positive cells should stain intensely as described above. Negative cells should stain red and should not exhibit specific apple-green fluorescence. Because the same antibody reagents are used for confirmation of tube cultures, shell vial cultures, and direct smears, culture confirmation testing can be used to demonstrate that the reagent is working properly. In laboratories that do not perform adenovirus isolations or laboratories where no isolations are ongoing, prepared slides (Fig. 3) should be stained whenever direct smears are stained. Control slides can be purchased commercially from a number of vendors or they can be prepared as described below.

Preparation of Control Slides

1. Add 20 µl of the positive control culture (described above) to 2 ml of sterile GLB. Hold the diluted virus on ice until it is used to inoculate the monolayer.
2. Remove the culture medium from one 75 cm^2 flask of newly confluent A549 cells.
3. Add 2.0 ml of the diluted virus and allow the virus to adsorb for 1-2 hours at 33-35°C. The flask should be rocked every 15 minutes to assure even virus distribution and to prevent monolayer desiccation.
4. Add 20 ml of complete medium containing MEM, 10% FCS, and antibiotics.
5. Incubate the flask at 35-37°C until the CPE involves 40-60% of the monolayer.

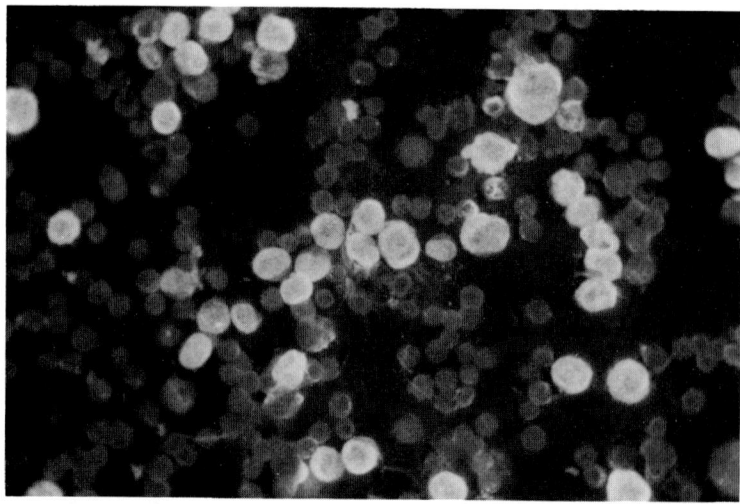

FIG. 3. Fluorescent antibody staining of positive control slide containing uninfected and adenovirus-infected cells.

6. Trypsinize the cells and remove them from the flask. Centrifuge the cells at 150 x g for 5 minutes.
7. Resuspend the cell pellet in 5 ml of PBS. Perform a cell count and adjust the cell concentration to 2-5 x 10^6 cells/ml.
8. Dispense 3-10 µl of the cell suspension onto each slide and allow the suspension to air dry.
9. Fix the slides in acetone at room temperature for 10 minutes and allow the slides to air dry.
10. Store the slides at -20°C or below for up to one year. Slides should be stored under desiccated conditions to minimize antigen degradation and background fluorescence.

NOTE: Slides prepared in this manner will contain both positive and negative cells (Fig. 3). Therefore, a single slide can be used for QC purposes. When slides are made from flasks with 100% CPE, two slides must be used - one containing a positive cell smear and a second slide containing an uninfected cell smear.

REFERENCES

1. Hierholzer JC, Wigand R, Anderson LJ, Adrian T, Gold JW. Adenoviruses from patients with AIDS: A plethora of serotypes and a description of five new serotypes of subgenus D (types 43-47). J Infect Dis 1988;158:804-813.
2. Christensen ML. Human viral gastroenteritis. *Clin Microbiol Rev* 1989;2:51-89.
3. Gary GW Jr, Hierholzer JC, Black RE. Characteristics of noncultivable adenoviruses associated with diarrhea in infants: a new subgroup of human adenoviruses. *J Clin Microbiol* 1979;10:96-103.
4. Horwitz MS. Adenoviruses. In: Fields BN, Knipe DM, Chanock RM, Hirsch MS, Melnick JL, Monath TP, Roizman B, eds. *Virology*, Second Edition. New York: Raven Press 1990;1723-1740.
5. Scott-Taylor T, Ahluwalia G, Klisko B, Hammond GW. Prevalent enteric adenovirus variant not detected by commercial monoclonal antibody enzyme immunoassay. *J Clin Microbiol* 1990;28:2797-2801.
6. Uhnoo, I, Wadell G, Svensson L, Johansson ME. Importance of enteric adenoviruses 40 and 41 in acute gastroenteritis in infants and young children. *J Clin Microbiol* 1984;20:365-372.
7. Shult PA, Polyak F, Dick EC, Warshauer DM, King LA, Mandel AD. Adenovirus 21 infection in an isolated antarctic station: Transmission of the virus and susceptibility of the population. *Am J Epidemiol* 1991;133:599-607.
8. Wadell G. *Adenoviridae*: The adenoviruses. In: Balows A, Hausler WJ Jr, Lennette EH, eds. *Laboratory diagnosis of infectious diseases - principles and practice*, Volume II. New York: Springer-Verlag 1988;284-300.
9. Hierholzer JC. Adenoviruses. In: Schmidt NJ, Emmons RW, eds. *Diagnostic procedures for viral, rickettsial, and chlamydial infections*, Sixth edition. Washington DC: American Public Health Association 1989;219-264.
10. Krisher KK, Menegus MA. Evaluation of three types of cell culture for recovery of adenovirus from clinical specimens. *J Clin Microbiol* 1987;25:1323-1324.
11. Espy MJ, Hierholzer JC, Smith TF. The effect of centrifugation on the rapid detection of adenovirus in shell vials. *Am J Clin Pathol* 1987;88:358-360.
12. Mahafzah AM, Landry ML. Evaluation of immunofluorescent reagents, centrifugation, and conventional cultures for the diagnosis of adenovirus infection. *Diagn Microbiol Infect Dis* 1989;12:407-411.
13. Ray CG, Minnich LL. Efficiency of immunofluorescence for rapid detection of common respiratory viruses. *J Clin Microbiol* 1987;25:355-357.

CHAPTER 9
Chlamydiae

INTRODUCTION

Chlamydiae are small intracellular bacteria that rely upon host cells for their energy requirements. In contrast to viruses, chlamydiae (a) contain both DNA and RNA, (b) reproduce by binary fission, (c) possess their own ribosomes, (d) are susceptible to several antibiotics, and (e) possess bacterial cell walls and lipopolysaccharides. Chlamydia elementary bodies (EBs) are infectious and enter susceptible cells through an endocytic process. Once inside the host cell, the infectious elementary body (EB) undergoes metabolic and structural changes to become the replicative reticulate body (RB). RBs replicate by binary fission producing microcolonies (inclusions) within endocytic vesicles (inclusion bodies), the hallmark of chlamydia-infected cells. After the RBs stop dividing, they condense to form infectious, metabolically inactive, EBs. The inclusion body bursts 48-72 hours after infection, destroying the host cell and releasing the newly formed EBs (1).

The order *Chlamydiales* consists of one genus, *Chlamydia*, and three species, *C. psittaci, C. trachomatis* and *C. pneumoniae* (formerly TWAR agents). *Chlamydia psittaci* is a common avian pathogen that causes sporadic cases of ornithosis and psittacosis in humans. The newly described *C. pneumoniae* has been associated with serious respiratory illness and has been identified as the cause of several pneumonia epidemics (2,3).

Chlamydia trachomatis

Chlamydia trachomatis is a human pathogen that causes a broad spectrum of clinical manifestations. There are 15 known serovars of *C. trachomatis*, A, B, Ba, C-K, and L1-L3. Four of these serovars (A, B, Ba, and C) cause trachoma - the most common, preventable form of blindness in the world. Today an estimated 500 million people in developing countries are affected with trachoma and 6-9 million of these people are already blind (4). Serovars L1-L3 cause lymphogranuloma venereum while serovars D-K are responsible for a broad spectrum of sexually transmitted diseases.

Chlamydia trachomatis is the leading cause of sexually transmitted disease in the United States with an estimated 3-5 million new infections occurring each year (5). These infections are of special concern to physicians and public health officials because up to two-thirds of all chlamydia-infected women are asymptomatic and do not seek medical treatment (6). Left untreated, chlamydia can ascend the female reproductive tract causing endometritis, salpingitis, and fallopian tube blockage (7). Chlamydia eventually spreads into the abdominal cavity causing a generalized pelvic inflammatory disease (PID) (8). Chlamydia infections are responsible for nearly 50% of all PID cases (9), 20% of which result in long-term complications including increased ectopic pregnancy rates and infertility (10). In males, chlamydia infections can cause urethritis, epididymitis, and rarely, proctitis.

Untreated chlamydia infections act as chlamydia reservoirs and contribute to the spread of these organisms to sexual partners and neonates. During pregnancy, untreated chlamydia infections have been shown to increase the risk of stillbirths and premature infants (11). When infants are born to chlamydia-infected mothers, the child's eyes, nose, throat, rectum, and vagina can be inoculated with organisms as the child passes through infected cervical secretions (12,13). Chlamydial pneumoni-

AT A GLANCE...

Chlamydia trachomatis

Detection Methods
Centrifugation-enhanced (shell vial) cultures using McCoy, HeLa-229, or Buffalo green monkey BGM) cells and an overlay medium containing cycloheximide.
Direct fluorescent antibody (DFA) staining of methanol or acetone-fixed smears.
Enzyme immunoassay screening tests are commercially available

Specimen Source
Swabbings or scrapings from endocervix, urethra or rectum.
Ocular swabbings or scrapings for trachoma or neonatal infections.
Oropharyngeal or nasopharyngeal swabs, washes, or aspirates for neonatal infections.

Time to Result
Culture: Positive culture - 48-72 hours. Negative culture - 72 hours.
DFA Method: 15-45 minutes.

***Chlamydia trachomatis* Serologies**
The usefulness of serological tests is limited because the antibody response depends upon the site of infection, duration of disease, the infecting serovars, and previous exposure to chlamydial antigens. Significant and predictable antibody responses generally occur in LGV or deep-seated chlamydial infections.

- **IgM** IgM determinations may be the method of choice in infants with chlamydial pneumonia. In adults, a single IgM antibody titer $\geq 1:32$ by microimmunofluorescence (micro IF) is suggestive of an active or recent chlamydial infection.

- **IgG** Diagnosis of recent infection is accomplished by demonstrating seroconversion or a four-fold rise in antibody titers in acute and convalescent sera collected 2-3 weeks apart. In salpingitis and LGV, micro IF titers may be greater than 1:512 while such titers are rare in active genital infections.

Epidemiology
Chlamydia trachomatis is the leading cause of sexually transmitted disease in the United States with 3-5 million new cases occurring each year. There is no seasonal variation in the incidence of *C. trachomatis* infections. There are 15 serovars of *C. trachomatis* (A, B, Ba, C-K, L1, L2, and L3). Reinfection is common.

Transmission/Incubation Period
Chlamydia trachomatis is transmitted through sexual contact with an infected person or by contact with, or aspiration of, chlamydia-infected cervical secretions during birth. Trachoma is spread by physical contact with chlamydia-infected secretions and through autoinoculation. The incubation period is usually 5-14 days.

Inactivators and Disinfectants
Chlamydia can be inactivated by heat and organic solvents. Effective disinfectants include 10% household bleach solutions, phenols, iodophor compounds, and glutaraldehyde compounds.

tis often occurs when chlamydia-infected secretions are aspirated with the baby's first breath (12,13).

Differential diagnosis of chlamydial infection is crucial for successful patient management because many chlamydia-infected individuals are also infected with *Neisseria gonorrhoeae* (14,15). Unfortunately, chlamydia infections and gonorrhea do not respond well to the same antibiotics. Laboratory diagnosis of *C. trachomatis* infection is usually based upon (a) the identification of chlamydia elementary bodies by direct fluorescent antibody (DFA) techniques, (b) detection of solubilized chlamydia antigens by enzyme immunoassays (EIAs), (c) detection of chlamydia-specific nucleic acid sequences, and/or (d) isolation and identification of the organism in McCoy, HeLa-229 or BGM cell cultures.

Chlamydia psittaci

Chlamydia psittaci is a common avian pathogen that can infect a variety of mammalian and reptilian species. *C. psittaci* causes sporadic cases of human ornithosis and psittacosis ("parrot fever") throughout the world. Although psittacine birds are considered the major reservoir for this agent, most human psittacosis cases have been associated with canaries, pigeons, sparrows, ducks, cockatiels, and occasionally, mammals. Occasionally, clusters of cases occur among workers at poultry (usually turkey) processing plants (15). Psittacosis is transmitted by the inhalation of infectious aerosols derived from feces, fecal dust, and secretions of infected animals. It should be noted that close or prolonged contact with birds is not necessary for the acquisition of infection. A number of cases have been reported where an infected person had only momentary contact with an environment where an infected bird had been. Percutaneous exposure (bites) provide an alternative, but less important, route of infection. Person-to-person transmission is rare but, when it occurs, the disease appears to be more severe.

After entry into the respiratory tract, *C. psittaci* is transported to the liver and spleen where it replicates in the reticuloendothelial cells. *C. psittaci* then spreads to the lung through hematogenous seeding. This two-phase infection process accounts for the relatively long incubation period of 4-14 days (16). Definitive diagnosis can only be established by the demonstration of a 4-fold rise in antibody titer in paired sera or by cell culture isolation and identification of the organism. *C psittaci* can be isolated in McCoy, HeLa-229, and BGM cells within 48-72 hours.

Chlamydia pneumoniae

C. pneumoniae is a respiratory tract pathogen that causes pneumonia, bronchitis, sinusitis, pharyngitis (often with hoarseness), myocarditis, endocarditis, and "influenza-like" febrile illnesses (2,3,17,18). *C. pneumoniae* is one of the 5 most common causes of pneumonia, causing approximately 10% of all pneumonias in the United States. *C. pneumoniae* infections occur throughout the world and seroepidemiologic surveys have shown that 30-50% of the general population have antibodies to *C. pneumoniae*. *Chlamydia pneumoniae* antibodies are rarely seen in children below the age of 2 years, but the prevalence increases after 5 years of age. At 10 years of age, approximately 10% of the population has antibodies to *C. pneumoniae* and the prevalence plateaus at 30-50% by 30 years of age. Approximately 70% of persons over the age of 60 have antibody to *C. pneumoniae*.

Chlamydia pneumoniae spreads slowly in the community with 1-2 month intervals between illnesses. Most *C. pneumoniae* infections are asymptomatic in young people. However, *C. pneumoniae* can cause severe pneumonia in older individuals. Reinfections can occur and serologic testing is the only method that can distinguish primary infections from reinfection. During primary infection, a group-specific CF antibody response is generally followed by a *C. pneumoniae*-specific IgM response that appears about 3 weeks after the onset of illness. IgM and CF antibody titers usually diminish 2-6 months after infection. *C. pneumoniae*-specific IgG antibodies usually appear 6-8 weeks after onset of symptoms and persist for varying lengths of time. Because antibody levels may be slow to develop in primary infection, convalescent sera should be obtained at 3 weeks (rather than 10-14 days) after the acute

AT A GLANCE...

Chlamydia psittaci

Detection Methods
Centrifugation-enhanced (shell vial) cultures using McCoy, HeLa-229, or Buffalo green monkey (BGM) cells and an overlay medium containing cycloheximide.

Specimen Source
Sputum, nasal aspirates, and throat swabs can be used for chlamydial pneumonia.
C. psittaci can be isolated from cloacal swabs, droppings from bird cages, and fecal pellets.

Time to Result
Culture: Positive culture - 48-72 hours. Negative culture - 72 hours.

Chlamydia psittaci Serologies

- **IgM**   IgM is usually detectable at the onset of symptoms and antibody levels remain elevated for 1-2 months. A single IgM titer of $\geq$ 1:32 is suggestive of an active or recent infection.

- **IgG**   First detectable 14 days after infection and antibody levels reach peak titers after 4-8 weeks. Diagnosis of recent infection is accomplished by demonstrating seroconversion or a four-fold increase in antibody titers in acute and convalescent specimens collected 2-3 weeks apart.

Epidemiology
Psittacosis has a worldwide distribution. *C. psittaci* causes sporadic disease and 40-60 psittacosis/ornithosis cases are reported annually in the United States. Owners of pet birds, pet shop employees, and workers in turkey processing plants are at higher risk of acquiring the disease than the general population. Re-infections have been described. There is no record of infection acquired by handling dressed, eviscerated birds or by eating poultry products.

Transmission/Incubation Period
The disease is transmitted by inhalation of infectious aerosols derived from feces, fecal dust, and secretions from infected animals. Animal bites provide a less important route of infection. Person-to-person transmission is rare but, when it occurs, the disease appears to be more severe. Incubation period is 7-15 days.

Inactivators and Disinfectants
Chlamydia can be inactivated by heat, and organic solvents. Effective disinfectants include 10% household bleach solutions, phenols, iodophor compounds, and glutaraldehyde compounds.

serum. Reinfections may not produce CF or IgM antibodies and IgG antibody levels usually rise more rapidly (within 1-2 weeks) than in primary infections. Serologic diagnosis is based upon seroconversion or demonstration of a fourfold rise in antibody titer. An IgM antibody titer of $\geq 1:16$ or an IgG antibody titer of $\geq 1:512$ in a single serum also suggests current or recent *C. pneumoniae* infection.

Although serological methods are often used for the laboratory diagnosis of *C. pneumoniae* infections, cell culture isolation also provides a sensitive and specific method for detecting these infections. *C. pneumoniae* can be isolated in embryonated chicken eggs, HeLa-229, HL (19), and HEp-2 cells. McCoy cells are not recommended for *C. pneumoniae* isolations. Two or three blind passages may be necessary when isolating *C. pneumoniae* from clinical specimens. Presumptive identification of *C. pneumoniae* inclusions can be made by immunofluorescent staining with genus-specific (anti-LPS) monoclonal antibodies. Specific identification is usually accomplished using specific monoclonal antibodies available from the University of Washington Research Foundation.

CELL CULTURE ISOLATION

Introduction

Although Chlamydiae can be isolated from embryonated chicken eggs, yolk sac isolations are rarely used for clinical specimens because they are time-consuming (14 days to result) and cumbersome. Most clinical laboratories use centrifugation-enhanced cell culture methods to isolate chlamydiae from clinical specimens. In this procedure, chlamydia specimens are inoculated onto cycloheximide-treated cells, centrifuged for 60 minutes at 1200-3000 x g, and allowed to replicate for 48-72 hours at 35°C. Cultures are considered positive if typical perinuclear intracytoplasmic inclusion bodies are observed within the infected cells.

Several methods have been used to detect chlamydia inclusion bodies including iodine, Giemsa staining, and fluorescein- (20) or peroxidase-labelled (21) monoclonal antibodies. Detection of chlamydia-infected cells has traditionally been accomplished with iodine staining. Iodine stains the glycogen in the inclusion bodies of cells infected with *C. trachomatis*. However, the usefulness of iodine staining is somewhat limited because cells infected with *C. trachomatis* are iodine-positive for only a short time. Furthermore, *C. psittaci* and *C. pneumoniae* inclusions do not stain with iodine.

Immunofluorescent staining is the most sensitive method for detecting chlamydial inclusions (20). Three types of chlamydial monoclonal antibodies are commercially available. Anti-lipopolysaccharide (anti-LPS) reagents are group specific and stain all chlamydiae species. Antibodies to the major outer membrane proteins (Anti-MOMP) of *C. trachomatis* stain only *C. trachomatis* strains while antibodies to *C. pneumoniae* (Washington Research Foundation) stain only *C. pneumoniae* strains. There are no commercially available *C. psittaci*-specific reagents. Fluorescent antibody staining should be done as directed by the manufacturer of the monoclonal antibodies or as described below.

Specimen Collection and Preparation

Specimens should be collected from the site of the suspected infection as soon as possible after onset of disease. Specimens should be transported to the laboratory in chlamydia transport medium containing gentamicin (10 μg/ml), vancomycin (100 μg/ml), and amphotericin B (4 μg/ml). *C. trachomatis* and *C. psittaci* specimens may be held at 2-8°C for up to 30 hours before inoculation onto cell cultures. *Chlamydia pneumoniae* is very sensitive to elevated temperatures. Kuo and Greyston (22) have reported that *C. pneumoniae* specimens retain only 1% of their original infectivity when stored at room temperature for 24 hours. In addition, *C. pneumoniae* specimens lose approximately 60% of their viability after rapid freezing. If specimens are first cooled to 4°C and then frozen at -70°C, they will usually retain 75% of their original viability (22). Therefore, *C. pneumoniae* specimens should be stored at 2-8°C for no more than 24 hours. If specimens cannot be inoculated within that time, they should be cooled

AT A GLANCE...

Chlamydia pneumoniae

Detection Methods
Centrifugation-enhanced (shell vial) cultures using HeLa-229, HL, or HEp-2 cells and an overlay medium containing cycloheximide.

Specimen Source
Sputum, nasal aspirates, broncheoalveolar lavages, and throat swabs can be used for the detection of chlamydial pneumonia.

Time to Result
C. pneumoniae grows slowly in culture and up to three subpassages may be required before the organism can be demonstrated. Positive culture - 2-10 days. Negative culture - 10 days.

***Chlamydia pneumoniae* Serologies**

IgM   *C. pneumoniae*-specific IgM is usually appears three weeks after the onset of symptoms and antibody levels generally persist for 2-6 months. A single *C. pneumoniae* IgM antibody titer of $\geq$1:16 suggests current or recent infection.

IgG   During primary infections, IgG levels are first detectable 6-8 weeks after the onset of symptoms and persist for varying lengths of time. After reinfection, IgG levels rise more quickly, usually within 1-2 weeks. Serologic diagnosis is based upon seroconversion or demonstration of a four-fold rise in antibody titer. An IgG titer of $\geq$1:512 is suggestive of a current or recent *C. pneumoniae* infection. A convalescent serum should be obtained 3 weeks after the first specimen.

Epidemiology
Chlamydia pneumoniae is a respiratory tract pathogen that causes sporadic cases of pneumonia throughout the world. Seroepidemiologic surveys have shown that 30-50% of the general population has antibodies to *C. pneumoniae*. These antibodies are rarely seen in children below the age of 2 years, but the prevalence increases after 5 years of age. At 10 years of age, approximately 10% of the population has antibodies to *C. pneumoniae* and the prevalence plateaus at 30-50% by 30 years of age. Approximately 70% of persons over the age of 60 have antibody to *C. pneumoniae*.

Transmission/Incubation Period
C. pneumoniae spreads slowly in the community and 1-2 month intervals between infections have been described. Infection does not produce lasting immunity and reinfections can occur. Little is known about the mechanism of transmission except that epidemics frequently occur outside the home.

Inactivators and Disinfectants
Chlamydia pneumoniae can be inactivated by heat, and organic solvents. Effective disinfectants include 10% household bleach solutions, phenols, iodophor compounds, and glutaraldehyde compounds.

to 2-8°C for at least 1 hour and then frozen at -70°C.

Urogenital Specimens

Urogenital specimens must be obtained by swabbing or scraping the transitional zone of the cervix or by swabbing the endourethra 2 to 4 cm from the meatus. Urine, cervical exudates and urethral discharges are not acceptable specimens. Calcium alginate and wooden-shafted swabs should not be used.

Ocular Specimens

Pus and other ocular exudates are not acceptable specimens. Purulent exudates must be carefully removed from the eye before collecting the specimen from the conjunctiva. For the best isolation results, the conjunctiva must be vigorously swabbed to obtain a representative sample of epithelial cells. The swab should then be placed into a tube containing chlamydia transport medium and sent to the laboratory.

Sputa or Throat Washings

Sputa and throat washings should be diluted 1:5 in chlamydia transport medium containing twice the normal concentration of antibiotics. Place the diluted specimen in a tightly stoppered sterile container and vortex the specimen vigorously with glass beads. Centrifuge the specimen for 15 minutes 600 x g and transfer the supernatant fluids to another sterile container. Incubate the supernatant fluids for 1 hour at room temperature to reduce the bacterial load.

Tissues

Tissue specimens should be weighed, minced with sterile scissors, and homogenized using a disposable homogenizer. Add enough chlamydia transport medium to the homogenizer to produce a 5% (w/v) suspension. Remove the suspension from the homogenizer and vortex vigorously to extract as much chlamydia antigen as possible. Centrifuge the specimen at 300-600 x g but do not separate the tissue from the supernatant fluids. Carefully remove two 100 μl and 10 μl aliquots for inoculation onto cell cultures.

Semen

Semen specimens are not recommended for the diagnosis of chlamydial infections. However, fertility clinics will occasionally submit semen specimens as part of their patient work-ups. Because semen specimens are extremely toxic to cell cultures, a small amount of semen should be added to 3 ml of chlamydia transport medium. Vortex the specimen to disperse the components and inoculate cultures with 100 μl and 10 μl aliquots of the specimen.

Fecal Samples

Cloacal swabs, droppings from bird cages, and/or fecal pellets are occasionally submitted to the laboratory to confirm the source of an ornithosis or psittacosis outbreak. Dust and aerosols from these specimens are often highly infectious to laboratory personnel and should only be handled in a biological safety cabinet. These specimens are also extremely toxic to cell cultures. Scrape a small amount of this material into a preweighed 50 ml centrifuge tube. Close the tube securely, wipe the outside with disinfectant, and reweigh the tube. Prepare a 5% (w/v) suspension in MEM containing 10% FCS and twice the normal concentration of antibiotics. Vortex the suspension vigorously and centrifuge at 600 x g to remove the solids. Transfer the supernatant fluids to a sterile tube and incubate for 1 hour at room temperature to further reduce the bacterial load. Inoculate each of two shell vial cultures with 100 μl, 50 μl and 10 μl of the specimen.

Cell Culture Procedure

1. Label two shell vials for each specimen.
2. Remove the maintenance medium from the cells and add 1 ml of MEM containing 2% FCS, 10 μg/ml gentamicin, 4 μg/ml amphotericin B, and 0.5-0.75 μg/ml cycloheximide.
3. Vortex the specimen vigorously. Remove the swab from the vial and firmly roll it against the inside of the vial to remove as much fluid as possible.
4. Add 0.1-0.2 ml of the specimen to each

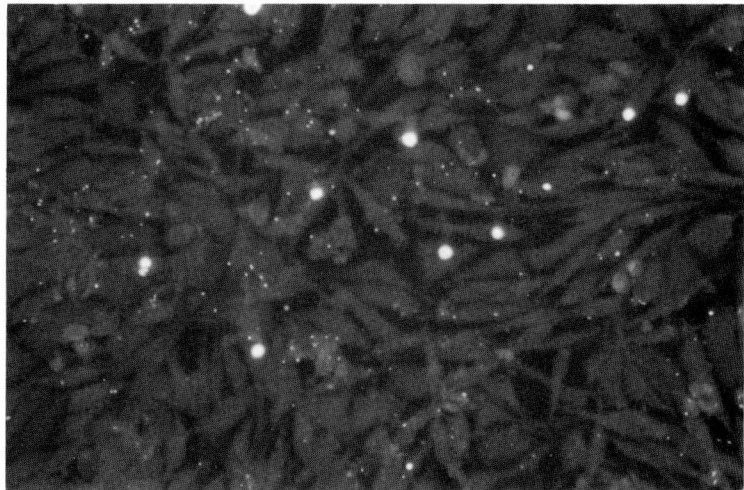

FIG. 1. Fluorescent antibody staining of *Chlamydia trachomatis*-infected McCoy cells showing inclusion bodies.

shell vial and replace the stoppers.
5. Centrifuge the vials at 1500-1700 x g for 1 hour at 25-35°C.
6. Incubate the vials at 35-37°C for 48-72 hours. Stain one coverslip after 48 hours and the second after 72 hours.

Staining of Coverslips
1. Carefully remove the culture fluids from the vials.
2. Carefully wash the coverslip twice by adding 1-2 ml of PBS to the vial and soaking for 5 minutes.
3. Remove the PBS and add 2-3 ml of methanol or acetone (as directed by the manufacturer of the reagent) to each vial. Fix the coverslips for 10 minutes at room temperature.
   NOTE: Fix *C. pneumoniae* coverslips with acetone. Methanol fixation will eliminate the antigenic reactivity (23).
4. Remove the fixative and allow the coverslip to air dry in the vial.
5. Rinse the coverslip briefly with PBS. (The moisture trapped between the coverslip and the vial will keep the stain from wicking under the coverslip.)
6. Add enough monoclonal antibody to the vial to cover the coverslip (100-150 µl) and incubate for 30 minutes at 35-37°C. The shell vials should be covered and incubated in a humidified chamber to prevent the antibody from drying on the coverslip.
7. Wash the coverslips twice as previously described (step 2).
8. Place a small drop of mounting fluid on a slide. Remove the coverslips from the vial and lay them, cell side down, on the mounting fluid.
9. Immediately examine the coverslips at 200-400X using a fluorescence microscope.

Interpretation of Results

Coverslips that contain brilliant apple-green cytoplasmic inclusions are considered positive for Chlamydiae (Fig. 1). Coverslips without the specific fluorescence described above are considered negative. Some specimens may nonspecifically trap the monoclonal antibodies between the cell monolayer and inoculum cells. Careful examination of the cell morphology can distinguish this type of reaction from chlamydia-specific reactions. Nonspecific staining can also occur if the monoclonal reagents are allowed to dry on the coverslip. In this case, the entire cell sheet may stain apple-green.

Chapter 9

Positive Test. The presence of chlamydia is indicated by the presence of brilliantly fluorescing cytoplasmic inclusion bodies (Fig. 1).
REPORT: Chlamydia species isolated (if using anti-LPS reagents).
Chlamydia trachomatis isolated (if using anti-MOMP reagents).
Chlamydia pneumoniae isolated if using specific monoclonal antibodies.

Negative Test. The absence of chlamydia is indicated by the lack of specific fluorescence described above. The entire cell sheet should be stained red by the counterstain. Lack of fluorescence does not preclude chlamydial infection.
REPORT: No Chlamydia species isolated (if using anti-LPS reagents).
No *Chlamydia trachomatis* isolated (if using anti-MOMP reagents).
No *Chlamydia pneumoniae* isolated (if using specific monoclonal antibodies).

QC Procedures.
A 1:10 dilution (in chlamydia transport medium) of recent chlamydiae isolates or ATCC cultures should be inoculated with each batch of shell vial cultures. Uninfected shell cultures can serve as negative controls. Infected and uninfected shell vial cultures should be processed and stained as described for the patient cultures. Positive controls must exhibit the typical fluorescence patterns. Negative controls must not exhibit specific fluorescent staining.

Preparation of Positive Control Cultures
After several cell culture passages, the *Chlamydia trachomatis* strain L2 from the ATCC can be propagated in flasks without centrifugation. Passage the initial inoculum as follows:
1. Dilute the ATCC cultures 1:1000 in chlamydia transport medium and use 0.2 ml aliquots to inoculate three McCoy shell vial cultures as described above.
2. Incubate the cultures at 35-37°C until 80-90% of each monolayer is destroyed (72-96 hours).
3. Pool the cell culture fluids. Freeze/thaw once to release any intracellular EBs.
4. Centrifuge the fluids at 600 x g for 20 minutes to pellet the cell debris.
5. Remove supernatant fluids, aliquot, and freeze at -70°C or below.
6. Dilute the (passage 1) chlamydia L2 serovar 1:1000 in chlamydia transport medium.
7. Infect three McCoy cell cultures as before. By the fourth passage, the L2 strain will completely destroy the shell vial monolayers within 72 hours. The chlamydia strain can then be used to infect cell monolayers as described below.

Positive Control Inocula
1. Add 20 µl of the chlamydia preparation (above) to 2 ml of chlamydia transport medium.
2. Remove the growth medium from one 75 cm^2 flask of 90% confluent McCoy cells.
3. Add the diluted chlamydia to the monolayer and allow the virus to adsorb for 1-2 hours at 33-35°C. Note: The flask should be rocked every 15 minutes to assure even distribution of the virus and to prevent monolayer desiccation.
4. Add 20 ml of chlamydia overlay medium containing cycloheximide.
5. Incubate the flask at 33-35°C until inclusions are present in 80-90% of the monolayer (usually 72 hours).
6. Freeze/thaw once to liberate any intracellular EBs.
7. Centrifuge the culture fluids at 600 x g for 20 minutes to pellet the cell debris.
8. Dilute the cell culture fluids to 100 ml with MEM containing 10% FCS.
9. Dispense 0.5 ml of the chlamydia suspension into each of 200 freezing vials.
10. Freeze the vials at -70°C or below. Vials may be stored at -70°C for up to 5 years and under liquid nitrogen for up to 15 years.

DIRECT SMEARS

Introduction

In the DFA procedure, the specimen is collected, smeared onto a glass slide, and fixed with acetone or methanol. The smears are then transported to the laboratory where they are allowed to react with fluorescein-labelled monoclonal antibodies. Once the unbound antibodies are washed away, the slide is dried, mounted and viewed under a fluorescence microscope. Positive specimens contain bright pinpoints of apple-green fluorescence. DFA methods are quite sensitive, exhibiting 60-90% sensitivity compared to cell culture. DFA methods are very rapid and staining and interpretation can be completed in 15-45 minutes. Because specimen adequacy can be evaluated microscopically, DFA procedures possess an advantage over culture or EIA. [Specimens without specific fluorescence or columnar epithelial cells can be reported as inadequate.]

Specimen Collection and Preparation

The success of the direct fluorescent antibody (DFA) procedure depends upon the preparation of a well made cell smear. Smears that are too thick or "lumpy" can cause nonspecific trapping of the antibody reagents. Smears that are grossly contaminated with red blood cells can cause interpretation difficulties due to red cell autofluorescence. Specimen adequacy and its definition are significant sources of intra-laboratory variation and an area of increasing regulatory concern. Many laboratories require at least 25 epithelial cells per smear while others require 1-2 cells/high power field. Whatever the number, too many cells can cause nonspecific trapping of antibody and with too few cells, the laboratory may not feel comfortable reporting a negative result.

Slide Preparation from Swab or Cytobrush Specimens
1. Swabs and cytobrushes should not be allowed to dry. Prepare smears immediately after specimen collection.
2. Rub the cytobrush or swab on the well of an acetone-cleaned slide making sure the specimen is evenly dispersed over the entire well.
3. Allow the slide to air dry.
4. Lay the slide flat and add 0.5 ml of methanol to the well. Let the methanol evaporate at room temperature.
5. Label the slide and send it to the laboratory in a petri dish or a slide mailer.
6. Slides may be stored at room (20-30°C) or refrigerator (2-8°C) temperature for up to 7 days before staining. If longer delays are anticipated, slides should be stored at or below -20°C.

Slide Preparation from Nasal Aspirates, Nasal Washes, and Throat Washes
1. Vortex the specimen gently to break up the mucus.
2. Place one drop of the specimen on the well of an acetone-cleaned slide and spread the specimen evenly over the well.
3. Allow the slide to air dry.
4. Lay the slide flat and add 0.5 ml of methanol to the well. Let the methanol evaporate at room temperature.
5. Slides may be stored at room (20-30°C) or refrigerator (2-8°C) temperature for up to 7 days before staining. If longer delays are anticipated, slides should be stored at or below -20°C.

Slide Preparation from Swabs in Viral Transport Medium
1. Vortex the specimen vigorously to release as many cells from the swab as possible.
2. Remove the swab from the viral transport medium and firmly roll it against the inside of the tube to remove as much fluid as possible.
3. Centrifuge the specimen at 10,000-15,000 x g to pellet the EBs.
4. Remove all of the supernatant fluids except for 10-20 μl.
5. Resuspend the cells in the remaining fluid.
6. Spread 5-10 μl of the cell suspension onto the well of an acetone cleaned slide.
7. Allow the smear to air dry at room temperature.
8. Lay the slide flat and add 0.5 ml of methanol to the well. Let the methanol

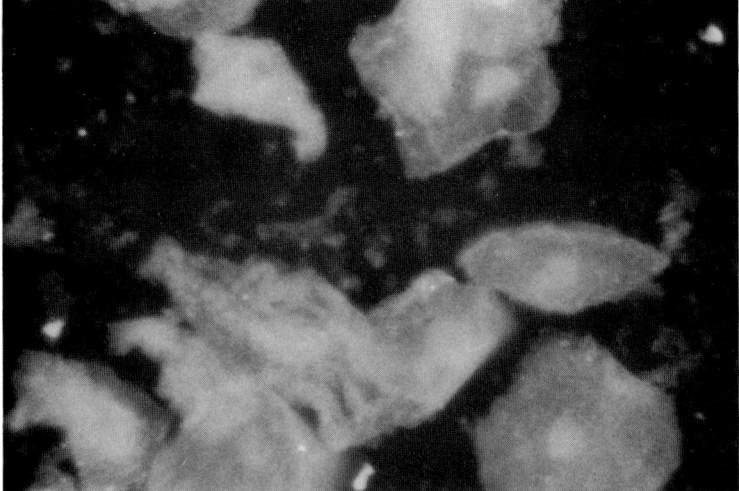

FIG. 2. Cervical smear stained with antibodies to *Chlamydia trachomatis*. Note the tiny fluorescing elementary bodies against a cellular background.

evaporate at room temperature.
9. Slides may be stored at room (20-30°C) or refrigerator (2-8°C) temperature for up to 7 days before staining. If longer delays are anticipated, slides should be stored at or below -20°C.

Test Procedure
1. Upon arrival in the laboratory, examine the slide wells at 100-300X magnification. Ideally, a minimum of 25 epithelial cells should be visible on each well before the specimen is considered adequate for further processing.
2. The slide should be processed as directed by the manufacturer of the fluorescein-labelled chlamydia antibody or as described below.
3. Place enough of the fluorescein-labelled monoclonal antibody on each well to cover the cell smear (approximately 15-30 µl).
4. Incubate the slide for 15-30 minutes at 33-35°C in a covered, humidified chamber. Do not allow the antibody to dry. Drying can cause nonspecific antibody binding.
5. Remove the excess antibody with a gentle stream of PBS. Do not direct the stream directly at the cell smear as this could dislodge the cells.
6. Soak the slide in PBS for 5 minutes, shake off the excess fluid, and allow the slide to air dry.
7. Carefully add a small drop of FA mounting fluid to the center of each well.
8. Place a number 1 coverslip on the mounting fluid and carefully remove all the air bubbles.
9. Immediately examine the slide using a fluorescence microscope. For optimum clarity, use 400-500X (oil) magnification for screening and 1000X (oil) for confirmation of morphology.

Interpretation of Results
Positive Test. The presence of chlamydia is indicated by the presence of at least 5 pinpoints of intense apple-green fluorescence in each well (Fig. 2; see Color Plate 3C following Chapter 7). At 1000X magnification, the elementary bodies should appear as a smooth-edged disc that are one one-hundredth the size of the epithelial cells. Elementary bodies should fluoresce evenly. Any irregular particles emitting an intense fluorescence or a duller yellow-green fluorescence should be disregarded.
REPORT: Chlamydia species detected (if using anti-LPS reagents).

Chlamydia trachomatis detected (if using anti-MOMP reagents).

Negative Test. The absence of chlamydiae is indicated by the lack of intensely fluorescing EBs. Occasionally, rectal specimens will contain bacteria that nonspecifically bind the antibody reagent. This type of staining can be distinguished from chlamydia-specific staining by cell size, morphology, and staining pattern. Fluorescing bacteria are generally 2-4 times larger than chlamydial EBs and staining usually appears only at the edge of the cells.
REPORT: No Chlamydia species detected (if using anti-LPS reagents).
No *Chlamydia trachomatis* detected (if using anti-MOMP reagents).

Inconclusive Test. Negative specimens with fewer than 25 cells on each well (or < 1-2 cells/hpf) may give inconclusive results.
REPORT: Unacceptable specimen - too few cells.

QC Procedures.
Positive and negative controls should be stained at least once each day to assure that the antibody reagent is performing properly and to provide a guide as to the size, shape, and staining pattern of EBs. Positive smears should stain intensely as described above. Negative cells should stain red and should not exhibit specific apple-green fluorescence.
Control slides can be purchased commercially from a number of vendors or they can be prepared in the laboratory as described below.

Preparation of Positive and Negative Controls
1. Trypsinize a 25 cm^2 flask of Vero cells and remove the cells from the flask.
2. Centrifuge the cells at 300 x g for 15 minutes at 2-8°C.
3. Resuspend the cells in 40 ml of cold PBS containing 2% FCS.
4. Centrifuge the cells as before (step 2) and resuspend in cold PBS without serum.
5. Perform a cell count and adjust the cell concentration to 1-3 x 10^6 cells/ml.
6. Divide the cell suspension and hold on ice.
7. Add 100-500 µl of the positive control inocula (above) to one tube of cells. The object is to add enough EBs to produce 2-20 EBs per 1000X oil immersion field.
8. Dispense 3-10 µl of the "positive" and "negative" cell suspensions onto the respective wells of an acetone-cleaned two-well slide (8 mm diameter wells). Allow the slides to air dry.
9. Fix the slides in methanol at room temperature for 10 minutes and allow the slides to air dry.
10. Store the slides at -20°C or below for up to one year. Slides should be stored under desiccated conditions to minimize antigen degradation and background fluorescence.

REFERENCES

1. Schacter J. Chlamydiaceae: The chlamydiae. In: Lennette, EH, Halonen P, Murphy FA, eds. *Laboratory Diagnosis of Infectious Diseases - Principles and Practice*, Volume II. New York: Springer-Verlag, 1988;847-863.
2. Grayston JT, Kuo CC, Wang SP, Altman J. 1986. A new *Chlamydia psittaci* strain, TWAR, isolated in acute respiratory tract infections. *N Eng J Med* 1986;315:161-168.
3. Marrie TJ, Grayston JT, Wang SP, and Kuo CC. Pneumonia associated with the TWAR strain of chlamydia. *Ann Intern Med* 1986;106:507-511.
4. Schacter, J. 1985. Chlamydiae (Psittacosis-Lymphogranuloma Venereum-Trachoma Group). Pp. 856-862. In: E.H. Lennette, A. Balows, W.J. Hausler, Jr., and H.J. Shadowmy (eds.), *Manual of Clinical Microbiology*, Fourth Edition. American Society for Microbiology, Washington, DC.
5. Judson FN. Assessing the number of genital chlamydia infections in the United States. *J Reprod Med* 1985;30:269-272.
6. Oriel JD, Johnson AL, Barlow D, Thomas BJ, Nayyar K, Reeve P. Infection of the uterine cervix with *Chlamydia trachomatis*. *J Infect Dis* 1978;137:443-451.
7. Schacter J. Chlamydial infections. *N Eng J Med* 1978;298:428-435.

8. Taylor-Robinson D, Thomas BJ. 1980. The role of *Chlamydia trachomatis* in genital tract and associated diseases. *J Clin Pathol* 1980;33:205-233.
9. Westrom L. Incidence, prevalence and trends of acute pelvic inflammatory disease and its consequences in industrialized countries. *Am J Obstet Gynecol* 1980;138:880-892.
10. Brunham RC. Pelvic inflammatory disease. *Med Clin North Am* 1983;6:515-524.
11. Thompson S., Lopez D, Wong KH, et al. A prospective study of chlamydia and mycoplasma infections during pregnancy: Relation to pregnancy outcome and maternal morbidity. In: Mardh PA, et al. *Chlamydia Infections*. New York: Elsevier Biomedical 1982;155-158.
12. Rettig PJ. Infections due to *Chlamydia trachomatis* from infancy to adolescence. *Pediatr Infect Dis* 1986;5:449-457.
13. Harrison HR, Alexander ER. *Chlamydia trachomatis* infections of the infant. In: Holmes KK, Mardh PA, Sparling PA, Wiesner PJ, eds. *Sexually Transmitted Diseases*. New York: McGraw-Hill 1984;270-280.
14. Stamm, WE, Guinan ME, Johnson C, et al. Effect of treatment regimens for *Neisseria gonorrhoeae* on simultaneous infection with *Chlamydia trachomatis*. *N Eng J Med* 1984;310:545-549.
15. Oriel JD, Reeve P, Thomas BJ, et al. Infection with Chlamydia group A in men with urethritis due to *Neisseria gonorrhoeae*. *J Infect Dis* 1975;131:376-382.
16. Centers for Disease Control. Psittacosis at a turkey processing plant - North Carolina, 1989. *MMWR* 1990;39:460-469.
17. Saikku P, Wang SP, Kleemola M, Brander E, Rusanen E, Grayston JT. An epidemic of mild pneumonia due to an unusual strain of *Chlamydia psittaci*. *J Infect Dis* 1985;151:832-839.
18. Kleemola M, Saikku P, Visakorpi R, Wang SP, Grayston JT. Epidemics of pneumonia caused by TWAR, a new *chlamydia* organism in military trainees in Finland. *J Infect Dis* 1988;157:230-236.
19. Cles LD, Stamm WE. Use of HL cells for improved isolation and passage of *Chlamydia pneumoniae*. *J Clin Microbiol* 1990;28:938-940.
20. Stamm WE, Tam M, Koester M, Cles L. Detection of *Chlamydia trachomatis* inclusions in McCoy cell cultures with fluorescein-conjugated monoclonal antibodies. *J Clin Microbiol* 1983;17:666-668.
21. Wiedbrauk DL, Nelhs S, Harris VA. Simplified procedure for large-scale chlamydiae isolation. *Am Clin Lab* 1989;8:27-30.
22. Kuo CC, Grayston JT. *Chlamydia* spp. strain TWAR: A newly recognized organism associated with atypical pneumonia and other respiratory infections. *Clin Microbiol Newsletter* 1988;10:137-140.
23. Wang SP, Grayston JT. *Chlamydia pneumoniae* elementary body antigenic reactivity with fluorescent antibody is destroyed by methanol. *J Clin Microbiol* 1991;29:1539-1541.

CHAPTER 10

Clostridium difficile Toxin Assay

INTRODUCTION

Clostridium difficile is not a virus, but rather, a Gram-positive anaerobic sporeforming bacillus. *Clostridium difficile* is the leading cause of pseudomembranous colitis (PMC) and is responsible for approximately one fifth of all cases of antibiotic-associated diarrhea (1). Although PMC had been described in patients prior to the start of the antibiotic era, the number of PMC cases increased dramatically after antibiotics began to be used. A wide variety of antibiotics have been implicated in *C. difficile* associated disease, including penicillins, cephalosporins, lincosamides, macrolides, tetracyclines, aminoglycosides, as well as some antineoplastic agents. In fact, any drug that disturbs the normal bacterial flora of the gut can make the host susceptible to colonization and overgrowth by the pathogen. In rare instances, PMC can occur without prior antibiotic therapy.

Because most adults do not carry *C. difficile* as part of their normal intestinal flora, it is thought that most cases of PMC result from *C. difficile* colonizations, presumably within the hospital environment (2). PMC in hospitals is especially difficult to control because the patients have diarrhea which, in turn, produces infectious aerosols. Once in the environment, *C. difficile* spores can persist for months because they are resistant to most hospital disinfectants. *Clostridium difficile* can be isolated from the clothing and room fixtures of colonized patients and from the hands of hospital personnel caring for them. Toxigenic *C. difficile* is also present as the "normal" flora of up to 60% of newborn infants in many hospital nurseries. In contrast with adults, infants with *C. difficile* in their stools are generally asymptomatic and do not acquire PMC. Because the rate of infant colonization parallels the incidence of outbreaks and cross-infections within hospitals, most infants probably acquire the organism nosocomially (3).

Toxigenic *C. difficile* strains produce two potent toxins, an enterotoxin (toxin A) and a cytotoxin (toxin B). The extensive tissue damage and fluid response that toxin A causes in experimental animals suggests that toxin A and not toxin B may be responsible for PMC. However, toxigenic strains of *C. difficile* produce both toxin A and toxin B and the virulence of *C. difficile* isolates correlates with the level of toxin production. Highly virulent strains produce high levels of both toxins and weakly virulent isolates produce low levels of both toxins. Nonpathogenic *C. difficile* isolates secrete neither toxin.

The most definitive diagnosis of PMC is by the endoscopic detection of pseudomembranes or microabscesses in antibiotic-treated patients with diarrhea who have *C. difficile* toxin in their stool (2). However, the diagnosis of *C. difficile* disease in patients who develop antibiotic-associated diarrhea is more difficult. In these cases, clinical diagnosis is often based upon clinical history (rather than invasive techniques), the presence of *C difficile* toxin in the stool, and/or stool cultures for *C. difficile*. Several methods are available for detecting *C. difficile* toxins including the cell culture neutralization test, a latex agglutination test for *C. difficile* toxin A, several enzyme immunoassays for toxin A, and one enzyme immunoassay for toxin A and B. An older latex agglutination test is also available that detects a 43,000 dalton protein from *C. difficile* (4). Although this test seems to correlate well with disease, it cannot distinguish toxigenic from nontoxigenic *C. difficile* isolates and the test cross-reacts with *Clostridium botulinum*, *Clostridium sporogenes*, *Peptostreptococcus anaerobius* and *Bacteroides asaccharolyticus* (2,5,6).

> **AT A GLANCE...**
>
> # Clostridium difficile TOXINS
>
> **Detection Methods**
> C. difficile toxins can be detected by EIA and latex agglutination assays. The gold standard for the detection of toxin is the cell culture neutralization assay. Cytotoxin assays can be performed in tube or microtiter plate cultures of HFF, Vero, CHO, or McCoy cells.
>
> **Specimen Source**
> Stool is the specimen of choice.
>
> **Epidemiology**
> Pseudomembranous colitis and antibiotic-associated diarrhea have been described in a wide variety of patients. These diseases have no seasonal incidence. A wide variety of antibiotics have been implicated in C. difficile associated disease, including penicillins, cephalosporins, lincosamides, macrolides, tetracyclines, aminoglycosides, as well as some antineoplastic agents. In fact, any drug that disturbs the normal bacterial flora of the gut can make the host susceptible to colonization and overgrowth by this pathogen.
>
> **Inactivators and Disinfectants**
> C. difficile spores are very hardy and can survive for months in the environment. Spores are resistant to most hospital disinfectants. Effective disinfectants include concentrated phenols and 10% household bleach solutions.

Bacterial cultures of *C. difficile* are not available in most microbiology laboratories. *Clostridium difficile* is grown anaerobically at 35-37°C on selective media, cycloserine-mannitol agar, cycloserine-mannitol-blood agar, or cycloserine-cefoxitin agar. The latter is the preferred medium (8). The bacterial culture takes 48-96 hours to grow and once the organism is isolated, a toxin assay must be performed to confirm the presence of toxins. Not all strains of *C. difficile* produce toxin and these nontoxigenic strains do not cause colitis or diarrhea. Compared with cytotoxin assays, the sensitivity of bacterial culture is 60-90% and the specificity is 85-99%. The limit of detectability of bacterial culture is about 100 colony forming units per gram of stool. Patients with antibiotic associated colitis generally have about 10^4-10^7 organisms per gram of stool. However, asymptomatic carriers (3-8% of the population) may have only a few organisms per gram of stool.

A number of tests are commercially available that can detect toxin A or a combination of toxin A and B. When compared with cytotoxin assays, the sensitivity of these kits range from 20-95% and the specificity from 90-98%. The advantages of using EIA methods over standard cytotoxin assays are (a) EIA tests are faster (3 hrs versus 24-48 hours), (b) cell cultures are not necessary, and (c) the level of required technical expertise is lower for EIA methods.

CELL CULTURE NEUTRALIZATION ASSAY

Toxigenic *C. difficile* strains produce nearly equal amounts of toxin A and toxin B. In addition to a potent enterotoxin activity, toxin A also has weak to moderate toxicity for cell cultures. Toxin B has no detectable enterotoxin activity but it is an extremely potent cytotoxin. As little as 1 pg of toxin B is sufficient to cause irreversible damage to the cytoskeletal system of cultured cells which, in turn, causes the "rounding" which is typical of *C. difficile* toxin activity (Fig. 1). The cell culture neutralization assay is primarily a toxin B test but the extreme sensitivity of this test and its excellent correlation with disease have made it the "gold" standard.

Although cells, antisera, and control reagents can be purchased commercially, there is little assay standardization and no external standards or CAP proficiency specimens routinely available. Therefore, the sensitivity of the cell culture neutralization assay can vary widely from laboratory to laboratory. Overall, the sensitivity of the cell culture neutralization assay has been estimated at 30-80% when compared with endoscopy or colonoscopy. Another disadvantage of cell culture neutralization is that it usually requires 24-48 hours before a result can be reported. Despite these disadvantages, the cell culture neutralization assay is the most sensitive and specific assay for the detection of *C. difficile* toxins.

Like other neutralization assays, the cell culture neutralization assay depends upon the ability of a specific antiserum to neutralize the cytotoxic effects of *C. difficile* toxins A and B. Neutralized and unneutralized fecal filtrates are compared in the same assay and, if the antitoxin neutralizes the cytotoxic effect of the filtrate, it is presumed that the cytotoxic activity is due to *C. difficile* toxins. Despite the fact that toxin B causes rounding in almost all mammalian cells, some cell lines are more sensitive to these toxins. The most sensitive cells include human foreskin fibroblasts (HFF), chinese hamster ovary (CHO), Vero, and McCoy cells (7). The neutralization assay can be done in tubes, shell vials, or in microtiter plates. Microtiter plates have an advantage because they require smaller reagent volumes and take up less incubator space.

Specimen Collection and Storage

At least 10 ml of diarrheal stool should be submitted in a leak-proof container without preservatives. Although *C. difficile* toxin has been detected in rectal swabs, these specimens are not recommended for *C. difficile* testing. Specimens should be held at 2-8°C until tested for the presence of toxin. Because the cytotoxic activity of toxin B is sensitive to proteases, changes in pH, and elevated temperatures, specimens should be processed and inoculated onto cell cultures immediately upon receipt in the laboratory. If immediate processing is not possible, the specimen can be stored at 2-8°C for up to 48 hours. If longer delays are anticipated, specimens should be stored at or below -70°C. Repeated freeze/thaw cycles should be avoided because they can adversely affect the activity of the cytotoxin. Do not freeze the specimens at -20°C.

Specimen Preparation

Prior to inoculation, specimens are usually treated to remove bacteria and fecal material. If the specimens are especially viscous, they can be diluted with saline, PBS, viral transport medium, or cell culture medium; vortexed vigorously; and centrifuged at 2-6,000 x g for 11 minutes to pellet the fecal material. The supernatant fluids can be forced through a 0.22-0.45 μm filter.

Microtiter Plate Procedure

1. Appropriately label two sterile tubes for each specimen. Note: Round or conical bottom microtiter plates can be used in place of tubes when handling a large number of specimens.
2. Place 25 μl of saline into the first tube and 25 μl of antitoxin into the second tube.
3. Add 25 μl of stool filtrate to each tube.
4. Vortex the tubes (or tap the plate) briefly to mix the contents. Toxin neutralization is nearly instantaneous and further incubations will not produce greater levels of toxin neutralization.
5. Add 50 μl of the diluted specimen to one well of a 96-well microtiter plate containing 90-100% confluent cell cultures and 0.2 ml of maintenance medium. Add 50 μl of the neutralized filtrate to an adjacent well.
6. Incubate the plates for 16-24 hours at 35-37°C in a humidified, 5% CO_2 incubator.
7. Examine the cells at 100-200X for the presence of typical CPE (Fig. 1).
8. If no CPE is present in the saline well, record the results and return the plate to the incubator.
9. Examine the plate again after 48 hours.

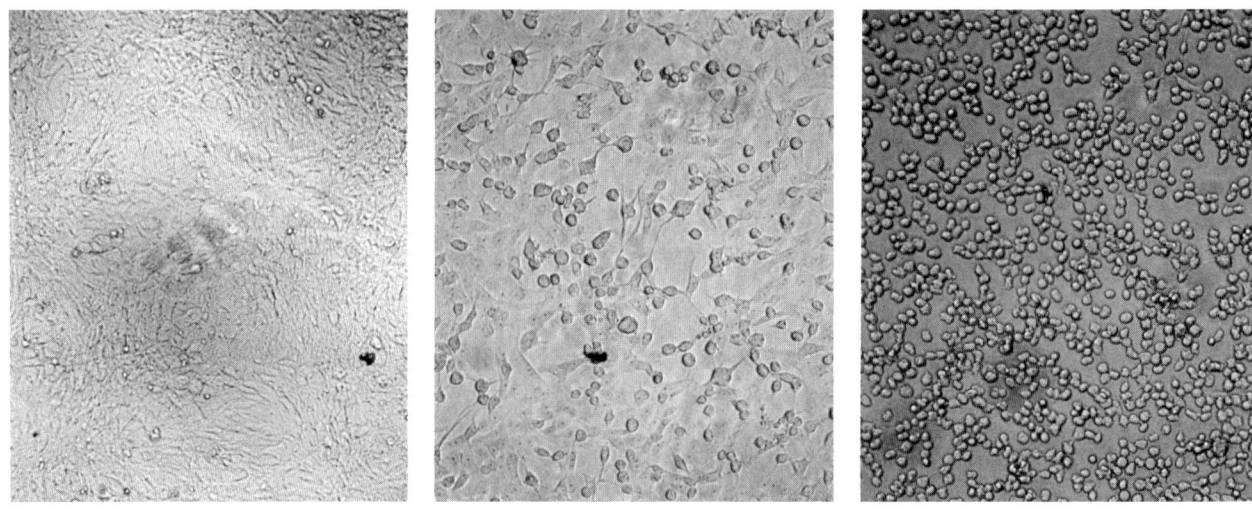

FIG. 1. Progression of cytotoxic effects caused by *Clostridium difficile* toxins. Normal Vero cell monolayer (left), 2+ CPE (middle), and 4+ CPE (right).

Tube/Vial Procedure
1. Appropriately label two cell culture tubes or vials for each specimen.
2. Aseptically remove the culture medium from the monolayers and add 1 ml of cell culture refeed media to each tube (0.5 ml to each vial).
3. Add 100 µl of saline to a separate test tube and 100 µl of antitoxin to a second tube.
4. Add 100 µl of stool filtrate to both tubes.
5. Tap the tubes gently to mix antisera and filtrate.
6. Add 200 µl of the diluted specimen to one cell culture tube and 200 µl of the neutralized filtrate to the other culture tube.
7. Incubate cells for 48 hours at 35-37°C.
8. Observe cells for cytopathic effect at 24 and 48 hours. Positive results can be observed as early as 4-6 hours after inoculation.

Interpretation of Results
Clostridium difficile toxin produces characteristic cell rounding with radiating arms of cytoplasm (Fig. 1) in most cell cultures. Approximately 70% of all cultures will be positive after 24 hours. Some cultures will produce CPE within 6 hours.

Positive Test. The presence of *C. difficile* toxin is indicated by the presence of characteristic CPE in the saline well and no CPE in the neutralization well.
REPORT: *Clostridium difficile* toxin detected.

Negative Test. The absence of *C. difficile* toxin is indicated by the lack of CPE in the saline well and the neutralization well.
REPORT: *Clostridium difficile* toxin not detected.
Inconclusive Test. Specimens that produce no CPE in the saline well and CPE in the neutralization well should be repeated because the specimens may have been reversed when they were added to the well. Extremely high levels of toxin can overwhelm the antitoxin, thereby producing CPE in the saline well and the neutralization well. Fecal filtrates from these specimens should be diluted 1:10, 1:100, and 1:1000 and retested as above. If no clear neutralization is demonstrable, the CPE may be due to a virus (e.g., enterovirus) or a different toxin.
REPORT: Cytotoxin detected. Negative for *C. difficile* toxin.

QC Procedures.
Clostridium difficile toxin and antitoxin sets are commercially available. Positive (*C. difficile* toxin)

and negative controls should be used in each neutralization assay to assure that the neutralizing antibody is working properly. The controls must be neutralized and inoculated onto cells in the same manner as the patient specimen. The positive (*C. difficile* toxin) control should produce rounding of the cells and radiating arms of cytoplasm within 24 hours. The cells may also appear vacuolated and granular as the CPE progresses. The negative control (cell control) and the neutralized toxin control and the antitoxin control should appear normal. Each microtiter plate well, cell culture tube or vial should be examined with a microscope prior to use to ensure that (a) a healthy, confluent monolayer is present, (b) there is no bacterial or fungal contamination, and (c) an adequate volume maintenance medium is present in the well.

REFERENCES

1. Woods GL, Iwen PC. Comparison of dot immunobinding assay, latex agglutination, and cytotoxin assay for laboratory diagnosis of *Clostridium difficile*-associated diarrhea. *J Clin Microbiol* 1990;28:855-857.
2. Lyerly DM, Krivan HC, Wilkins TD. *Clostridium difficile*: Its disease and toxins. *Clin Microbiol Rev.* 1988;1:1-18.
3. Bolton RP, Tait SK, Dear PRF, Losowsky MS. Asymptomatic neonatal colonization by *Clostridium difficile*. *Arch Dis Child*. 1984;59:466-472.
4. Lyerly DM, Ball DW, Toth J, Wilkins TD. Characterization of cross-reactive proteins detected by culturette brand rapid latex test for *Clostridium difficile*. *J Clin Microbiol.* 1988;26:397-400.
5. Borrello SP, Barclay FE, Reed PJ, Welch AR, Brown JD, Burdon DW. Analysis of latex agglutination test for *Clostridium difficile* toxin A (D-1) and differentiation between *C. difficile* toxins A and B and latex reactive protein. *J Clin Pathol.* 1987;40:573-580.
6. Kamiya S, Nakamura S, Yamakawa K, Nishida S. Evaluation of a commercially available latex immunoagglutination test kit for detection of *Clostridium difficile* D-1 toxin. *Microbiol Immunol* 1986;30:177-181.
7. Maniar AC, Williams TW, Hammond GW. Detection of *Clostridium difficile* toxin in various tissue culture monolayers. *J Clin Microbiol.* 1987;25:1999-2000.
8. Iwen PC, Booth SJ, Woods GL. Comparison of media for screening of diarrheic stools for the recovery of *Clostridium difficile*. *J Clin Microbiol.* 1989;27:2105-2106.

CHAPTER 11

Cytomegalovirus

INTRODUCTION

Cytomegalovirus (CMV) is a human herpesvirus and a member of the betaherpesvirinae subfamily. CMV is the largest member of the herpesviruses. In electron micrographs, CMV virions appear as a 64 nm core enclosed by a 110 nm icosahedral capsid. The capsid is surrounded by an amorphous tegument and a 200 nm diameter lipid-containing envelope. The CMV genome consists of a linear, double-stranded (240 kilobase pair) DNA molecule that codes for more than 30 polypeptides. Although CMV is relatively unstable, it can persist for several hours on surfaces and skin (1). CMV is sensitive to drying and can be inactivated by heat, alcohols, detergents, and organic solvents.

Cytomegalovirus is found throughout the world and 40-100% of adults have antibodies to the virus (2). CMV antibody prevalence is highest in developing countries and in areas where people are living in crowded, intimate conditions. The number of CMV-positive individuals in a population increases slowly through childhood then rises rapidly during adolescence and early adulthood. There appears to be no seasonal variation in CMV prevalence.

CMV is principally transmitted through close contact with saliva, urine, breast milk, cervical secretions, blood, and semen of infected individuals. Transmission can also occur through blood transfusions and organ transplantation. The incubation period has not been well defined but it is assumed to be 3-8 weeks following transfusion and 3-12 weeks after perinatal infections. Once the infection is established, patients frequently shed virus from several sites for weeks to years before the infection becomes latent (3). Episodes of recurrent infection with renewed viral shedding are common.

The vast majority of CMV infections are asymptomatic. However, several serious, often life-threatening syndromes have been associated with CMV infection including (a) heterophile-negative mononucleosis-like syndrome in young adults, (b) cytomegalic inclusion disease in congenitally infected infants, (c) pneumonitis in young infants, (d) interstitial pneumonia, retinitis, and febrile illness among organ transplant and other immunocompromised patients, and (e) a post-transfusion syndrome characterized by pneumonitis, hepatitis, and atypical lymphocytosis, especially in premature infants (3,4).

The best means of diagnosing productive CMV infection is by virus isolation. However, virus isolations cannot distinguish primary infections from reactivations. Virus isolation plus demonstration of CMV IgM, seroconversion, or a four-fold rise in antibody titer are the best procedures for identifying recent or active infections in normal persons. Demonstration of virus and CMV IgM is perhaps the best method for detecting primary infection.

In immunocompromised or immuno-suppressed individuals, it is important to distinguish primary from recurrent CMV infections because primary CMV infections have increased virulence in this population. Virus isolation and the presence of CMV IgM are not sufficient to establish a diagnosis of primary CMV infection in these individuals because IgM responses can occur during CMV reactivation. Therefore, demonstrating seroconversion is essential. Establishing the immune status of an organ donor and the recipient is also important.

A number of different assays including complement fixation, virus neutralization, indirect immunofluorescence, anticomplement immunofluorescence, latex agglutination, indirect hemagglu-

AT A GLANCE...

CYTOMEGALOVIRUS

Virus Detection Methods
Tube Cultures of human diploid fibroblasts, HFF, WI-38, and MRC-5 cells.
Centrifugation-enhanced (shell vial) cultures using MRC-5, HFF, HEL, or ML cells.
Direct fluorescent antibody staining of bronchial cell smears.

Specimen Source
Specimens of choice are urine, throat swabs or washings, bronchial washings or blood.

Time to Result
Standard Culture: Positive Culture - 1 day to 4 weeks. Negative Culture - 4 weeks.
Shell Vial Method: Positive Culture - 1-2 days. Negative Culture - 2 days.
Direct Fluorescent Antibody Method: 30-45 minutes.

CMV Serology

- **IgM** Usually appears 3-4 days after the appearance of symptoms and is present for several months. IgM may or may not be present during CMV reactivation. Because IgM does not pass through the placenta, the presence of CMV-IgM in cord blood is evidence of congenital infection.

- **IgG** Usually present 7-14 days after symptoms appear and antibody levels fluctuate over the patient's lifetime. Demonstration of a four-fold rise in antibody titer in acute and convalescent specimens is suggestive of an active primary or recurrent infection. IgG1 and IgG3 are the predominant subclasses seen in healthy donors. IgG1 may be higher during CMV reactivations.

- **IgA** Usually present at the onset of symptoms. IgA may be present for long periods of time in normal, healthy individuals. Serum containing CMV IgA contain both IgA1 and IgA2.

Epidemiology
Cytomegalovirus infections are common throughout the world. By puberty, 40-80% of children are infected and by age 35, almost 100% of the population is infected. In pregnancy, 1-2% of seropositive individuals shed CMV from the cervix. Cervical shedding increases to 10-15% during the last trimester and the resulting risk of perinatal infection is 0.4-2.6%.

Transmission/Incubation Period
CMV is transmitted by contact with body fluids, through blood transfusion, or organ transplantation. The most common routes of infection are from close contact with infected oropharyngeal secretions, vaginal or cervical secretions, semen, urine, breast milk or blood. Congenitally or perinatally infected children often shed large amounts of virus in urine for 5 or more years and in saliva for 2-4 years. The incubation period is 3-8 weeks following transfusion and 3-12 weeks following perinatal infection.

Inactivators and Disinfectants
CMV is very labile and is quickly inactivated by heat, alcohols, organic solvents, and repeated freeze/thaw cycles. Effective disinfectants include 70% ethanol, quaternary ammonium compounds, phenols, iodophor compounds, and glutaraldehyde compounds.

tination, radioimmunoassay, and enzyme immunoassays have been used for the detection of CMV antibodies. Following CMV infection, IgM usually appears 3-4 days after the appearance of symptoms and remains detectable for several months. IgM may or may not be present during CMV reactivation. Because IgM does not pass through the placenta, the presence of CMV-IgM in cord blood is evidence of congenital infection. CMV IgG is usually present 7-14 days after the onset of symptoms. IgG levels will fluctuate over the patient's lifetime but once established, antibody levels will usually remain detectable. Demonstration of a four-fold rise in antibody titer in acute specimens collected early in disease and convalescent specimens collected 3-4 weeks later is suggestive of an active primary or recurrent infection. During primary infection, CMV IgA is present at the onset of symptoms. However, IgA may be present for long periods of time in normal, healthy individuals.

VIRUS ISOLATION

Introduction

Virus isolation provides the best method for the diagnosis of productive CMV infections. However, traditional cell culture methods have limited clinical utility because physicians may not receive isolation results for 3-4 weeks. To decrease isolation times, many laboratories have begun to use the centrifugation-enhanced (shell vial) culture methods originally described by Gleaves, et al. (5). Shell vial methods can accommodate the same clinical specimens as traditional tube cultures and can reduce CMV isolation time from 28 days to 1-2 days.

CMV grows readily in human diploid fibroblast cell lines and virus isolations are most often accomplished in MRC-5, human foreskin fibroblasts (HFF), human embryonic lung (HEL), or WI-38 cells. Mink lung (ML) cells may be less susceptible to the toxic effects of blood specimens and seem to produce more fluorescent foci sites when used in shell vial procedures (6,7). Isolation times vary from lab-to-lab and depend upon a number of factors including (a) virus concentration, (b) volume of inoculum, and (c) culture conditions. Some laboratories perform blind passages on negative cultures after 14 days while others incubate the original culture for 14-28 days without subpassage. In tube cultures, CPE may be visible the first day after inoculation and rapidly spread throughout the cell monolayer. However, most isolates require 5-7 days of cultivation to produce visible CPE. The best virus recovery is obtained with 90% confluent monolayers that are actively growing (8). To accomplish this, cells should be grown and maintained in growth medium containing 10% fetal calf serum (FCS). For best isolation results, two diploid fibroblast culture tubes (e.g., MRC-5 and HFF) and at least two MRC-5 shell vials should be inoculated.

Specimen Collection and Storage.

Urine, blood (buffy coats), throat washings, and broncheoalveolar lavages are the specimens of choice. However, CMV has also been isolated from breast milk, semen, cervical secretions, gastrointestinal specimens, saliva amniotic fluid, and tissues. CMV is relatively labile and specimens should be inoculated onto cell cultures immediately upon receipt in the laboratory. If this is not possible, specimens can be stored at 4°C for up to 48 hours. **DO NOT FREEZE SPECIMENS AT -20°C** as CMV is rapidly inactivated at this temperature. Alternatively, specimens can be frozen at -70°C or below in the presence of 30% sorbitol (8)

Specimen Preparation
Blood (Buffy Coats)

White cells are separated from red cells by centrifugation through a ficoll-hypaque density gradient. A number of ready-made gradient materials can be purchased commercially (Ficol-Paque from Pharmacia Fine Chemicals, Histopaque from Sigma Chemical Company or Lymphocyte Separating Medium from Organon-Teknika) for this purpose or gradient materials can be prepared in the laboratory (Appendix C). Separation should be done as directed by the manufacturer or as described below.

1. Aseptically collect 10 ml of blood in green

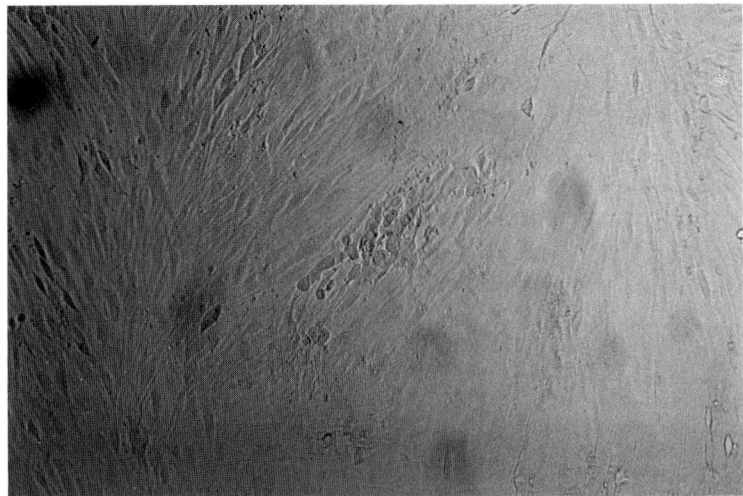

FIG. 1. Focal CMV CPE in a human foreskin fibroblast tube culture 5 days after infection.

top vacutainer tubes containing preservative-free heparin.
2. Place 3 ml of ficoll-hypaque in clear, sterile 15 ml centrifuge tubes.
3. Dilute the heparinized blood with an equal volume of sterile saline.
4. Carefully overlay 5 ml of heparinized blood onto the ficoll-hypaque.
5. Centrifuge the tubes at 400 x g in a swinging bucket rotor for 30 minutes at room temperature.
6. Lymphocytes will band at the plasma-ficoll interface.
7. Carefully remove and discard the clear upper (plasma) layer.
8. Remove the lymphocyte band and the ficoll layer down to the red blood cells. Pool the buffy coat preparations from a single patient in a sterile 50 ml centrifuge tube.
9. Add PBS to the 30 ml mark.
10. Centrifuge the cells at 200 x g for 15 minutes and remove the supernatant fluids.
11. Resuspend the cell pellet in 1 ml of viral transport medium.

Tissues

Tissue specimens should be weighed, minced with sterile scissors, and homogenized using a disposable tissue homogenizer. Add enough viral transport medium to the homogenizer to produce a 10-20% (w/v) suspension. Remove the suspension from the homogenizer and vortex vigorously to extract as much antigen as possible. Centrifuge the specimen at 300-600 x g for 10 minutes at 4°C but do not separate the tissue from the supernatant fluids. Carefully remove the required amount of specimen for inoculation.

Standard Tube Culture
Procedure
1. Appropriately label two actively growing human fibroblast cell culture tubes.
2. Add 0.2-0.5 ml of specimen to each tube.
3. Incubate the tubes at 35-37°C in a roller drum (10-15 rph) for 28 days.
4. Examine cultures at least every other day for 14 days and weekly thereafter.
5. Replace the culture medium weekly or whenever it becomes acidic. Toxic cultures should be transferred to new culture tubes.

Interpretation of Results
Positive Test. Presumptive identification of CMV is based upon the presence of focal areas of large, rounded, refractile cells in an otherwise undisturbed cell sheet (Fig. 1). CMV is cell-associated and spreads slowly from cell-to-cell until the entire cell sheet is involved. Positive cultures should be

scraped and stained with monoclonal antibodies to cytomegalovirus.
REPORT: No report should be made based solely upon the presence of CPE.

Negative Test. The absence of characteristic CPE after 28 days indicates the absence of viable CMV.
REPORT: Cytomegalovirus not detected.

Culture Confirmation Procedure
1. Once CPE develops, remove all but a few drops of culture medium from the tube.
2. Scrape the cells from the surface of the tube and suspend the cells in the remaining fluid.
3. Using a pipette, place a drop of the cell suspension onto an acetone-cleaned slide. Retain the remainder of the cell suspension for passage into new culture tubes.
4. Allow the slide to air dry at room temperature.
5. The slide should be processed as directed by the manufacturer of the monoclonal antibody to CMV or as described below.
6. Fix the slide in cold acetone for 10-15 minutes and allow the slide to air dry. Slides should be stained immediately after fixation. Alternatively, fixed slides may be stored for up to 1 week at 2-8°C or one year at -20°C or below. Slides should be stored under desiccated conditions in order to minimize antigen degradation and background staining.
7. Add enough fluorescein-labelled CMV antibody to cover the cell smear (15-30 μl).
8. Incubate the slide for 30 minutes at 35-37°C in a covered, humidified chamber. Do not allow the antibody to dry as this could cause nonspecific antibody binding.
9. Remove the excess antibody with a gentle stream of PBS. Do not direct the stream directly at the cell smear as this could dislodge the cells.
10. Soak the slide in PBS for 5 minutes, shake off the excess fluid, and allow the slide to air dry.
11. Carefully add a small drop of FA mounting fluid to the center of the smear.
12. Place a number 1 coverslip on the mounting fluid and carefully remove all the air bubbles.
13. Immediately examine the slide using a fluorescence microscope. For optimum clarity, use 200-300X magnification for screening and 400X for confirmation.

Interpretation of Results
Positive Test. The pattern of specific fluorescence will vary depending upon the type of monoclonal antibody used. If antibodies to CMV early antigens are used, a positive result is indicated by the presence of characteristic CPE in the tube culture **and** intense apple green fluorescence in the nucleus of the infected cells. If antibodies to CMV late antigens are used, a positive result is indicated by the presence of nuclear and diffuse cytoplasmic fluorescence. Only experienced individuals should interpret FA patterns because cell debris and cell clumps may exhibit a dull fluorescence which can be interpreted as a specific, positive reaction by inexperienced technologists.
REPORT: Cytomegalovirus isolated.

Negative Test. The absence of CMV virus is indicated by the uniform red coloration of the cells and/or the absence of specific staining.
REPORT: Cytomegalovirus not isolated.

QC Procedures.
Subpassages of CMV clinical isolates or ATCC cultures should be inoculated with each batch of CMV isolations. Uninfected cell cultures can serve as negative controls. Infected and uninfected cell monolayers should be scraped and stained as described above. Positive controls must produce focal CPE and the cell smears must exhibit a typical staining pattern. Negative controls must stain with the red counterstain and should not exhibit any CPE or specific fluorescent staining.

Preparation of Positive Control Inocula
1. Prepare a 1:1000 dilution of a vigorously growing CMV isolate or an ATCC culture in GLB or MEM containing 10% FCS. Hold the dilution on ice until ready for inoculation.
2. Remove the growth medium from one 75 cm^2

flask containing an 80-90% confluent HFF or MRC-5 cell monolayer.
3. Add 2 ml of the inoculum to the monolayer and allow the virus to adsorb for 1-2 hours at 35-37°C. The flask should be rocked every 15 minutes to assure even distribution of the inoculum and to prevent monolayer desiccation.
4. Add 10 ml of MEM containing high glucose, L- glutamine, and 10% FCS.
5. Incubate the flask at 35-37°C until the CPE involves 80-100% of the monolayer (usually 5-7 days). Replace the medium after 3-5 days or if the medium becomes acidic.
6. Scrape the cells from the flask into the growth medium.
7. Transfer the cell suspension to a 50 ml polypropylene centrifuge tube and freeze the suspension quickly at -70°C.
8. Thaw the suspension quickly in a 37°C water bath.
9. Centrifuge the cells at 600 x g for 15 minutes to pellet the cell debris.
10. Transfer the supernatant fluids to a sterile 50 ml centrifuge tube and add MEM containing 20% FCS to the 50 ml mark.
11. Dispense 1 ml of the cell suspension into each of 50 freezing vials and freeze the cells at -70°C or below. Store the vials for up 2 years at -70°C or 5 years in liquid nitrogen.

Centrifugation Enhanced (Shell Vial) Method

Rapid detection of cytomegalovirus is essential for the timely administration of antiviral therapeutic measures and to identify at-risk neonates and transplant patients. The centrifugation-enhanced (shell vial) method described by Gleaves, et al. (5) has been almost universally adopted for CMV isolations in a wide variety of clinical specimens (10). The use of monoclonal antibodies to CMV early antigens allows this procedure to detect CMV before CPE appears in the cell monolayers. Since its introduction in 1984, a number of procedural modifications have been described including the use of dexamethasone and dimethylsulfoxide (11-13) and mink lung cells (6,7). Although HFF and MRC-5 cells have been used for CMV shell vial isolations, MRC-5 cells appear to be more sensitive than HFF cells (14). Centrifugation enhanced methods are rapid and specific but they may not detect slow-growing CMV isolates (15).

Procedure
1. Appropriately label two or three shell vial cultures for each specimen.
2. Aseptically remove the culture medium from the vials.
3. Inoculate each vial with 0.2-0.5 ml of the specimen.
4. Centrifuge the vials at 700 x g for 40 minutes at 36°C. If maintaining this temperature is not possible, cultures can be centrifuged at room temperature.
5. Add 1 ml of MEM containing 2% FCS and antibiotics to each vial.
6. Incubate the vials at 35-37°C. Stain one vial after 16-24 hours and the other vial after 40 hours. Some laboratories inoculate a third vial and stain it after 96 hours.

Staining of Coverslips

Identification of CMV in shell vial systems is accomplished by staining the coverslips with monoclonal antibodies to the 72 kilodalton immediate-early proteins of CMV. Most specimens will produce 20-2000 stained nuclei after 48 hours. Fixation and staining should be done as directed by the manufacturer of the monoclonal antibodies or as described below.
1. Carefully remove the culture fluids from the vial and wash the coverslip twice by adding 1-2 ml of PBS and soaking for 5 minutes.
2. Remove the PBS and add 2-3 ml of acetone to each vial. Fix the coverslips for 10 minutes at room temperature.
3. Remove the acetone and allow the coverslips to air dry.
4. Rinse the coverslip briefly with PBS. The moisture trapped between the coverslip and the shell vial will keep the stain from wicking under the coverslip.
5. Add enough monoclonal antibody to the vial to cover the coverslip (100-150 μl) and incubate 30 minutes at 35-37°C. The shell

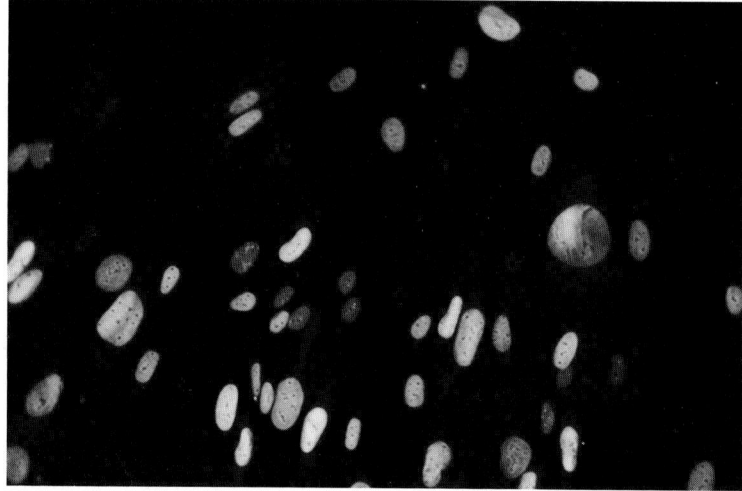

FIG. 2. Immunofluorescent staining CMV-infected MRC-5 shell vial culture 24 hours after inoculation.

vials should be covered or incubated in a humidified chamber to prevent the antibody from drying.
6. Wash the coverslips twice as previously described (step 1) and remove the PBS.
7. Add enough FITC-labelled conjugate to the vial to cover the coverslip (100-150 µl) and incubate 30 minutes at 35-37°C as before.
8. Wash the coverslips twice as previously described (step 1) and remove the PBS.
9. Place a small drop of mounting fluid on a slide
10. Remove the coverslip from the vial and lay it cell side down on the mounting fluid.
11. Immediately examine the coverslips at 100-300X using a fluorescence microscope.

Interpretation of Results

Coverslips that have individual cells or small groups of cells that exhibit brilliant apple-green nuclear fluorescence (Fig. 2) are positive for cytomegalovirus. Coverslips without specific fluorescence are considered negative. Some specimens may trap the monoclonal antibody or the conjugate between the cell monolayer and inoculum cells. Careful examination of the cell morphologies can distinguish this type of reaction from specific CMV reactions. In addition, nonspecific staining can occur if the monoclonal reagents are allowed to dry on the coverslip. In this case, the entire cell sheet may stain apple-green.

Positive Test. The presence of CMV is indicated by brilliant, apple green fluorescence in the nucleus of infected cells (see Color Plate 4C following Chapter 7).
REPORT: Cytomegalovirus isolated.

Negative Test. The absence of viable CMV is indicated by the lack of specific fluorescence described above. The entire cell sheet should be stained red by the counterstain.
REPORT: Cytomegalovirus not detected.

QC Procedures.

A 1:10 dilution of a recent CMV isolate or an ATCC culture should be inoculated with each batch of CMV shell vial cultures. Uninfected shell vial cultures serve as negative controls. Infected and uninfected shell vial cultures should be processed and stained as described for the patient cultures. Positive controls must exhibit the typical fluorescence patterns described above. Negative controls must not exhibit specific fluorescent staining.

DIRECT SMEARS

Introduction

Although virus isolation is the gold standard for the laboratory diagnosis of CMV infections, virus lability and long isolation times can limit the diagnostic utility of most CMV culture systems. Direct fluorescent antibody (DFA) staining of broncheoalveolar lavage specimens can provide a rapid adjunct test to cell culture isolations (16,17).

In contrast with 1-28 day isolation times, DFA staining of cell smears can usually be accomplished in 1-2 hours. Although rapid, the DFA methodology is not without problems and the sensitivity of this method has yet to be determined with other types of specimens. When used alone, DFA methods can give false negative results when testing specimens containing only free CMV and with specimens containing viruses other than CMV. Exclusive use of DFA reagents could also miss any new infectious agent that might appear in the community. For these reasons, backup cultures should be inoculated for all DFA-stained specimens.

Specimen Preparation

The success of the DFA procedure depends upon the preparation of an adequate cell smear. Smears that are too thick or "lumpy" can cause nonspecific trapping of the DFA reagent. Smears that are grossly contaminated with red blood cells can also cause interpretation difficulties due to red cell autofluorescence. Finally, cell smears containing too few cells could cause the laboratory to report a false-negative result because no positive cells were observed.

Test Procedure for BAL Samples

1. Centrifuge the BAL specimen at 600 x g for 10 minutes at 4°C to pellet the cellular fraction.
2. Suspend the cells in 10 ml of PBS and mix gently.
3. Centrifuge the cells as before and remove all but a few drops of fluid.
4. Suspend the cells in the remaining fluid.
5. Using a pipette, place a drop of the cell suspension on an acetone-cleaned slide.
6. Allow the smear to air dry at room temperature.
7. Fix and stain the slide as directed by the manufacturer of the labeled monoclonal antibody to CMV late antigens or as described below.
8. Fix the cells by immersing the slides in acetone (room temperature) for 10 minutes. Allow the slides to air dry.
9. Place enough of the fluorescein-labelled monoclonal antibody on the slide to cover the cell smear (approximately 15-30 μl).
10. Incubate the slide for 30 minutes at 35-37°C in a covered, humidified chamber. Do not allow the antibody to dry. Drying could cause nonspecific antibody binding.
11. Remove the excess antibody with a gentle stream of PBS. Do not direct the stream directly at the cell smear as this could dislodge the cells.
12. Soak the slide in PBS for 5 minutes, shake off the excess fluid, and allow the slide to air dry.
13. Carefully add a small drop of FA mounting fluid to the center of the smear.
14. Place a number 1 coverslip on the mounting fluid and carefully remove all the air bubbles.
15. Immediately examine the slide using a fluorescence microscope. For optimum clarity, use 200-300X magnification for screening and 400X for confirmation.

Interpretation of Results

Positive Test. The presence of CMV is indicated by apple green fluorescence that covers the entire cell (Fig. 3). At least three positive cells should be present in the smear before reporting a positive result. An indeterminate report should be given if only 1 or 2 fluorescing cells are observed. Only intact cells should be examined because cell debris and clumps of normal cells may exhibit a dull fluorescence which could be misinterpreted as a specific, positive reaction.

REPORT: Cytomegalovirus detected.

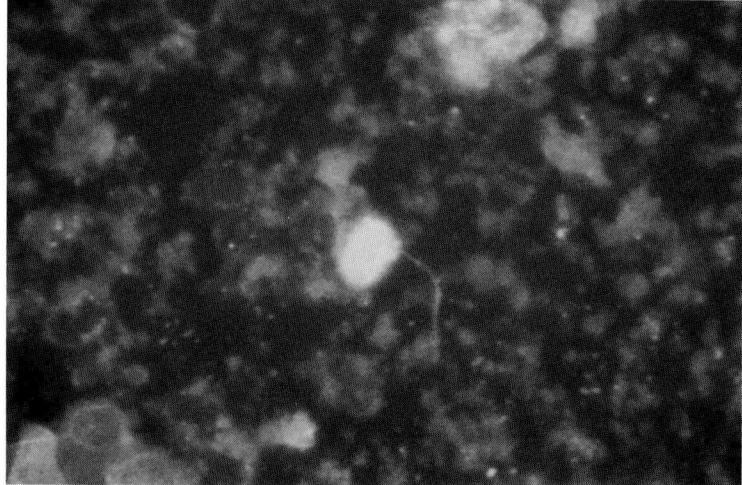

FIG. 3. Direct immunofluorescent antibody staining of a broncheoalveolar lavage specimen with monoclonal antibodies to CMV.

Negative Test. The absence of CMV is indicated by the lack of intense apple green fluorescence in the cells.
REPORT: Cytomegalovirus not detected by direct smear. This report does not preclude the possibility of CMV.

Indeterminate Test. Specimens with only 1 or 2 fluorescing cells are considered indeterminate and an additional specimen should be requested.
REPORT: Unacceptable specimen - too few cells.

QC Procedures.
Positive and negative controls should be stained at least once each day to assure that the antibody reagent is performing properly. Positive cells should stain intensely as described above. Negative cells should be red and should not exhibit specific apple green fluorescence. Prepared slides should be stained whenever direct smears are stained. Control slides can be purchased commercially from a number of vendors or they can be prepared in the laboratory as described below.

Preparation of Control Slides
1. Prepare a 1:1000 dilution of the positive control culture (described above) in GLB or MEM containing 10% FCS. Hold the dilution on ice until ready for inoculation.
2. Remove the growth medium from one 75 cm^2 flask containing an 80-90% confluent HFF cell monolayer.
3. Add 2 ml of the inoculum to the monolayer and allow the virus to adsorb for 1-2 hours at 35-37°C. The flask should be rocked every 15 minutes to assure even distribution of the inoculum and to prevent monolayer desiccation.
4. Add 10 ml of MEM containing high glucose, L- glutamine, and 10% FCS.
5. Incubate the flask at 35-37°C until the CPE involves 30-60% of the monolayer (usually 3-5 days). Replace the medium if it becomes acidic.
6. Trypsinize the cells and remove them from the flask.
7. Centrifuge the cells at 150 x g for 5 minutes to pellet the cells.
8. Suspend the cell pellet in 5 ml of PBS containing 2% FCS.
9. Perform a cell count and adjust the cell concentration to 2-5 x 10^6 cells/ml.
10. Dispense 3-10 µl of the cell suspension onto each slide and allow the suspension to air dry.
11. Fix the slides in acetone at room temperature for 10 minutes and allow the slides to air dry.

12. Store the slides at -20°C or below for up to one year. Slides should be stored under desiccated conditions to minimize antigen degradation and background fluorescence.

NOTE: Slides prepared in this manner will contain both positive and negative cells. Therefore, a single slide can be used for most QC purposes. When slides are made from flasks with 100% CPE, two separate slides must be used - one containing a positive cell smear and a second slide containing an uninfected cell smear.

REFERENCES

1. Hutto C, Little EA, Lee JD, Pass RF. Isolation of cytomegalovirus from toys and hands in a day care center. *J Infect Dis* 1986;154:527-530.
2. Krech U. Complement-fixing antibodies against cytomegalovirus in different parts of the world. *Bull WHO* 1973;49:103-106.
3. Stagno S, Britt WJ, Pass RF. Cytomegalovirus. In: Schmidt NJ, Emmons RW, eds. *Diagnostic procedures for viral, rickettsial, and chlamydial infections.* Sixth Edition. Washington DC: American Public Health Association, 1989;321-378.
4. Alford CA, Britt WJ. Cytomegalovirus. In: Fields BN, Knipe DM, Chanock RM, Hirsch MS, Melnick JL, Monath TP, Roizman B, eds. *Virology*, Second Edition. New York: Raven Press, 1990;1981-2010.
5. Gleaves CA, Smith TF, Shuster EA, Pearson GR. Rapid detection of cytomegalovirus in MRC-5 cells inoculated with urine specimens by using low-speed centrifugation and monoclonal antibody to an early antigen. *J Clin Microbiol* 1984;19:917-919.
6. MacKenzie D, McLaren LC. 1989. Increased sensitivity for rapid detection of cytomegalovirus by shell vial centrifugation assay using mink lung cell cultures. *J. Virol Methods* 1989;26:183-188.
7. Gleaves CA, Hursh DA, Meyers JD. Detection of human cytomegalovirus in clinical specimens by centrifugation culture with a nonhuman cell line. *J Clin Microbiol* 1992;30:1045-1048.
8. Fedorko DP, Ilstrup DM, Smith TF. Effect of age of shell vial monolayers on detection of cytomegalovirus from urine specimens. *J Clin Microbiol* 1989;27:2107-2109.
9. Weller TH, Hanshaw JB. Virologic and clinical observations on cytomegalic inclusion disease. *N Eng J Med* 1962;266:1233-1244.
10. Paya CV, Wold AD, Smith TF. Detection of cytomegalovirus infections in specimens other than urine by the shell vial assay and conventional tube cell cultures. *J Clin Microbiol* 1987;25:755-757.
11. West PG, Aldrich B, Hartwig R, Haller GJ. Enhanced detection of cytomegalovirus in confluent MRC-5 cells treated with dexamethasone and dimethyl sulfoxide. *J Clin Microbiol* 1988;26:2510-2514.
12. Espy MJ, Wold AD, Ilstrup DM, Smith TF. Effect of treatment of shell vial cultures with dimethyl sulfoxide and dexamethasone for detection of cytomegalovirus. *J Clin Microbiol* 1988;26:1091-1093.
13. Thiele GM, Woods GL. The effect of dexamethasone on the detection of cytomegalovirus in tissue culture and by immunofluorescence. *J Virol Methods* 1988;22:319-328.
14. Gleaves CA, Meyers JD. Comparison of MRC-5 and HFF cells for identification of cytomegalovirus in centrifugation culture. *Diagn Microbiol Infect Dis* 1987;6:179-182.
15. Ashley R, Peterson E, Abbo H, Gold D, Corey L. Comparison of monoclonal antibodies for rapid detection of cytomegalovirus in spin-amplified plate cultures. *J Clin Microbiol* 1989;27:2858-2860.
16. Emanuel D, Peppard J, Stover D, Gold J, Armstrong D, Hammerling U. Rapid immunodiagnosis of cytomegalovirus pneumonia by broncheoalveolar lavage using human and murine monoclonal antibodies. *Ann Intern Med* 1986;104:476-481.
17. Gleaves CA, Meyers JD. Rapid detection of cytomegalovirus in broncheoalveolar lavage specimens from marrow transplant patients: Evaluation of a direct fluorescein-conjugated monoclonal antibody reagent. *J Virol Methods* 1986;26:345-350.

CHAPTER 12

Enteroviruses

INTRODUCTION

The genus enterovirus comprises one of the four genera belonging to the family *Picornaviridae*. Enteroviruses include polioviruses (3 serotypes), coxsackie A viruses (23 serotypes: types 1-17, 19-22 and 24), coxsackie B viruses (6 serotypes), echoviruses (30 serotypes: 1-7, 9, 11-27, 29-33), and the newer enteroviruses (serotypes 68-71). Human hepatitis A virus (Chapter 14) is classified as enterovirus 72. Enteroviruses are small (20-30 nm) non-enveloped icosahederal viruses. Their 7.5 kilobase, positive sense, single-stranded RNA genome codes for a single polycystronic message that is post-translationally modified into four structural proteins (VP1-4). Enteroviruses can be distinguished from their rhinovirus cousins by their characteristic CPE, ability to replicate at 35-37°C, and their stability at pH 3. Enteroviruses are extremely hardy and persist for long periods of time in water milk, food, feces and sewage. Thermostability is significantly enhanced in the presence of $MgCl_2$. Enteroviruses are resistant to chloroform, ether, many household disinfectants (70% alcohol, 5% Lysol or Amphyl, and 1% quaternary ammonium compounds) and most detergents. The presence of organic material can further protect enteroviruses from inactivation.

Enteroviruses are responsible for a wide variety of diseases. Polioviruses are the prototypic enteroviruses and clinically, they are the most significant members of the enterovirus genus. In developing countries, polioviruses cause paralytic disease in 4 of every 1,000 school-age children (1). Nonpolio enteroviruses are responsible for 5-10 million symptomatic infections each year and they are the leading cause of aseptic meningitis, viral myocarditis, and nonspecific febrile exanthematous illnesses (2-4). Although 50-80% of enterovirus infections are asymptomatic, these viruses still cause hepatitis, pleurodynia, stomatitis, and neonatal sepsis in a significant number of patients each year.

Coxsackie viruses produce a variety of syndromes ranging from aseptic meningitis, herpangina (coxsackie A), epidemic myalgia and pleurodynia (coxsackie B), hand, foot and mouth disease (coxsackie A16), myocarditis and pericarditis (coxsackie B), generalized disease in the newborn (coxsackie B), pneumonia, rashes and common colds. Echoviruses are typically associated with febrile illnesses which may include rashes and common colds. Acute hemorrhagic conjunctivitis has been associated with enterovirus 70 and coxsackie A24.

Approximately 75% of enteroviral infections occur in children under 15 years of age and attack rates are highest in children under 1 year of age. Aseptic meningitis is most commonly recognized in very young infants while pleurodynia and myocarditis are found predominantly in adolescents and young adults. Symptomatic enteroviral infections in elderly individuals are unusual. Among children, boys are at greater risk of illness, acquiring aseptic meningitis and poliomyelitis almost twice as often as girls. After puberty, females are more likely to acquire enteroviral infections, perhaps because adult women have greater exposure to children who are shedding virus. Pregnancy and vigorous exercise prior to infection also appear to enhance the severity of enteroviral infections.

Enteroviruses are most efficiently transmitted through fecal-oral contact. However, virus is shed from the oropharynx and transmission can occur through inhalation of virus-laden aerosols. The incubation period ranges from 2 to 25 days with an average of 3-5 days. The virus initially gains entry

AT A GLANCE...

ENTEROVIRUSES

Virus Detection Methods
No single cell culture system is suitable for recovery of all enteroviruses. Most laboratories utilize a combination of at least three cell cultures for enterovirus isolation including a primary monkey kidney culture (PMK, RMK, or CMK), a human cell culture system (HFF, MRC-5, HEK, HeLa, HEp-2, or WI-38), and RD cells. Virus typing is accomplished via virus neutralization.

Specimen Sources
Enteroviruses may be isolated from stool, throat swabs, CNS and spinal fluids, blood, urine, skin, and conjunctiva. Virus may be isolated from throat swabs during the first week of illness and from stool for 3 weeks to several months. Two specimens (rectal and throat swabs) should be collected for optimal enteroviral isolation. Exceptions are enteroviruses that cause acute hemorrhagic conjunctivitis (enterovirus 70 and Coxsackie A24).

Time to Results
Positive culture - 1 to 7 days.   Negative culture - 2 weeks.

Enterovirus Serologies

IgM First appears 1-3 days after infection and persists for 2-3 months.

IgG First appears 5-7 days after infection and antibody levels persist for years. The antibody response appears to be predominantly of the IgG1 and IgG3 subtypes.

IgA Serum IgA first appears 2-6 weeks after infection and antibody levels are very low. Not all patients will have a serum IgA response.

Epidemiology
Enteroviruses are responsible for sporadic infections and epidemic disease outbreaks throughout the world. In temperate climates, enterovirus infections have a sharp seasonal incidence (May-November) with peak incidence in July-September. Children are infected more frequently than adults, and males more frequently than females.

Transmission/Incubation Period
Enteroviruses are spread mainly through the fecal-oral route and by inhalation of infectious aerosols. Virus is shed from the oropharynx from 2-3 days before, to 1-2 weeks after, the onset of symptoms and in feces for several weeks to months. Asymptomatic virus excretion is common. The incubation period for enteric or systemic infections is 2-25 days with an average of 3-5 days. Ocular discharges from patients with acute hemorrhagic conjunctivitis (AHC) are highly infectious. Person-to-person transmission of AHC can occur through contamination of smooth surfaces and subsequent autoinoculation of the eyes. The incubation period for acute hemorrhagic conjunctivitis is approximately 24 hours.

Inactivators and Disinfectants
Enteroviruses are very stable and can survive for several days at room temperature. Enteroviruses are resistant to organic solvents, detergents, and many household disinfectants. Effective disinfectants include 0.3% formaldehyde solutions, 10% household bleach solutions 0.1 N HCl, 2% sodium hydroxide, and 2% glutaraldehyde solutions. The presence of organic material can protect enteroviruses from inactivation.

through the oro- or nasopharynx and infects susceptible epithelial cells of the mouth, nose, throat, and occasionally, the eye. Infection reaches the regional lymph nodes within 24-48 hours. Primary viremia occurs three days after infection and is responsible for the spread of virus to secondary sites. A secondary (or major) viremia occurs 5-7 days postinfection and results in clinical symptoms (6) and dissemination of virus to target organs such as the meninges, pancreas, myocardium, and skin. Necrosis and inflammatory lesions are often seen in the target organs but lesions are generally not seen in the gut and the lymphoreticular tissues associated with earlier replicative events (7).

With the exception of some coxsackie A and hepatitis A viruses, laboratory diagnosis of enterovirus infections depends upon cell culture isolation and identification. Because asymptomatic patients can shed virus for prolonged periods of time, isolation of virus from rectal or throat swabs provides strong, but not conclusive proof that a particular enterovirus is responsible for the disease symptomatology. The most definitive evidence is provided by enterovirus isolation from CSF, pericardial fluid, or skin lesions. Serological testing has only limited usefulness in the diagnosis of enteroviral infection because it is not feasible to test for antibodies to each of the virus serotypes. Nucleic acid testing protocols (10) have shown some promise but they are not yet commercially available.

VIRUS ISOLATION

Introduction

No single cell culture system is suitable for recovery of all enteroviruses and most laboratories utilize multiple cell types to isolate enteroviruses. The most efficient combination of cell cultures includes a primary monkey kidney culture (PMK, RMK, or CMK), a human cell culture system (HFF, MRC-5, HEK, HeLa, HEp-2, or WI-38), and RD cells (5). The RD cell line was derived from a rhabdomyosarcoma and has been shown to support the replication of many of the coxsackie A viruses which had previously been grown only in newborn mice. With the addition of RD cells, only coxsackie A1, A19, and A22 must be grown in suckling mice.

Many wild-type enteroviruses and vaccine strains of polioviruses produce CPE in a variety of cell lines within days of inoculation. The rapid nature of CPE production requires frequent observation of cell cultures during the first week of culture. Following isolation, preliminary identification can be made on the basis of CPE in cell lines that support enterovirus growth, and local epidemiology (8). For instance, coxsackie B viruses produce characteristic teardrop-shaped CPE in RMK, HEp-2, and BGM cells but do not grow well in RD or human fibroblast cells. Echoviruses and the culturable coxsackie A viruses produce CPE in RMK, RD and human fibroblast cells but not in BGM or HEp-2 cells. These viruses typically produce CPE later than polioviruses. Definitive identification and virus typing can be accomplished by neutralization assays.

Coxsackie A1, A19, and A22 can only be isolated in suckling mice. In this procedure, a litter of newborn mice not older than 1 day is inoculated by three routes: intracerebrally with 20 μl; subcutaneously with 30 μl; and intraperitoneally with 50 μl of specimen. Mice are observed daily for 14 days. The disease caused by coxsackie A virus is a polymyocytis and animals show a progressive flaccid paralysis due to muscle degeneration. Coxsackie B virus infections produce encephalitis, spastic paralysis, and tremors in mice. Virus passages are made by harvesting brain, heart, skin, or skeletal muscle after the onset of paralysis and inoculating tissue homogenates into cell culture or newborn mice.

Specimen Collection

Stool, throat, rectal, or conjunctival swabs are the specimens of choice. Specimens should be collected as early in the disease as possible and placed in suitable viral transport media to prevent drying and to inhibit bacterial growth. At least 1 ml of spinal or pericardial fluid should be collected and sent to the laboratory in a sterile tube. Do not add viral transport medium to CSF specimens.

Specimens from multiple sites will often improve the isolation frequency. Although enteroviruses are stable at room temperature for several days, specimens should be held on wet ice or at 4°C until they are shipped to the laboratory.

Specimen Preparation
Swabs
1. Vortex the specimen with several glass beads for 20-30 seconds to release any bound cells or virus.
2. Remove the swab from the viral transport medium and firmly roll it against the inside of the tube to remove as much fluid as possible.
3. Viral transport medium containing rectal swabs or other grossly contaminated specimens should be filtered through a 0.45 μm syringe filter before inoculation onto cell cultures.

Stool
1. Combine approximately 1 gram of stool in 3 ml of viral transport medium. Add an additional 10-20 ml of viral transport medium, BSS, or MEM.
2. Vortex the stool with glass beads to emulsify the specimen and extract as much virus as possible.
3. Centrifuge the suspension at high speed (1000-3,000 x g) for 10-15 minutes.
4. Filter the supernatant fluids through a 0.45 μm syringe filter.

Standard Tube Culture
Procedure
1. Appropriately label 1 RMK, 1 HEp-2 and 1 RD cell culture tube. Many laboratories substitute 1 Vero and 1 HFF tube for the HEp-2 culture.
2. Add 0.2-0.5 ml of specimen supernatant or CSF to each tube. If CSF a specimen does not have sufficient volume to inoculate 4 tubes, a single RMK and RD tube should be inoculated.
3. Incubate the tubes on a roller drum (10-15 rph) at 35-37°C. Conjunctival specimens from patients with suspected acute hemorrhagic conjunctivitis should be incubated at 33°C.
4. Observe the cells daily for CPE for the first 5-7 days. Thereafter, the tubes should be examined every other day until 12-14 days postinoculation.
5. Tubes displaying characteristic teardrop CPE should be frozen, thawed, and passed to fresh RMK, HEp-2, and RD cells for presumptive identification. Definitive typing should be accomplished by neutralization.

Interpretation
Positive test. The presence of characteristic teardrop CPE (Fig. 1) that develops within 1-3 days in RMK, HEp-2 and RD tubes can be presumptively identified as poliovirus. Characteristic CPE that develops more slowly (4-7 days) in only RMK and HEp-2 tubes can be presumptively identified as coxsackie B virus. Characteristic CPE developing only in RMK and RD tubes can be presumptively identified as echovirus or coxsackie A virus. CPE developing in only the RMK tube should be identified as enterovirus.
REPORT: Enterovirus isolated, CPE suggestive of (Coxsackie B, Echovirus or Coxsackie A virus, or Poliovirus).

Negative Test. The absence of characteristic CPE in any of the tubes after 14 days indicates the absence of viable and culturable enterovirus.
REPORT: Enterovirus not detected.

QC Procedures.
Subpassages of clinical isolates or ATCC cultures should be inoculated with each batch of enterovirus isolations. Uninfected cell cultures can serve as negative controls. QC cultures should produce enteroviral CPE in appropriate tubes within 1-7 days. CPE should not be present in uninoculated control tubes. Representative polioviruses, coxsackie A and B viruses, echoviruses, and enteroviruses should be tested periodically to assure that the presumptive identification scheme is performing properly.

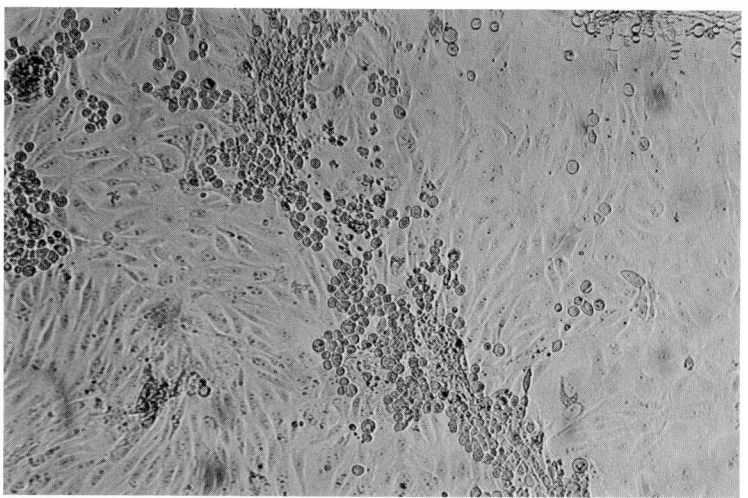

FIG. 1. Enterovirus CPE in primary rhesus monkey kidney cells.

Preparation of Positive Control Cultures
1. Add 20 µl of a recent enterovirus isolate or an (ATCC culture) to 2 ml of EMEM-2% FCS.
2. Remove the growth medium from one 75 cm^2 flask of freshly confluent RMK (or other appropriate cells)
3. Add 2 ml of the inoculum to the flask and allow the virus to adsorb for 1-2 hours at 35-37°C. The flask should be rocked every 15 minutes to assure even virus distribution and to prevent monolayer desiccation.
4. Add 20 ml of MEM containing L-glutamine, antibiotics and 2% FCS.
5. Incubate the flask at 35-37°C until CPE involves 80-100% of the cell monolayer.
6. Scrape the cells from the flask into the cell culture medium
7. Transfer the cell suspension to a 50 ml polypropylene tube and freeze at -70°C.
8. Thaw the medium quickly in a 37°C water bath.
9. Centrifuge the tube at 600-800 x g to pellet the cell debris.
10. Transfer the supernatant fluids to a sterile vessel containing 80 ml of EMEM-10% FCS.
11. Dispense 0.5 ml of the diluted supernatant fluids into each of 200 freezing vials and freeze the vials at -70°C or below. Vials can be stored in liquid nitrogen for up to 15 years or -70°C for up to 10 years.

ENTEROVIRUS TYPING USING THE LIM BENYESCH-MELNICK POOLS
Introduction

The diversity of enterovirus types make typing by neutralization with monospecific antisera extremely tedious. In order to facilitate enterovirus typing, Lim and Benyesch-Melnick (9) developed 15 pools of type-specific antisera. The first 8 pools (pools A-H) have the most clinical utility because they can identify 42 enteroviruses that can be readily isolated in cell culture. These pools include most of the echoviruses, all coxsackie B viruses, and some coxsackie A viruses. The next seven pools are directed against 19 coxsackie A viruses, some of which can be isolated only in mice.

Procedure
1. Prepare a freshly confluent microtiter plate containing an appropriate cell line for propagating the enterovirus isolate.
2. Grow the isolate in a tube culture until the CPE involves 80-100% of the cells in the monolayer.
3. Freeze and thaw the isolate.
4. Centrifuge the isolate at 600-1000 x g to pellet the cell debris.

5. Prepare 10^{-1} through 10^{-6} dilutions of the virus isolate in MEM containing antibiotics and 2% FCS.
6. Add 30 μl of each antibody neutralization pool (working titer of 50 antibody units/0.1 ml) to duplicate wells of a sterile 96 well microtiter plate.
7. Add 30 μl of MEM 2% FCS to seven additional (virus titration) wells.
8. Add 30 μl of the 10^{-1} virus dilution to one set of virus neutralization wells.
9. Add 30 μl of the 10^{-3} virus dilution to the second set of virus neutralization wells.
10. Add 30 μl of the 10^{-1} through 10^{-3} virus dilutions to the virus titration wells (step 7).
11. Cover the dilution plate, tap the plate gently to mix, and incubate for 1 hour at 35-37°C.
12. Transfer 50 μl of the virus suspensions to a cell culture microwell (step 1) containing a confluent monolayer and 200 μl of maintenance medium. **NOTE:** Be sure to use separate pipette tips for each transfer.
13. Cover the plate and incubate at 35-37°C for 3-5 days. Record CPE of each well daily.
10. Compare the neutralization pattern with the chart provided with the antiserum pools to determine a presumptive identification. Neutralization using the homologous serum provides a definitive identification.
11. Repeat the procedure using the homologous (serotype-specific) antiserum for definitive identification.

Interpretation of Results

Occasionally, the infectivity of some isolates will not be neutralized by any of the antisera pools. If the virus titer is greater than 10^{-5}, the amount of antiserum will not be sufficient to neutralize the virus. In this case, the procedure must be repeated using an appropriate virus dilution. Occasionally, the patient isolation tube will contain more than one enterovirus. Because dual infections usually do not produce viruses with exactly the same virus titers, the 10^{-1} and 10^{-3} dilutions may produce a different neutralization pattern. Plaque purification or limiting dilution assays can be used to separate these isolates for typing. Failure of the antiserum to neutralize the enterovirus isolate may be due to the presence of a new enteroviral strain or the presence of a nonenterovirus isolate.

REPORT: Coxsackievirus B5 (etc.) isolated.

REFERENCES

1. Assaad F, Ljungars-Esteves K. World overview of poliomyelitis: regional patterns and trends. *Rev Infect Dis* 1984;6:S302-S307
2. Rotbart HA. Human enterovirus infections: Molecular approaches to diagnosis and pathogenesis. In: Semler BL, Ehrenfeld E. *Molecular aspects of picornavirus infection and detection.* American Society of Microbiology, Washington, DC. 1989;243-264.
3. Connolly KJ, Hammer SM. The acute aseptic meningitis syndrome. *Infect Dis Clinics N Am* 1990;4(4)599-622).
4. See DM, Tilles JG. Viral Myocarditis. *Rev Infect Dis* 1991;13:951-956.
5. Melnick, JL. Enteroviruses. In: Fields BN, Knipe DM, Chanock RM, Hirsch MS, eds. *Virology*, Second Edition. New York: Raven Press 1990;549-605.
6. Cherry JD. Enteroviruses: polioviruses (poliomyelitis), coxsackieviruses, echoviruses, and enteroviruses. In: Field RD, Cherry JD, eds. *Textbook of pediatric infectious diseases* Part III, Infections with specific microorganisms, 2nd ed. Philadelphia: W.B. Saunders, 1987;1729-1790.
7. Moldlin JF. *Picornaviridae*: introduction. In: Mandell GL, Douglas RG Jr, Bennett JE, eds. *Principles and practice of infectious diseases*, Third Edition. New York: Churchill Livingstone 1990;1352-1359.
8. Johnston SLG, Siegel CS. Presumptive identification of enteroviruses with RD, HEp-2, and RMK cell lines. *J Clin Microbiol* 1990;28:1049-1050.
9. Lim KA, Benyesch-Melnick M. Typing of viruses by combinations of antiserum pools. Application to typing of enteroviruses (coxsackie and ECHO). *J Immunol* 1960;84:309-317.
10. Hyypia T, Stalhandske P, Vainionpaa R, Pettersson U. Detection of enteroviruses by spot hybridization. *J Clin Microbiol* 1984;19:436-438.

CHAPTER 13

Epstein-Barr Virus

INTRODUCTION

Epstein-Barr virus (EBV) is a human herpesvirus and a member of the gammaherpesvirinae subfamily. In electron micrographs, EBV virions appear as 45 nm hexagonal nucleocapsids surrounded by a complex, 120 nm diameter envelope. The EBV genome consists of a linear, double-stranded (172 kilobase pair) DNA molecule that codes for more than 30 polypeptides. A relatively unstable virus, EBV has not been recovered from environmental surfaces or fomites. EBV is rapidly inactivated by heat, alcohols and organic solvents.

EBV is the etiological agent of infectious mononucleosis and has been implicated in African Burkitt's lymphoma and nasopharyngeal carcinoma. EBV infections are ubiquitous and show no seasonal variability, even in temperate climates. Most individuals acquire EBV infections early in life. Seroepidemiologic surveys have shown that 50% of children have antibodies to the virus by the age of five. Childhood infections may be asymptomatic or produce mild "flu-like" illness. However, adolescents and adults who escape infection during childhood experience infectious mononucleosis (IM) upon primary infection. IM is characterized by irregular fever, pharyngitis, and lymphadenopathy lasting 1 to 4 weeks. Hematological abnormalities including an absolute increase in lymphocytes and monocytes exceeding 50% and more than 15% atypical lymphocytes, last for at least 2 weeks (1). Liver function tests generally reveal a mild to moderate increase in SGGT, SGOT, bilirubin, and LDH levels. Although IM is generally a benign and self-limited disease, it may be complicated by splenomegaly and splenic rupture, hepatitis, pericarditis, myocarditis, or central nervous system involvement (Guillain-Barre syndrome, Bell's palsy, transverse myelitis, and meningoencephalitis). Patients with X-linked lymphoproliferative syndrome and ataxia telangiectasis often succumb to overwhelming fatal infections upon primary infection.

Epstein-Barr virus is transmitted via salivary contact through kissing or exposure to contaminated eating utensils. The main portal of entry is the oropharynx and the incubation period is estimated at 4-7 weeks (2). Once transmitted, the virus infects the epithelial cells of the oropharynx, the salivary gland ducts, and possibly, the tongue (3,4). About 10-20% of normal, seropositive individuals have been shown to secrete EBV in the saliva. However, if the saliva is concentrated, up to 100% of these individuals can be shown to secrete the virus (5). It is not clear, however, whether EBV latently infects the epithelial cells or whether EBV merely produces a persistent, productive infection.

The oropharynx is the principal site of EBV replication and provides the source of the virus that infects B lymphocytes (6). Like most other herpesviruses, EBV can cause latent infections and once infected, the virus persists for the remainder of the host's life. In the case of EBV, latent infections are usually established in resting B cells similar to the ones found in germinal centers. EBV nuclear antigens can be detected in B cells from nearly all EBV-seropositive individuals (7). EBV reactivations can occur and are heralded by increases in antibody levels to EBV early antigens, increased virus titers in the saliva, and occasionally, by clinical symptoms (6).

In adolescents and young adults, a preliminary diagnosis of IM can often be made based upon classic symptomatology. However, cytomegalovirus and Toxoplasma infections can also produce mononucleosis-like symptoms. Differential diagnosis usually requires serological testing. Paul-

AT A GLANCE...

EPSTEIN-BARR VIRUS

Virus Detection Methods
Virus isolation is difficult and rarely provides relevant clinical information because many asymptomatic, seropositive individuals shed virus from the oropharynx.

Specimen Source
Throat washes or oropharyngeal swabs are the specimens of choice. EBV can be isolated from nearly all heterophile-positive patients and virus shedding can persist for up to 18 months after clinical recovery.

Time to Result
Positive culture - 2-4 weeks.
Negative Culture - 4 weeks.

EBV Serology

- **VCA IgM** — First detectable at 2-4 weeks after primary infection and peak levels are present at the onset of clinical symptoms. Antibody titers usually decline to undetectable levels 2-3 months after the symptoms subside.

- **VCA IgG** — First detectable within 1-2 weeks of onset of symptoms. Demonstration of a four-fold increase in antibody titer is usually not possible. Antibody levels decline somewhat after clinical illness but remain detectable throughout the patient's lifetime. IgG1 is the predominate subclass and other subclasses are found only infrequently. IgG4 is prevalent in nasopharyngeal carcinoma (NPC) while IgG3 is found during EBV reactivations.

- **VCA IgA** — First detectable 2-4 weeks after primary infection. Titers decline quickly thereafter and may not be present at the onset of symptoms. NPC patients have a prolonged and elevated IgA response that is predictive of disease.

- **EBV-EA IgG** — Most patients show a transient rise to the diffuse component of the early antigen complex [EA(D)] during the acute infection and EA(D) antibody is usually undetectable 3-6 months after clinical illness. Antibodies to the restricted component [EA(R)] may be present in asymptomatically infected individuals. Antibodies to either/both EA antigens can reappear during EBV reactivation. In African Burkitt's lymphoma, EA(R) antibodies are present at moderate to high levels while patients with NPC have high EA(D).

- **EBNA IgG** — EBNA antibodies are rarely present in acute-phase sera and are first detectable 3-4 weeks after the onset of symptoms. Antibody levels persist throughout the patient's lifetime EBNA antibodies may not be detectable in immunodeficient patients.

Epidemiology
EBV occurs throughout the world and more than 90% of adults have antibodies to the virus. Most individuals acquire EBV early in life. Seroepidemiologic studies have indicated that 50% of children have antibodies to the virus by the time they are 5 years of age. No clear seasonal incidence has been noted. Only one serotype is known.

Transmission/Incubation Period
EBV is poorly contagious and transmission is via salivary contact, either through kissing or by exposure to contaminated eating implements. The incubation period is 4-7 weeks.

Inactivators and Disinfectants
EBV is relatively labile and is rapidly inactivated by heat, alcohols and organic solvents. Effective disinfectants include 70% alcohol, 10% household bleach solutions, quaternary ammonium compounds, phenols, iodophor compounds, and glutaraldehyde compounds.

Bunnell-Davidsohn heterophile test kits are commonly used for the laboratory diagnosis of infectious mononucleosis and a number of rapid and relatively inexpensive heterophile tests are commercially available. As the name implies, heterophile antibody tests are not specific for EBV, but rather, they detect heterotypic antibodies to ox, horse, and sheep red blood cells that are produced during primary EBV infections. Heterophile antibodies are usually detectable at the onset of symptoms and reach peak levels within 2 weeks. Heterophile antibody levels decline rapidly thereafter and are usually not detectable after 3 months. Heterophile tests are not suitable for testing sera from children because children do not always produce heterophile antibodies during primary EBV infection. In children, demonstration of EBV IgM antibodies directed to the viral capsid antigens (VCA) is diagnostic of primary infection with EBV.

Other antibodies including those to the nuclear antigens (EBNA) and early antigens (EBV-EA) can be used to stage EBV disease and to detect EBV reactivations. In nasopharyngeal carcinoma, patients have a prolonged and elevated IgA response to the viral capsid antigens that is predictive of disease. EBV isolations are usually of little clinical value because many asymptomatic, seropositive individuals shed the virus.

VIRUS ISOLATION

Introduction

The presence of EBV in throat swabs and washings can be determined by the ability of the virus to transform cord blood lymphocytes. Although this assay has been an important research tool, EBV isolations are generally not useful for clinical diagnosis because many seropositive individuals shed virus from the oropharynx. In addition, this assay is cumbersome, requiring freshly fractionated cord blood lymphocytes, and results are generally not available for 4-5 weeks. Once transformed cells are observed, they are tested for the presence of EBV nuclear antigen using the anti-complement immunofluorescence test.

Specimen Collection

During infectious mononucleosis, virus can be recovered from the saliva of nearly all heterophile-positive individuals. Virus shedding can persist for up to 18 months after clinical recovery. Virus also can be recovered from 10-20% of normal healthy seropositive individuals, from 50% of renal transplant recipients, and from 70-90% of critically ill leukemia or lymphoma patients. The specimen of choice is a throat wash (Chapter 5). If the patient cannot gargle, a pharyngeal swab in viral transport medium will often yield EBV.

Specimen Preparation and Storage

1. Upon receipt in the laboratory, vortex the specimen briefly and centrifuge at 1500 x g for 10 minutes to pellet any cellular material.
2. Filter the supernatant fluids through a 0.45 μm filter.
3. Specimens should be stored at or below -70°C until they can be tested.

Test Procedure
Cord Blood Cells

Cord blood lymphocytes must be separated from red cells and granulocytes by centrifugation through a ficoll-hypaque density gradient. A number of ready-made gradient materials can be purchased commercially (Ficol-Paque from Pharmacia Fine Chemicals or Histopaque from Sigma Chemical Company, or Lymphocyte Separating Medium from Organon-Teknika) for this purpose or gradient materials can be prepared in the laboratory (Appendix C). Lymphocyte separation should be done as directed by the manufacturer or as described below.

1. Aseptically collect 10 ml of human umbilical cord blood with 5-10 units of preservative-free heparin per milliliter.
2. Place 3 ml of the ficoll-hypaque in a clear sterile 15 ml centrifuge tube.
3. Dilute the heparinized blood with an equal volume of sterile saline.
4. Carefully overlay 5 ml of heparinized blood onto the ficoll-hypaque.

5. Centrifuge the tubes at 400 x g in a swinging bucket rotor for 30 minutes at room temperature.
6. Lymphocytes will band at the plasma-ficoll interface.
7. Carefully remove and discard the clear upper plasma layer.
8. Transfer the lymphocyte band to a sterile centrifuge tube and estimate the fluid volume.
9. Add 4 volumes of Dulbecco's PBS to the tube to produce a 1:5 dilution.
10. Perform a cell count.
11. Centrifuge the cells at 600 x g for 15 minutes and remove the supernatant fluids.
12. Resuspend the cell pellet at 5×10^6 cells/ml in RPMI-1640 containing 10% FCS.
13. Transfer the cells to a flask and incubate overnight at 35-37°C to assure sterility.
14. Cells are ready for use and should be inoculated within 24 hours. Alternatively, the cells can be diluted 1:2 in 2X Freezing Medium (Appendix C) and frozen in liquid nitrogen (Chapter 6) for up to 5 years. Prior to use, cells are washed twice to remove the cryoprotectants and used as described below.

Inoculation
1. Place 5×10^6 cord blood leukocytes into each of four sterile 16 x 125 mm culture tubes.
2. Centrifuge the tubes at 600 x g to pellet the cells.
3. Resuspend two of the cell pellets with 0.5 ml of the filtered throat wash specimen.
4. **(Negative Control).** Resuspend one cell pellet with 0.5 ml of RPMI-1640 containing 10% FCS.
5. **(Positive Control).** Resuspend one cell pellet with 0.5 ml of RPMI-1640 containing 10% FCS and 1-30 transforming units of EBV.
6. Incubate the cultures at 33-35°C in an upright position for 1-2 hours to allow the virus to adsorb to the cells.
7. Add 1.5 ml of RPMI-1640 containing 10% FCS and incubate the cultures at 33-35°C in an upright position for 4 weeks.
8. Replace the cell culture medium each week and examine the cultures for evidence of virus transformation.

Interpretation of Results
Positive Test. Cell transformation is signalled by the appearance of clusters of large proliferating lymphoblastoid cells and an increase in metabolic activity (more acidic pH). No report should be made based upon the transformation test alone. All positive cultures should be tested for the presence of EBV nuclear antigens using the ACIF test.
REPORT: No report should be issued.

Negative Test. The absence of transforming agent is characterized by cell degeneration and the lack of cell aggregates that increase in numbers.
REPORT: Transforming agent not detected.

QC Procedures.
Positive and negative controls must be inoculated with each batch of EBV isolations. The positive control consists of 10-30 transforming units of EBV from either a positive throat washing or from spent cell culture fluids from the EBV-transformed B95-8 cell line. Transformation is usually recognized in the positive control tube after 2-3 weeks. Some lymphocyte cultures cannot be readily transformed. Therefore positive control tubes must exhibit the transformation described above. If transformation is not present, the test is invalid and must be repeated with a different virus preparation and/or a new batch of cord blood lymphocytes.

Negative controls must not contain transformed lymphocytes and uninfected cell cultures will usually degenerate after two weeks. Some lymphocyte preparations contain "spontaneous" transformants. Therefore, any transformed lymphocytes in the negative control invalidates the test. Invalid tests must be repeated using a new batch of cord blood lymphocytes.

Culture Confirmation
Preparation of Cell Smears

1. Centrifuge the cultures at 600-800 x g to pellet the cells.
2. Remove all but a few drops of cell culture medium from the tubes making sure to use separate pipettes for each culture.
3. Resuspend the cells in the remaining fluid. Using a Pasteur pipette, place one drop (10-20 µl) of the suspension on an acetone-cleaned slide.
4. Repeat the process with the other tubes and controls.
5. Allow the smears to air dry.
6. Fix the slides for 10 minutes in cold (-20°C) acetone/methanol (equal volumes).
7. Slides prepared in this manner can be stored desiccated at -20°C for up to 5 years or they can be immediately stained as described below.

ACIF Procedure

In the anticomplement immunofluorescence (ACIF) procedure, complement-fixing antibodies (human serum or certain types of monoclonal antibodies) to the nuclear antigens of EBV are added to the smears. If EBNA antigens are present, the antibodies will bind to the cells. The unbound antibody is washed away and guinea pig complement is added. If complement fixing antibodies are bound to the cells, complement fraction C3' will bind to the cells at the location of the antibody. Another wash is performed to remove any unbound complement and fluorescein-labelled antiserum to guinea pig C3' is added. An Evans blue counterstain is usually included in the final antiserum to provide staining contrast. Because activation of the complement cascade causes a large number of C3' molecules to be bound for each antibody molecule, this procedure is much more sensitive than the standard indirect immunofluorescence test. This amplified method is necessary because EBV nuclear antigens are present in very low concentrations. If antiserum or mouse ascites fluids are used, they must be heated to 56°C for 30 minutes to remove endogenous complement activity. Complement and antiserum concentrations must be matched for optimal sensitivity.

1. Remove a control slide from the freezer and allow it to come to room temperature.
2. Place one drop (10-30 µl) of the EBNA antiserum on each smear and on the control smear(s).
3. Incubate the slides for 45 minutes at 35-37°C in a covered, humidified chamber. Do not allow the antibody to dry on the slide as this could cause nonspecific antibody binding.
4. Remove the excess antibody with a gentle stream of PBS. Do not direct the stream directly at the cell smear as this could dislodge the cells.
5. Soak the slide in PBS for 5 minutes, shake off the excess fluid. Do not allow the slides to dry.
6. Add one drop of guinea pig complement (15-30 µl) to each cell smear.
7. Incubate the slide for 45 minutes at 33-35°C in a covered, humidified chamber. Do not allow the complement to dry on the slide.
8. Remove the excess complement with a gentle stream of PBS as before.
9. Soak the slide in PBS for 5 minutes and shake off the excess fluid. Do not allow the slide to dry.
10. Add one drop (15-30 µl) of fluorescein-labelled antibody to guinea pig C3'.
11. Incubate the slide for 30 minutes at 33-35°C in a covered, humidified chamber. Do not allow the antibody to dry on the slide.
12. Remove the excess antibody with a gentle stream of PBS as before.
13. Soak the slide in PBS for 5 minutes, shake off the excess fluid, and allow the slide to air dry.
14. Carefully add a small drop of FA mounting fluid to the center of each well.
15. Place a number 1 coverslip on the mounting fluid and carefully remove all the air bubbles.
16. Immediately examine the slide using a

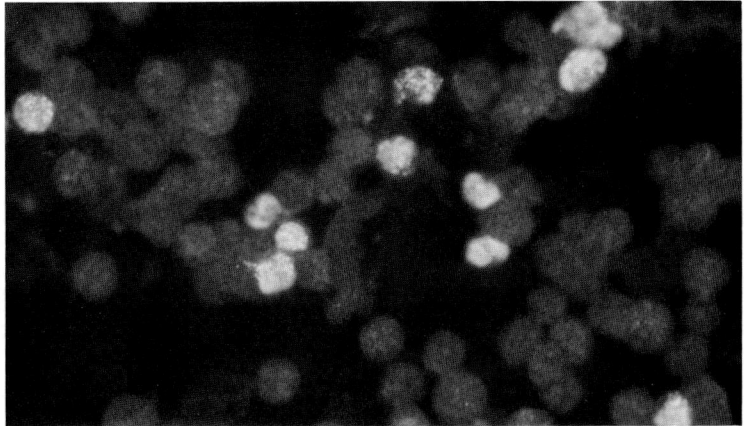

FIG. 1. Anticomplement immunofluorescence staining of a mixture of Raji and BJAB cells..

fluorescence microscope. For optimum clarity, use 200-300X magnification for screening and 400X for confirmation of nuclear staining.

Interpretation of Results
Positive Test. The presence of Epstein-Barr virus nuclear antigen is indicated by the presence of speckled apple-green fluorescence in the nucleus of the transformed cells. Other cells will stain red due to the counterstain (Fig. 1). Only individuals experienced in reading FA reactions should examine the cell smears because cell debris and cell clumps may exhibit a dull fluorescence which can be misinterpreted as a specific, positive reaction by inexperienced technologists.
REPORT: Epstein-Barr virus isolated.

Negative Test. If all control reactions function properly, the absence of nuclear fluorescence in transformed cells indicates that another agent or set of circumstances are responsible for the lymphocyte transformation.
REPORT: Transforming agent isolated, not Epstein-Barr Virus.

QC Procedures.
Positive control smears must exhibit typical speckled apple-green fluorescence in the nucleus of an appropriate number of cells (see below). If no fluorescence is observed, the test is invalid and must be repeated with fresh guinea pig complement and/or fluorescein-labeled antibody to guinea pig C3'.

Preparation of Control Slides
Positive cell smears are prepared from Raji cells, EBV-transformed lymphocytes that express only EBV nuclear antigens (EBNA) when cultivated under normal conditions. Negative control smears are prepared from BJAB cells, a continuous lymphocyte line that does not express EBNA. Both cell lines are propagated as suspension cultures in RPMI-1640 containing 10% FCS. Cell smears (see below) are prepared in the same manner for both cells. Control slides can be prepared on 2-well slides with Raji cells (EBNA positive) in one well and BJAB (EBNA negative) cells in the other well. An alternative format is to mix 1 part Raji cells with 3 parts BJAB to produce the concentration listed below. In this format, both positive and negative cells are present in the same well and only one well need be stained.

1. Perform viable cell counts on rapidly growing BJAB and Raji cell cultures.
2. Add enough RPMI-1640 containing 10% FCS to each culture to produce a cell concentration of 3×10^5 viable cells/ml.
3. Incubate the cells in a cell culture flask for 7 days and perform cell counts. Cell

densities should be at least 1×10^6 cells/ml.
4. Centrifuge the culture at 600-800 x g for 15 minutes to pellet the cells.
5. Resuspend the cells in cold (2-8°C) PBS at a final cell concentration of 1×10^7 cells/ml.
6. Dispense 3-5 µl of the cell suspension onto a 5-8 mm slide well. Use larger volumes to cover larger areas.
7. Allow the cell smears to dry at room temperature.
8. Fix the slides in cold (-20°C) acetone/methanol (equal volumes) for 10 minutes.
9. Store the slides at -20°C for up to 5 years. Slides should be stored under desiccated conditions to minimize antigen degradation and background fluorescence.

REFERENCES

1. Henle W, Henle G, Horowitz CA. Infectious mononucleosis and Epstein-Barr virus-associated malignancies. In: Lennette EH, Schmidt N, eds. *Diagnostic procedures for viral, rickettsial, and chlamydial infections.* Fifth edition. Washington DC: American Public Health Association 1985;441-470

2. Lennette ET. Herpesviridae: Epstein-Barr virus. In: Lennette EH, Halonen P, Murphy FA, eds. *Laboratory diagnosis of infectious diseases - Principles and practice.* New York: Springer-Verlag 1988;230-246.

3. Sixbey JW, Nedrud JG, Raab-Traub N, Hanes RA, Pagano JS. Epstein-Barr virus replication in oropharyngeal epithelial cells. *N Eng J Med* 1984;310:1225-1230.

4. Wolf H, Haus M, Wilmes E. Persistence of Epstein-Barr virus in the parotid gland. *J Virol* 1984;51:795-798.

5. Yao QY, Rickinson AB, Epstein MA. A re-examination of the Epstein-Barr virus carrier state in healthy seropositive individuals. *Int J Cancer* 1985;35:35-42.

6. Miller G. Epstein-Barr Virus - Biology pathogenesis and medical aspects. In: Fields BN, Knipe DM, Chanock RM, Hirsch MS, Melnick JL, Monath TP, Roizman B, eds. *Virology*, Second Edition. New York: Raven Press 1990;1921-1958.

7. Gerber P, Monroe JH. Studies on leukocytes growing in continuous culture derived from normal human donors. *J Natl Cancer Inst* 1968;40:855-866.

CHAPTER 14

Hepatitis A Virus

INTRODUCTION

Hepatitis A virus (HAV or enterovirus 72) is a nonenveloped positive-stranded RNA virus belonging to the *Picornaviridae*. HAV virions have icosahedral symmetry with an average diameter of 27-32 nm. The HAV genome consists of a 7.48 kilobase, single-stranded RNA genome that codes for a single polyprotein that is subsequently processed into four viral proteins. HAV can survive for up to one month under ambient conditions and is only partially inactivated after 12 hours at 60°C. In addition, HAV is resistant to drying, low pH, and to organic solvents.

HAV is primarily transmitted through the fecal-oral route and through contaminated food and water. Fomites and person-to-person spread are principal means of transmission, especially in day care centers handling children under two years of age. This is not surprising because in one study, two year old children placed an object or a hand in their mouth every two minutes (1).

The mean incubation time is one month with a range of 10-50 days. The sequence of events following infection are not well understood. Like poliovirus, HAV is thought to establish an early infection in the oropharynx because virus has been found in saliva during the early stages of infection. Viremia usually occurs 7-10 days before the onset of jaundice and HAV is rarely demonstrable in blood at the onset of illness. HAV infects the liver and replicates in hepatocytes and Kupffer cells. Because there is little evidence of virus replication in the intestinal mucosa, it is assumed that HAV in the feces originates in the liver (2). High concentrations (up to 10^8 infectious units/ml) of virus are excreted in the stool from two weeks before, to one week after, the onset of jaundice. The hepatic symptomatology may be due to host immune responses rather than to a direct cytolytic effect (2).

The severity of HAV disease varies markedly with age. Children are often asymptomatic or have only mild illness lasting 10-14 days. Malaise, nausea, fever, and diarrhea are reported in about half of those infected while joint pain, abdominal pain, or vomiting are reported in 20-30% of infected persons. Infected children under three years of age rarely develop jaundice. Jaundice occurs in only 10% of HAV-infected 4-6 year olds. In contrast, approximately 75% of HAV-infected adults develop jaundice. The duration of illness is longer in adults (2-8 weeks) and symptoms may persist for 4 months or longer (3). Fulminant hepatitis is a rare but potentially lethal complication of infection. Other rare complications include myocarditis and arthritis.

Acute illnesses caused by the various hepatitis viruses cannot be distinguished clinically. Differential diagnosis is made by demonstrating hepatitis A virus in the stool, hepatitis A IgM antibodies, or by seroconversion. Antigen capture assays and immune electron microscopy have been used to demonstrate the presence of virus. However, virus isolation has little clinical value because culture methods require 4-8 weeks and are frequently unrewarding. Laboratory diagnosis is most often made by demonstrating the presence of IgM antibodies in the early phase of clinical disease HAV IgM antibodies are present at the onset of symptoms and peak approximately 4 weeks later. IgM antibodies usually disappear about 3-6 months after onset of disease. The presence of HAV-specific IgM in serum indicates a current or recent hepatitis A infection. IgG can be detected within 2 weeks of the onset of symptoms and antibody levels persist for life. There is only one HAV serotype and the presence of antibody indicates

AT A GLANCE...

HEPATITIS A VIRUS

Virus Detection Methods
Wild hepatitis A virus (HAV) strains replicate poorly in cell culture. Therefore, HAV isolation from human specimens is an uncertain, difficult, and prolonged process. HAV has been isolated in FRhK-4, BSC-1, LLC-MK2 cells. Virus can also be detected by antigen capture assays and via electron microscopy.

Specimen Source
Stool is the specimen of choice. HAV is present in stool 2-3 weeks before, and up to 8 days after, the onset of jaundice. Although low-level viremia is present 7-10 days before the onset of symptoms, HAV is rarely isolated from blood when the patient is symptomatic.

Time to Result
Positive culture - 4-8 weeks
Negative culture - 8 weeks.

Hepatitis A Virus Serology

IgM   Laboratory diagnosis of HAV infection is most often made by demonstrating the presence of IgM antibodies. HAV IgM antibodies are present at the onset of symptoms and peak approximately 4 weeks later. IgM antibodies usually disappear 3-6 months after onset of disease. The presence of HAV-specific IgM in serum indicates a current or recent hepatitis A infection.

IgG   IgG can be detected within 2 weeks of the onset of symptoms and antibody levels persist for life. The presence of antibody indicates immunity to reinfection.

IgA   IgA has been detected in stool during the early stages of infection.

Epidemiology
Hepatitis A has a worldwide distribution and man is the only known reservoir for the virus. Recurrent epidemics are a prominent feature of the disease. In the U.S., the disease appears to peak in the fall and winter months. HAV is endemic in many developing countries and childhood infection is common. In developing countries, up to 90% of adults have antibodies to the virus. In industrialized countries HAV antibodies are uncommon in young children (<5%) but they are present in 5-20% of those under 20 years of age and in 30-50% of older adults.

Transmission/Incubation Period
Transmission is predominantly by the fecal-oral route or through contaminated water or food. Patients are infectious 2-3 weeks before the onset of symptoms and for approximately 8 days thereafter. The incubation period is 10-50 days with a mean incubation time of 1 month. Only one type of HAV serotype is known.

Inactivators and Disinfectants
HAV is resistant to organic solvents and phenols and is relatively stable at low pH. Effective disinfectants include 10% household bleach solutions, 0.3% formaldehyde solutions, and 2% glutaraldehyde compounds.

immunity to subsequent infections. Antibodies are detected using EIA or RIA methods.

VIRUS ISOLATION

Introduction

Hepatitis A virus (HAV) was first propagated in cell cultures by Provost and Hilleman in 1979 (4). However, isolation of HAV has little clinical efficacy because isolation of HAV from human specimens is a difficult, lengthy, and uncertain procedure (5,6). A variety of cells have been used for the propagation of culture-adapted HAV strains. However, primary African green monkey kidney and FRhK-4 cells appear to be the best cells for primary isolation. The chemical 5,6-dichloro-1-beta-D-ribofuranosylbenzimidazole (DRB) has also been shown to enhance the propagation of HAV in FRhK-4 cells (7). HAV does not produce reproducible cytopathology and virus-infected cultures must be stained blindly with fluorescein-labelled antibodies. In addition, one to three passages may be required before some isolates can be detected.

Specimen Collection

Stool is the specimen of choice and specimens should be collected as soon after onset of disease as possible. Hepatitis A virus is present in stool 2-3 weeks before and up to 8 days after the onset of jaundice. Low-level viremia precedes the onset of symptoms by 7-10 days and virus is rarely detected in blood when symptoms appear.

Specimen Preparation

1. Place approximately 1 gram of stool in a centrifuge tube containing 5 ml of HBSS.
2. Vortex the tube vigorously for at least 30 seconds to emulsify the specimen and extract as much virus as possible.
3. Centrifuge the tube at 5-7,000 x g for 30 minutes.
4. Collect the supernatant fluid and pass it through a 0.45 μm filter.

Standard Tube Culture

Procedure

1. Appropriately label 2 tubes containing freshly confluent AgMK or FRhK-4 cells.
2. Remove the medium from the cells and add 0.2 ml of the sterile filtrate.
3. Incubate the tubes in a roller rack (10-15 rph) for 2 hours at 33-35°C and 10-15 rph to allow the virus to adsorb.
4. Add 1 ml of MEM containing 2% FCS and antibiotics to each tube.
5. Place the tubes in a roller drum and incubate for 2 weeks at 33-35°C as before.
6. After one week, replace the medium with fresh cell culture medium.
7. After 2 weeks, scrape the cells from one tube and stain them with fluorescein-labelled antibodies to hepatitis A virus. The second tube should be passaged into two fresh monolayers.
8. Repeat steps 5, 6, and 7 for 8 weeks or until the culture is positive.

Culture Confirmation

1. Remove all but a few drops of medium from the cell culture tube.
2. Scrape the cells from the tube surface and resuspend the cells in the remaining medium.
3. Resuspend the cells in the medium left in the tube.
4. Use a pipet to spot cells onto an acetone cleaned glass slide.
5. The slide should be processed as directed by the manufacturer of the fluorescein-labelled hepatitis A antibody or as described below.
6. Allow the slide to air dry slide then fix it in acetone for 5 min. Fixed slides may be stored held for up to 12 months at -20°C in a moisture free container.
7. Add enough of the fluorescein-labelled HAV antibody on each well to cover the cell smear (approximately 15-30 μl).
8. Incubate the slide for 15-30 minutes at 33-35°C in a covered, humidified chamber. Do not allow the antibody to dry on the slide as this could cause nonspecific antibody binding.

9. Remove the excess antibody with a gentle stream of PBS. Do not direct the stream directly at the cell smear as this could dislodge the cells.
10. Soak the slide in PBS for 5 minutes, shake off the excess fluid, and allow the slide to air dry.
11. Carefully add a small drop of FA mounting fluid to the center of each well.
12. Place a number 1 coverslip on the mounting fluid and carefully remove all the air bubbles.
13. Immediately examine the slide using a fluorescence microscope. For optimum clarity, use 200-300X magnification for screening and 400X for confirmation of cell morphology.

Interpretation of Results
Positive Test. The presence of hepatitis A virus is indicated by characteristic granular apple-green fluorescence in the cytoplasm of infected cells on the slide. Only individuals experienced in reading FA reactions should examine the cell smears because cell debris and cell clumps may exhibit a dull fluorescence which can be misinterpreted as a specific, positive reaction by inexperienced technologists.
REPORT: Hepatitis A virus isolated.

Negative Test. The absence of hepatitis A is indicated by the lack of intense apple-green granular cytoplasmic fluorescence in the infected cells. Cultures should be passaged four times before they are considered negative. Because wild-type hepatitis A virus does not grow well in cell culture, a negative result does not preclude hepatitis A infection.
REPORT: Hepatitis A virus not detected.

QC Procedures.
Subpassages of recent hepatitis A or ATCC cultures should be inoculated into cell culture tubes with each batch of isolations. Uninfected cell cultures serve as negative controls. Infected and uninfected cell monolayers are scraped and stained as described above. Positive controls must contain typical cells exhibiting intense apple-green granular cytoplasmic fluorescence. Negative controls must not exhibit specific staining.

REFERENCES

1. Hadler SC, Webster HM, Erben JJ, Swanson JE, Maynard JE. Hepatitis A in day-care centers: a community-side assessment. *N Eng J Med* 1980;302:1222-1227.
2. Hollinger FB, Ticehurst. Hepatitis A virus. In: Fields BN, Knipe DM, Chanock RM, Hirsch MS, eds. *Virology*, Second Edition. New York: Raven Press, 1990;631-667.
3. Hadler SC, McFarland L. Hepatitis in day care centers: epidemiology and prevention. *Rev Infect Dis* 1986;8:548-557.
4. Provost PJ, Hilleman MR. Propagation of human hepatitis A virus in cell culture in vitro. *Proc Soc Exp Biol Med* 1979;160:213-221.
5. Provost PJ, Giesa PA, McAleer WJ, Hilleman MR. Isolation of hepatitis A virus in vitro in cell cultures directly from human specimens. *Proc Soc Exp Biol Med* 1981;167:201-206.
6. Binn LN, Lemon SM, Marchwicki RH, Redfeld RR, Gates NL, Bancroft WH. Primary isolation and serial passage of hepatitis A virus strains in primate cell cultures. *J Clin Microbiol* 1984;20:28-33.
7. Widell A, Hansson BG, Nordenfeldt E, Oberg B. Enhancement of hepatitis A virus propagation in tissue culture with 5,6-dichloro-1-beta-D-ribofuran-osyl-benzimidazole (DRB). *J Med Virol* 1988;24:369-376.

CHAPTER 15

Herpes Simplex Virus

INTRODUCTION

Herpes simplex virus (HSV) is a human herpesvirus and a member of the alphaherpesvirinae subfamily. HSV is large virus and electron micrographs reveal an ultrastructure consisting of a 75 nm core surrounded by a 95-105 nm icosahedral nucleocapsid. The nucleocapsid is in turn, surrounded by an amorphus fibrillar tegument and a 110-120 nm, lipid-containing envelope. The HSV envelope is composed of a lipid-containing, trilaminar membrane into which 24 nm and 8-10 nm glycoprotein spikes are embedded. The HSV genome consists of a linear double-stranded (150 kilobase pair) DNA molecule that codes for more than 70 polypeptides (1), 30 of which are found in the virion. HSV is thermolabile and can be inactivated by heat, alcohols, detergents, and organic solvents.

Two HSV serotypes have been identified: HSV type 1 (HSV-1) and HSV type 2 (HSV-2). Differentiation of HSV-1 and HSV-2 has been based upon cell culture range (HSV-2 replicates in chicken embryo cells); pock-size studies on chorioallantoic membranes (HSV-2 produces larger lesions); monoclonal antibody-based assays; and restriction endonuclease analysis. Monoclonal antibody reagents to the gC and gE glycoproteins are widely used to distinguish HSV-1 from HSV-2. However, some laboratory strains and clinical isolates appear to have an intermediate reaction to these reagents. Typing of these isolates requires restriction endonuclease analysis .

Recent serological studies indicate that HSV-1 infections are generally acquired during childhood and by 60 years of age, up to 90% of the population has antibodies to HSV-1 (2). In contrast, HSV-2 infections are usually not acquired until puberty and antibody prevalence rates seem to correlate with past sexual activity (2,3). Although HSV-1 antibody prevalence rates have declined steadily over the past decade, HSV-2 antibody prevalence rates appear to be rising in many populations (3). This rise in seroprevalence appears to correlate with the increased number of HSV infections seen in sexually transmitted disease clinics.

Acquisition of primary HSV-1 infection usually occurs after contact with infected saliva or a person with oral lesions. Although most infections are asymptomatic, HSV-1 infections can cause gingivostomatitis, conjunctivitis, keratitis, and herpetic whitlow. Gingivostomatitis is particularly common in children under 5 years of age. This disease is characterized by the presence painful vesicular lesions on the palate, buccal mucosa, pharynx, tongue and the floor of the mouth. These lesions rapidly ulcerate and may be accompanied by a low-grade fever and submandibular lymphadenopathy. Lesions usually resolve within 2-3 weeks after primary infection and 4-7 days after recurrent infection (4). Autoinoculation of other sites, particularly the fingers, is common among young children.

HSV-1 infections are also responsible for more than 95% of herpes simplex virus encephalitis cases. HSV encephalitis is the most commonly reported viral infection of the central nervous system, accounting for 10-20% of all viral encephalitides in the United States (3). Left untreated, HSV encephalitis is a vicious, often fatal neurologic infection that eventually destroys the tissues of the temporal and frontal lobes of the brain. Epidemiologic studies indicate that HSV encephalitis may have a biphasic distribution with increased incidence of disease occurring in patients who are 5-30 years of age and in patients greater than 50 years of age (3).

AT A GLANCE...

HERPES SIMPLEX VIRUS

Virus Detection Methods
Tube cultures of human diploid fibroblasts, RD, Hep-2, A549, HFF, MRC-5, RK, ML, Vero cells.
Centrifugation-enhanced (shell vial) cultures using MRC-5 cells.
Direct fluorescent antibody staining of vesicular cell smears.

Specimen Source
Specimen of choice is vesicle fluid up to 3 days after vesicle appears. HSV can also be isolated from blood, cerebrospinal fluid, urine, and lesion swabs until the lesion becomes crusted.

Time to Result
Standard Culture: Positive culture - 16 hours to 7 days. Negative culture - 7 days.
Shell Vial Culture: Positive culture - 16-48 hours. Negative culture - 48 hours.
Direct Fluorescent Antibody Stain: 45 minutes - 2 hours.

HSV Serology

- **IgM** First appears 3-10 days after infection and persists for 6-8 weeks. IgM may or may not be produced during HSV reactivation. In CNS disease, HSV IgM detection in CSF may be of value, however, IgM is not produced until rather late in the disease.

- **IgG** First appears 7-14 days after infection and antibody levels peak 4-6 weeks thereafter. Antibody levels remain relatively stable over the lifetime of the patient. Demonstration of seroconversion or a four-fold increase in IgG antibody titer can help to document recent or active infections. During reactivation, IgG titers may or may not increase significantly. Demonstration of a four-fold rise in antibody titer in paired CSF specimens has diagnostic import. Subclass prevalence during primary infection IgG1 >> IgG2, IgG3, and IgG4. During reactivation IgG1, IgG2, IgG3 >> IgG4

- **IgA** First appears 7-14 days after infection and antibody levels persist for life.

Epidemiology
HSV infections are ubiquitous in man and most humans are infected in childhood. By adulthood, 80-90% of all humans are infected. Infections occur year-round and there is little variation in seasonal or annual incidence rates.

Transmission/Incubation Period
Transmission of HSV typically occurs through close personal contact (kissing, sharing eating utensils, etc.) or through some form of sexual contact. Virus is shed during primary infection, during episodes of recurrent herpes, and periodically in the absence of any clinically apparent disease. Asymptomatic shedding is a significant source of virus transmitted to susceptible hosts. The incubation period ranges from 1-26 days and with a mean of 6-8 days.

Inactivators and Disinfectants
Herpes simplex virus is relatively unstable and does not survive for long periods in the environment. HSV is rapidly inactivated by organic solvents, low pH (<4) conditions, and at elevated temperatures. Effective disinfectants include 70% alcohol, quaternary ammonium compounds, phenols, iodophor compounds, and glutaraldehyde compounds.

Historically, HSV-1 has been associated with oral infections while HSV-2 has been associated with genital infections. This distinction is no longer valid and today, 30-50% of genital herpes infections are caused by HSV-1 and 5-20% of oral infections are caused by HSV-2. It is interesting to note that the frequency of HSV reactivation depends upon the virus type and the anatomic site of infection (5). Genital HSV-2 infections recur 8-10 times more frequently than HSV-1 genital infections (5,6). In addition, more than 80% of patients with primary HSV-2 genital infections have a recurrence within 12 months compared with 55% of patients with HSV-1 genital infections (5). Conversely, oral HSV-1 infections recur much more frequently than oral HSV-2 infections (5,6).

Primary HSV-2 infections typically present as herpes genitalis and are characterized by extensive, bilaterally distributed papules or vesicles that merge to form large pustular or ulcerative lesions. Lesions often crust after 10-15 days and resolve within 2-4 weeks. Primary infections may be accompanied by fever, inguinal lymphadenopathy and dysuria. Patients with preexisting HSV-1 antibodies often have less severe symptoms during primary HSV-2 infection.

In recurrent herpes genitalis, lesions generally persist for 7-8 days. Unlike primary infections, recurrent HSV-2 episodes are usually of shorter duration, they have less associated morbidity, and they usually produce unilateral lesions. In about 50% of infected individuals, recurrent HSV-2 infections are heralded by burning or tingling in the affected area. These prodromal symptoms usually precede the eruption of vesicles by 1-2 days (7). As in primary infection, the symptoms are less severe and of shorter duration in men than in women.

The increased incidence of genital herpes has caused a concomitant increase in the incidence of neonatal herpes (8). Neonates have the highest incidence of visceral and CNS infections of any patient population with more than 70% of untreated cases producing disseminated or CNS infections. These infections can be devastating in neonates and they are associated with an overall mortality rate of 65%. In addition, less than 10% of these neonates have normal development after infection.

Patients compromised by immunodeficiency, immunosuppression, malnutrition, or by burns are at greater risk of developing severe HSV infections (4) In transplant patients, HSV is often isolated from throat washings in the first few weeks after transplantation. While these infections are often asymptomatic, they can occasionally cause severe tracheobronchitis, pneumonia, and esophagitis. Severe HSV infections are a prominent feature of AIDS and these patients often have HSV perianal ulcers, colitis, esophagitis, pneumonia, and a variety of neurological disorders.

Laboratory diagnosis of HSV infections can be accomplished by isolation and identification of HSV in cell cultures, by direct fluorescent antibody staining of vesicular smears, histochemical staining of biopsies, and serological methods. A number of serological tests including serum neutralization, complement fixation, passive hemagglutination, radioimmunoassay, EIA, and indirect fluorescent antibody methods have been used to detect HSV antibodies. In general, seroconversion or a four-fold increase in HSV antibody levels is indicative of recent infection. However, four-fold increases should be interpreted carefully because heterotypic immune responses can occur with primary varicella-zoster virus (VZV) infection. In this situation, primary VZV infection in a HSV-positive individual often produces a four-fold rise in HSV antibody titers.

HSV IgM may be helpful in diagnosing neonatal infections. In neonatal HSV infections, IgM is usually present 4 weeks after birth and persists for many months. In herpes encephalitis, detection of HSV IgM in cerebral spinal fluid may be of value. However, intrathecal IgM production is usually not detectable until late in disease. IgM is not useful for separating primary from recurrent HSV infections.

VIRUS ISOLATION

Introduction

Tube culture isolation is the gold standard for HSV detection and the reference method against which all other HSV detection methods are measured. Although centrifugation-enhanced (shell vial) and direct fluorescent antibody methods are used increasingly for HSV detection, these methods are not as sensitive as tube cultures. Therefore, tube cultures should be inoculated whenever these other methods are used.

HSV grows readily in a wide variety of cell lines (9,10) including HFF, MRC-5, A549, RD, ML, primary rabbit kidney, CV-1, Vero, and HEp-2 cells. Of these cells, HFF and MRC-5 cells are used most often while Vero and HEp-2 cells are used least often due to decreased sensitivity. RMK cells are generally resistant to HSV infection.

HSV isolation times vary widely from lab-to-lab and depends largely upon the sensitivity of the cell cultures used for isolations (11,12). The length of time that cultures are held can also influence the isolation rates. Some laboratories hold cultures for 10 days before reporting a negative result while other laboratories hold their cultures for only 5-7 days. In sensitive culture systems, HSV CPE may be visible 1 day after inoculation. However, most isolates require 2-3 days of cultivation to produce visible foci. Isolation times depend upon the (a) virus concentration and specimen type, (b) volume of inoculum, (c) culture conditions, and (d) cell line sensitivity. Maximum virus recovery is obtained when newly confluent cell monolayers are used.

Herpes simplex virus CPE is characterized by groups of enlarged refractile cells scattered throughout the culture. In addition, HSV-2 will occasionally produce a number of small syncytia within a monolayer. HSV CPE begins with cytoplasmic granularity that rapidly progresses to cellular enlargement and rounding (Fig. 1). Infected cells become refractile and may combine to form multinucleated giant cells. CPE spreads rapidly through the cell monolayer causing complete destruction of the monolayer within just a few days.

Once CPE forms, cultures should be stained with antibody reagents before reporting a positive result. Serotyping of HSV isolates is an area of considerable controversy. Although clinicians do not need to know the HSV type to administer antiviral therapies, they have come to expect this data from the laboratory. In addition, HSV typing can provide some prognostic information in cases of HSV encephalitis. HSV typing can also help the physician predict the probability of recurrences. These predictions are often used when counseling patients and helping them to cope with their disease. HSV typing should be available in all virology laboratories and most laboratories routinely type all HSV isolates.

For routine isolation, HSV specimens should be inoculated into two different cell lines to minimize any variations in cell line sensitivities. One human diploid fibroblast culture (e.g., MRC-5 or HFF) and one continuous heteroploid culture are recommended. The choice of heteroploid culture can be varied according to the time of year to provide the maximum virus isolation capability with the minimum number of cell lines. In this scenario, RD cells could be used during enterovirus season and A549 cultures could be used during respiratory season. MRC-5 or HFF cells should be used at a low passage number (<30) to have the best sensitivity.

Specimen Collection and Storage.

Although HSV has been isolated from nearly all visceral and mucocutaneous sites, fresh vesicular fluids from nonpurulent lesions are the specimens of choice. Selecting a lesion containing clear vesicle fluid is important because viral recovery rates decline as lesions become pustular or ulcerate. The efficiency of cell culture isolation is highest when specimens are inoculated onto cell cultures immediately after collection. If immediate inoculation is not possible, specimens may be stored for up to 48 hours at 2-8°C. If longer delays are anticipated, specimens should be frozen rapidly on dry ice and stored at -70°C. **DO NOT FREEZE SPECIMENS AT -20°C.** Storage at -20°C and repeated freeze/thaw cycles will destroy viral infectivity.

Specimen Preparation
Swabs
1. Vortex the specimen with several glass beads for 20-30 seconds to release any bound cells or virus.
2. Remove the swab from the transport medium

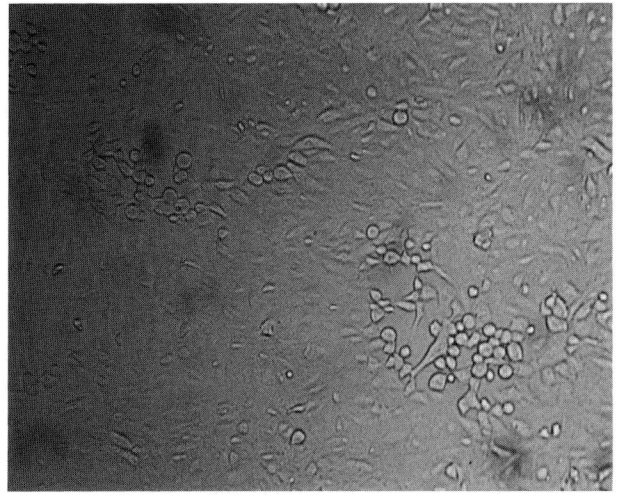 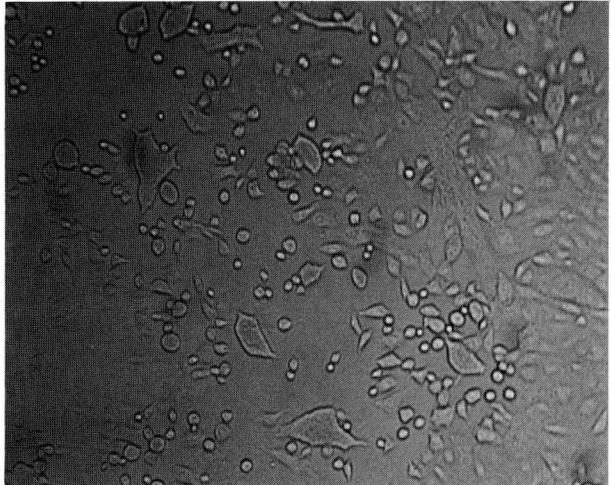

FIG. 1. HSV-1 CPE (left) and HSV-2 CPE (right) in CV-1 cells. Note the presence of small syncytia in the HSV-2 culture.

and firmly roll it against the inside of the tube to remove as much fluid as possible.

Tissues

Tissue specimens should be weighed, minced with sterile scissors, and homogenized using a disposable tissue homogenizer. Add enough viral transport medium to the homogenizer to produce a 10-20% (w/v) suspension. Remove the suspension from the homogenizer and vortex vigorously to extract as much antigen as possible. Centrifuge the specimen at 300-600 x g for 10 minutes at 4°C but do not separate the tissue from the supernatant fluids. Carefully remove the required amount of specimen for inoculation.

Standard Tube Culture

Procedure
1. Appropriately label one human fibroblast and one heteroploid cell culture tube.
2. Remove the medium from the cells.
3. Add 2 ml of maintenance medium to each tube.
4. Add 0.2-0.5 ml of the specimen to each tube.
5. Place the tubes in a roller rack (10-15 rph) and incubate at 35-37°C for 7 days. Some studies have indicated that rolling specimens at high speeds can enhance HSV isolation (13).
6. Cell cultures should be examined daily for evidence of CPE (Fig. 1). CPE is typically present 1-5 days after inoculation. Cultures that become toxic should be inoculated onto fresh cell cultures.
7. Perform culture confirmation testing on all positive cultures.

Interpretation of Results
Positive Test.

HSV CPE is characterized by enlarged, refractile cells within the monolayer (Fig. 1). In addition, HSV-2 isolates sometimes produce multinucleated giant cells within the cell sheet. HSV CPE begins as increased cytoplasmic granularity that is quickly followed by enlargement and rounding of the cells. The cells become refractile and may combine to form multinucleated giant cells. CPE spreads rapidly through the cell monolayer causing complete monolayer destruction within just a few days. Cultures exhibiting herpes-like CPE must be tested by the culture confirmation procedure.

REPORT: No report should be made based solely upon the presence of CPE.

Negative Test. The absence of characteristic CPE after 7 days indicates the absence of viable virus.
REPORT: Herpes simplex virus not isolated.

Culture Confirmation Procedure

1. Once CPE develops in 30-75% of the cell monolayer, remove all but a few drops of culture medium from the tube.
2. Scrape the cells from the tube surface.
3. Suspend the cells in the remaining culture medium.
4. Using a pipette, place a small drop of the cell suspension onto two wells of an acetone cleaned slide. Retain the remainder of the cell suspension for passage into new culture tubes if necessary.
5. Allow the slide to air dry.
6. The slide should be processed as directed by the manufacturer of the HSV antibody reagents or as described below.
7. Fix the slide in acetone for 10 minutes and allow the slide to air dry. Fixed slides may be stored for up to one week at 2-8°C or for one year at -20°C or below under desiccated conditions.
8. Add 15-50 µl of fluorescein-labeled HSV-1 monoclonal antibody to the first smear and HSV-2 monoclonal antibody to the second smear. When using bivalent (HSV-1 and HSV-2) HSV reagents, the extra well can be stained for VZV.
9. Incubate the slide for 30 minutes at 35-37°C in a covered, humidified chamber. Do not allow the antibody to dry as this could cause nonspecific antibody binding.
10. Remove the excess antibody with a gentle stream of PBS. Do not direct the stream directly at the cell smear as this could dislodge the cells.
11. Soak the slide in PBS for 5 minutes, shake off the excess fluid, and allow the slide to air dry.
12. Carefully add a small drop of FA mounting fluid to the center of each smear.
13. Place a number 1 coverslip on the mounting fluid and carefully remove all the air bubbles.
14. Immediately examine the slide using a fluorescence microscope. For optimum clarity, use 200-300X magnification for screening and 400X for confirmation of cell morphology.

Interpretation of Results
Positive Test. The presence of herpes simplex virus is indicated by the presence of characteristic CPE in the tube culture **and** intense apple-green fluorescence in the cytoplasm of the infected cells on the slide (Fig. 2) (see Color Plate 2A following Chapter 7). Only individuals experienced in reading FA reactions should examine the cell smears because cell debris and cell clumps may exhibit a dull fluorescence which can be misinterpreted as a positive reaction.
REPORT: Herpes simplex virus (type 1 or type 2 if appropriate) isolated.

Negative Test. The absence of herpes simplex virus is indicated by the uniform red coloration of the cells and a lack of intense apple-green fluorescence. CPE-positive, antibody negative specimens should be passaged and stained with antibodies to VZV.
REPORT: Herpes simplex virus not detected.

QC Procedures.
Subpassages of HSV clinical isolates or ATCC cultures should be inoculated with each batch of HSV isolations. Uninfected cell cultures serve as negative controls. Infected and uninfected cell monolayers are scraped and stained as described above. Positive controls must exhibit typical CPE and the cell smears must contain cells exhibiting intense apple-green fluorescence. Negative controls must stain with the red counterstain and should not exhibit any CPE or specific fluorescent staining.

Preparation of Positive Control Cultures
1. Add 20 µl of a recent HSV isolate (or an ATCC culture) to 2 ml of sterile GLB. Hold the diluted virus on ice until used to inoculate the monolayer.
2. Remove the growth medium from one 75

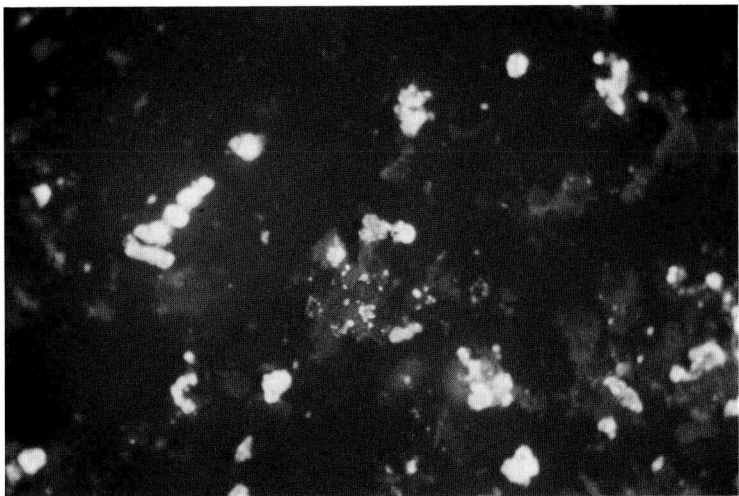

FIG. 2 HSV-infected CV-1 cells. Cells were scraped from the culture tubes and stain with monoclonal antibodies to HSV-2.

cm² flask of 80-90% confluent RD, A549, or Vero cells.

3. Add 2 ml of the diluted virus to the monolayer and allow the virus to adsorb for 1-2 hours at 35-37°C. Note: The flask should be rocked every 15 minutes to assure even virus distribution and to prevent monolayer desiccation.
4. Add 20 ml of MEM containing 2% FCS.
5. Incubate the flask at 35-37°C until the CPE involves 80-100% of the monolayer.
6. Scrape the cells from the flask into the cell culture medium.
7. Transfer the medium to a 50 ml polypropylene centrifuge tube and freeze at -70°C.
8. Thaw the medium quickly in a 37°C water bath.
9. Centrifuge the tube at 600-800 x g for 15 minutes to pellet the cell debris.
10. Transfer the supernatant fluids to a sterile vessel containing 80 ml of sterile GLB.
11. Dispense 1.0 ml of the diluted supernatant fluids into each of 100 freezing vials and freeze the vials at or below -70°C. Store the vials in liquid nitrogen for up to 7 years or at -70°C for up to 3 years.

Centrifugation Culture (Shell Vial) Method

Many laboratories have begun using centrifugation-enhanced (shell vial) culture methods to reduce virus isolation times. Shell vial methods can accommodate the same clinical specimens as traditional tube cultures and can reduce HSV isolation time from 1-7 days to 16-48 hours. Although shell vial methods are rapid and specific, they are slightly less sensitive than traditional tube cultures (14). Shorter turn-around times are usually not necessary for most HSV cultures. However, they may be warranted in cases involving women in active labor, in neonates, and for immunocompromised patients.

A number of different cell types including HFF, MRC-5, ML, CV-1, RD, and A549 cells can be used for this procedure. However, MRC-5 cells are used most often. Because of the reduced sensitivity of the shell vial method, at least one heteroploid tube culture should also be inoculated for each specimen.

Identification of HSV in shell vials is accomplished by staining the coverslips with bivalent or type-specific HSV antibodies to HSV. The number of fluorescent foci on the coverslips will vary depending upon the specimen type and the length of incubation.

FIG. 3. HSV-2 in a CV-1 shell vial culture. The coverslip culture was stained with antibodies to HSV-2.

Procedure
1. Appropriately label 4 MRC-5 shell vials.
2. Remove the medium from the shell vials.
3. Inoculate each vial with 0.2-0.5 ml of the specimen.
4. Centrifuge the vials at 700 x g for 1 hour at room temperature.
5. Add 1 ml of MEM containing 2% FCS to each vial.
6. Incubate the vials at 35-37°C.
7. Fix two vials after 24 hours and the remaining vials after 48 hours. One coverslip is stained with HSV-1 antibody reagents and the other with HSV-2 reagents. Only two shell vials are needed if bivalent reagents are used.

Staining of Coverslips
1. Coverslips should be fixed and stained as directed by the manufacturer of the antibody reagents or as described below.
2. Carefully remove the culture fluid from the vials and wash the coverslips twice by adding 1-2 ml of PBS and soaking for 5 minutes.
3. Remove the PBS and add 2-3 ml of acetone to each vial. Fix the coverslips for 10 minutes at room temperature.
4. Remove the acetone and allow the coverslips to air dry.
5. Rinse the coverslips briefly with PBS. (The moisture trapped between the coverslip and the vial will keep the stain from wicking under the coverslip.)
6. Add enough fluorescein-labeled monoclonal antibody to the vial to cover the coverslip (100-150 μl) and incubate for 30 minutes at 35-37°C in a humidified chamber.
7. Wash the coverslips twice as described above (step 2).
8. Place a small amount of mounting fluid on a slide. Remove the coverslips from the vial and lay them, cell side down, on the mounting fluid.
9. Immediately examine the coverslips at 100-300X using a fluorescence microscope.

Interpretation of Results

Coverslips that have individual or small groups of cells exhibiting brilliant apple-green fluorescence (Fig. 3) are positive for HSV. Coverslips without specific fluorescence are considered negative. Some specimens may trap the monoclonal antibodies between the cell monolayer and inoculum cells. These cultures may be interpreted as positive by an inexperienced technician. However, careful examination of the cell morphologies can distinguish this type of reaction from specific, HSV reactions. In addition, nonspecific staining can occur if the antibody reagents are allowed to dry on the coverslip. In this case, the entire cell sheet may stain apple-green.

Positive Test. The presence of HSV is indicated by the brilliant apple-green cytoplasmic and nuclear fluorescence.

REPORT: Herpes simplex virus (type 1 or type 2 if appropriate) isolated.

Negative Test. The absence of HSV isolation is indicated by the lack of specific fluorescence described above. The entire cell sheet should be stained red by the counterstain. A negative result should be reported after 48 hours of culture.

REPORT: Herpes simplex virus not isolated.

QC Procedures.

A 1:10 and 1:100 dilutions of recent herpes simplex virus isolates or ATCC cultures (see Preparation of Positive Control Cultures, above) should be inoculated with each batch of herpes shell vial cultures. Uninfected shell vial cultures serve as negative controls. Infected and uninfected shell vial cultures should be processed and stained as described for the patient cultures. Positive controls must exhibit the typical fluorescence patterns. Negative controls must not exhibit specific fluorescent staining. The 1:100 dilution may be needed to preclude monolayer destruction after 48 hours in culture.

DIRECT SMEARS

Introduction

Although virus isolation is the gold standard for HSV detection, the relatively long isolation times can limit the diagnostic utility of culture methods. This is especially true when isolating HSV from women in active labor. In these instances, direct fluorescent antibody (DFA) staining of smears can provide a rapid adjunct to cell culture isolations. Although rapid, DFA tests are not without problems. The sensitivity of this method is less than culture and depends heavily upon the submission of a high quality specimen. Given an adequate specimen, DFA testing has a sensitivity of up to 88% compared with virus isolation. Despite the relative lack of sensitivity, DFA methods are very specific and the false-positive rate is low.

DFA methods must be backed up by cultures. When used alone, DFA methods can give false negative results when testing specimens containing only free HSV and in specimens containing viruses other than HSV. Exclusive use of DFA reagents could also miss any new infectious agent that might appear in the community.

Specimen Collection

The success of the DFA procedure depends upon the submission of a well made cell smear. Smears that are too thick or "lumpy" can cause nonspecific trapping of the DFA reagent. Smears that are grossly contaminated with red blood cells can cause interpretation difficulties because of red cell autofluorescence. Finally, cell smears containing too few cells could cause the laboratory to report a false-negative result because no positive cells were observed. Specimen adequacy and its definition are significant sources of laboratory--to-laboratory variation and an area of increasing regulatory concern. Many laboratories require at least 50 cells per smear while other laboratories require 1-2 cells/high power (400X) field. Whatever the number, too many cells can cause nonspecific trapping of the conjugate while with too few cells, the laboratory may not be able to find an infected cell.

Procedure
1. Upon arrival in the laboratory, examine the slide wells at 100-300X magnification. Ideally, a minimum of 50 cells should be visible on each well before the specimen is considered adequate for further processing.
2. The slide should be processed as directed by

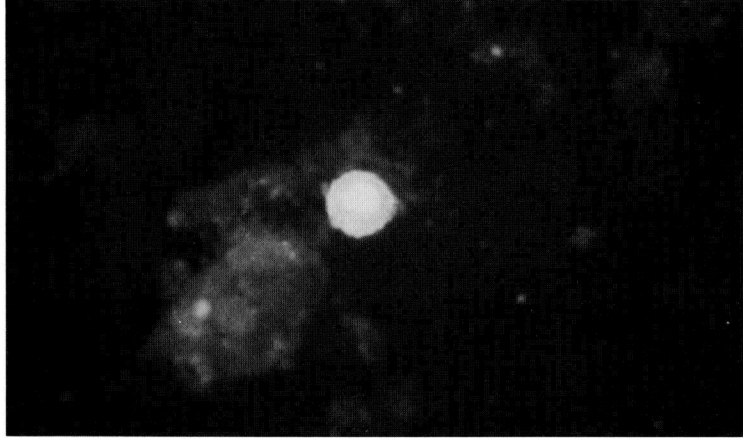

FIG. 4. Direct fluorescent antibody staining of a vesicular cell smear using bivalent HSV antibodies.

the manufacturer of the fluorescein-labelled antibody to herpes simplex virus or as described below.
3. Fix the cells by immersing the slides in acetone (room temperature) for 10 minutes. Allow the slides to air dry.
4. Place enough of the fluorescein-labelled monoclonal antibody on each well to cover the cell smear (approximately 15-30 μl). If more than one smear is received, one smear can be stained with HSV-1 reagents and the other well with HSV-2 reagents. If only one smear is submitted, a bivalent stain must be used.
5. Incubate the slide for 30 minutes at 35-37°C in a covered, humidified chamber. Do not allow the antibody to dry. Drying could cause nonspecific antibody binding.
6. Remove the excess antibody with a gentle stream of PBS. Do not direct the stream directly at the cell smear as this could dislodge the cells.
7. Soak the slide in PBS for 5 minutes, shake off the excess fluid, and allow the slide to air dry.
8. Carefully add a small drop of FA mounting fluid to the center of each well.
9. Place a number 1 coverslip on the mounting fluid and carefully remove all the air bubbles.
10. Immediately examine the slide using a fluorescence microscope. For optimum clarity, use 200-300X magnification for screening and 400X for confirmation of cell morphology.

Interpretation of Results

Positive Test. The presence of herpes simplex virus is indicated by intense apple-green fluorescence in the nucleus (and cytoplasm) of the basal or parabasal cells (Fig. 4; see Color Plate 1B following Chapter 7). There is typically a greater concentration of fluorescence around the periphery of the cell. Only intact cells should be examined because cell debris and clumps of normal cells may exhibit a dull fluorescence which could be misinterpreted as a positive reaction.
REPORT: Herpes simplex virus (type 1 or type 2 if appropriate) detected.

Negative Test. The absence of herpes simplex virus is indicated by the lack of intense apple green fluorescence in the basal or parabasal cells.
REPORT: Herpes simplex virus not detected.

Inconclusive Test. Specimens with fewer than 50 cells on each well may give erroneous results.
REPORT: Unacceptable specimen - too few cells.

QC Procedures.

Positive and negative controls should be stained

at least once each day to assure that the antibody reagent is performing properly. Positive cells should stain intensely as described above. Negative cells should be red and should not exhibit specific apple-green fluorescence. Because the same antibody reagent is used for confirmation of tube cultures, shell vial cultures, and direct smears, these tests can be used to demonstrate that the reagent is working properly. In laboratories that do not perform HSV isolations or laboratories where no isolations are ongoing, prepared slides should be stained whenever direct smears are stained. Control slides can be purchased commercially from a number of vendors or they can be prepared in the laboratory as described below.

Preparation of Positive and Negative Controls
1. Infect one 75 cm^2 flask of newly confluent Vero cells with 1.0 ml of a HSV isolate or the ATCC subculture.
2. Allow the virus to adsorb for 1-2 hours at 35-37°C.
3. Add 12 ml of complete maintenance medium containing 2% FCS.
4. Incubate the flask at 35-37°C until CPE involves 40-60% of the monolayer.
5. Trypsinize the cells and remove them from the flask. Centrifuge the cells at 150 x g for 5 minutes.
6. Resuspend the cell pellet in 5 ml of PBS. Perform a cell count and adjust the cell concentration to 2-5 x 10^6 cells/ml.
7. Dispense 3-10 µl of the cell suspension onto each slide and allow the suspension to air dry.
8. Fix the slides in acetone at room temperature for 10 minutes and allow the slides to air dry.
9. Store the slides at -20°C or below for up to one year. Slides should be stored under desiccated conditions to minimize antigen degradation and background fluorescence.

NOTE: Slides prepared in this manner will contain both positive and negative cells. Therefore, a single slide can be used for QC purposes. When slides are made from flasks with 100% CPE, two slides must be used - one containing a positive cell smear and a second slide containing an uninfected cell smear.

REFERENCES

1. Whitley R. Herpes simplex virus. In: Fields BN, Knipe DM, Chanock RM, Hirsch MS, Melnick JL, Monath TP, Roizman B, eds. *Virology*, Second edition. New York: Raven Press, 1990;1843-1887.
2. Nahmias AJ, Josey WE, Naib ZM, Luce CF, Duffey A. Antibodies to herpesvirus hominis types 1 and 2 in humans. *Am J Epidemiol* 1970;92:539-546.
3. Ashley RL, Corey L. Herpes simplex viruses. In: Schmidt NJ, Emmons RW, eds. *Diagnostic procedures for viral, rickettsial and chlamydial infections*. Sixth edition. Washington DC: American Public Health Association, 1989;256-317.
4. Hirsch MS. Herpes simplex virus. In: Mandell GL, Douglas RG, Bennett JE. *Principles and practice of infectious diseases*. Third edition. New York: Churchill Livingstone, 1990;1144-1153.
5. Lafferty WE, Coombs RW, Benedetti J, Critchlow C, Corey L. Recurrences after oral and genital herpes simplex virus infection: Influence of anatomic site and viral type. *N Eng J Med* 1987;316:1444-1449.
6. Corey L, Adams HG, Brown ZA, Holmes KK. Genital herpes simplex virus infection: Clinical manifestations, course and complications. *Ann Intern Med* 1983;98:958-972.
7. Lycke E, Jeansson S. *Herpesviridae*: Herpes simplex virus. In: Balows A, Hausler WJ, Lennette EH, eds. *Laboratory diagnosis of infectious diseases - principles and practice*. New York: Springer-Verlag, 1988;211-229.
8. Sullivan-Bolyai J, Hull HF, Wilson C, Corey L. Neonatal herpes simplex virus infection in King County, Washington: Increasing incidence and epidemiologic correlates. *JAMA* 1983;250:3059-3062.
9. Johnston SLG, Wellens K, Siegel CS. Rapid isolation of herpes simplex virus by using mink lung and rhabdomyosarcoma cell cultures. *J Clin Microbiol* 1990;28:2806-2807.
10. Peterson EM, Hughes BL, Aarnaes SL, de la

Maza LM. Comparison of primary rabbit kidney and MRC-5 cells and two stain procedures for herpes simplex virus detection by a shell vial centrifugation method. *J Clin Microbiol* 1988;26:22-224.

11. Hsiung GD. The impact of cell culture sensitivity on rapid viral diagnosis: A historical perspective. *Yale J Biol Med* 1989;62:79-88.

12. Zhao L, Landry ML, Balkovic ES, Hsiung GD. Impact of cell culture sensitivity and virus concentration on rapid detection of herpes simplex virus by cytopathic effects and immunoperoxidase staining. *J Clin Microbiol* 1987;25:1401-1405.

13. Hughes JH, Hamparian VV, Mavromoustakis CT. Continuous high-speed rolling versus centrifugation for detection of herpes simplex virus. *J Clin Microbiol* 1989;27:2884-2886.

14. Johnston SLG, Siegel CS. Comparison of enzyme immunoassay, shell vial culture, and conventional cell culture for the rapid detection of herpes simplex virus. *Diag Microbiol Infect Dis* 1990;13:241-244.

CHAPTER 16

Human Herpesvirus 6

INTRODUCTION

Human herpesvirus 6 (HHV-6) is the newest member of the *Herpesviridae*. The HHV-6 virion is large (160-200 nm) and consists of lipoprotein envelope surrounding a 100 nm icosahederal capsid (1). The HHV-6 genome consists of a double stranded (161.5 kilobase pair) linear DNA molecule that codes for more than 20 polypeptides. The genomic structure, organization, and homology among specific genes suggests that HHV-6 most closely resembles CMV at the genetic level. HHV-6 is relatively unstable and is quickly inactivated at 56°C, by alcohols, detergents, and organic solvents.

HHV-6 was originally designated human B-cell lymphotropic virus (HBLV) by Salahudden, et al. (1) because the virus was thought to preferentially infect B-lymphocytes and because most of the patients from whom the virus was isolated, suffered from B-lymphoproliferative disorders or malignancy. It was subsequently found that HHV-6 has a tropism for CD4 lymphocytes (2-4) *in vitro*. In addition, HHV-6 has been found in a variety of other cells including monocytic, megakaryocytic, and neuronal cell lines.

In 1988, Asano and coworkers (5) discovered that HHV-6 causes roseola infantum (exanthem subitum or fourth disease) in infants and small children. Roseola is characterized by the abrupt onset of a high fever which is followed in 2-4 days by erythematous maculopapular rash that lasts for 1-2 days. Roseola lesions are macular or papular and never become vesicular like the lesions associated with VZV or HSV. Lesions typically appear on the neck and back and then spread to the abdomen and thighs. The syndrome is self-limiting but convulsions associated with high fever can cause permanent neurologic injury. Atypical roseola can occur in the absence of rash and disease is accompanied by seroconversion, four-fold rise in HHV-6 antibody titers, and the presence of virus in saliva and blood (6).

Despite attempts to link HHV-6 infections with HIV-1-related disease, Kawasaki disease, lymphoproliferative disease, hepatitis, mononucleosis, and chronic fatigue syndrome, roseola is the only disease clearly associated with HHV-6 infection. Seroepidemiologic studies have demonstrated that most children acquire antibodies to HHV-6 between 6 and 18 months of age. Seroconversion rates appear to increase as maternal antibody levels diminish in the neonate. By four years of age, three-quarters of the population has antibodies to HHV-6 and infection is nearly universal in adults (7,8). Transmission of HHV-6 is poorly understood. However, the high frequency of virus isolation from adult saliva suggests that transmission may occur after contact with oral secretions. The incubation period appears to be 5-15 days.

Most cases of roseola can be diagnosed clinically. However, atypical (rashless) roseola can be documented either through isolation of the virus from blood or by serological methods. Serological diagnosis of primary infections can be made by demonstrating seroconversion, a four-fold increase in antibody titer in acute and convalescent sera, or by demonstration of HHV-6 specific IgM. An IgM antibody response may or may not occur during HHV-6 reactivation. Therefore, serological testing may not be useful for detecting reactivation.

VIRUS ISOLATION

Routine laboratory confirmation is usually not required for children who present with typical roseola. However, HHV-6 isolation is required in

> *AT A GLANCE...*
>
> # HUMAN HERPESVIRUS-6 (HHV-6)
>
> **Virus Detection Methods**
> HHV-6 can be isolated by co-cultivation with phytohemagglutinin (PHA) stimulated cord blood or peripheral blood leukocytes.
>
> **Specimen Source**
> Heparinized peripheral blood is the specimen of choice. However, HHV-6 has frequently been isolated from saliva. All specimens should be obtained as early as possible after the onset of symptoms.
>
> **Time to Result**
> Positive Culture - 1-2 weeks.
> Negative Culture - 2 weeks.
>
> **HHV-6 Serology**
>
> IgM — First detectable 2-5 days after the onset of rash and antibody levels peak after 7-14 days. IgM antibody levels generally persist for 4-8 weeks. IgM antibody levels may be detectable during HHV-6 reactivation.
>
> IgG — First detectable 7-8 days after onset of symptoms, IgG antibody levels usually peak 2-3 weeks later. IgG titers generally decline for several months but usually remain detectable for the lifetime of the patient.
>
> **Epidemiology**
> HHV-6 infections are common and the virus has been found throughout the world. HHV-6 infections can occur in all age groups. However, infection is most often acquired before three years of age. HHV-6 is the causative agent of roseola (exanthem subitum) in children. Roseola is a common infection of infancy and the peak incidence of roseola occurs between the ages of 6 and 18 months. Infants less than 6 months are thought to be protected by maternal antibodies. Adults who become infected are often only mildly symptomatic.
>
> **Transmission/Incubation Period**
> Transmission of HHV-6 is poorly understood. However, the high frequency of virus isolation from the saliva of adults suggests that transmission may occur after contact with oral secretions. The incubation period is 5-15 days.
>
> **Inactivators and Disinfectants**
> HHV-6 is readily inactivated at elevated temperatures (56°C) and by organic solvents. Effective disinfectants include 10% household bleach solutions, phenols, glutaraldehyde compounds, and formalin solutions.

order to determine the frequency of virus reactivation and to determine the spectrum of HHV-6 disease in children and adults. Unfortunately, methods for isolation and detection of HHV-6 are still in development and the method listed below may change radically in the next few years. This procedure is a modification of the methods used to isolate HIV-1 from peripheral blood leukocytes. In this procedure, leukocytes are isolated from the patient's blood and co-cultivated with peripheral blood or cord blood leukocytes that have been stimulated for 2-3 days with PHA, IL-2, and hydrocortisone to increase the number of HHV-6 susceptible cells. The co-cultivation medium contains IL-2, polybrene, and antibodies to interferon as described in Chapter 25. Although HHV-6 has been isolated from saliva, peripheral blood leukocytes are the specimen of choice.

Specimen Collection
Saliva
1. Dilute the saliva with an equal volume of RPMI-1640 containing 10% FCS and twice the normal concentration of antibiotics.
2. Centrifuge the specimen at 1000-3,000 x g to pellet any cells or debris.
3. Filter the supernatant fluids through a 0.45 μm syringe filter before inoculation.
4. Specimens should be inoculated onto PHA-stimulated PBL cultures as soon as possible.

Blood Collection
1. Aseptically collect 10 ml of venous blood in green top vacutainer tubes containing preservative-free heparin.
2. Once collected, the tubes should be inverted several times to assure that the heparin is evenly dispersed.
3. The specimens may be shipped to the laboratory at room temperature. However, virus is rarely isolated if the specimen is older than 24 hours old.
4. Freezing of mononuclear cells for later isolation is not recommended.

Lymphocyte Separation
Lymphocytes are separated from red cells and granulocytes by centrifugation through a ficoll-hypaque density gradient. A number of ready-made gradient materials can be purchased commercially (Ficol-Paque from Pharmacia Fine Chemicals, Histopaque from Sigma Chemical Company or Lymphocyte Separating Medium from Organon-Teknika) for this purpose or gradient materials can be prepared in the laboratory (Appendix C). Lymphocyte separation should be done as directed by the manufacturer or as described below.

1. Aseptically collect 10-20 ml of blood in green top vacutainer tubes containing heparin.
2. Place 3 ml of ficoll-hypaque in clear sterile 15 ml centrifuge tubes.
3. Dilute the heparinized blood with an equal volume of sterile saline.
4. Carefully overlay 5 ml of heparinized blood onto the ficoll-hypaque.
5. Centrifuge the tubes at 1000 x g in a swinging bucket rotor for 30 minutes at room temperature.
6. Lymphocytes will band at the plasma-ficoll interface.
7. Carefully remove and discard the clear upper (plasma) layer.
8. Remove the lymphocyte bands and pool them in a sterile 50 ml centrifuge tube.
9. Add PBS to the 30 ml mark.
10. Centrifuge the cells at 600 x g for 15 minutes and remove the supernatant fluids.
11. Resuspend the cell pellet in 5 ml of PBS.
12. Perform a cell count and resuspend the cells at 2×10^7 cells/ml.

PHA-Stimulated Peripheral Blood Leukocytes
1. Separate the lymphocytes from peripheral or cord blood as described above (steps 1-12).
2. Resuspend the cells at 1×10^6 cells/ml in PHA-H medium (Appendix C) and distribute into an appropriate number of flasks (25 cm² flask = 10 ml; 75 cm² flask = 30 ml; 150 cm² flask = 100 ml).
3. Incubate the flasks in an upright position for 2-3 days at 35-37°C. Contaminated cultures should be discarded.
4. Cells are ready for cocultivation or feeding. PHA-stimulated cells can be frozen in liquid nitrogen (Chapter 6) in RPMI-1640 containing 20% FCS and 10% DMSO for 3-4 years. Prior to use, frozen cells should be thawed quickly in a 35-37°C water bath and washed twice to remove the cryoprotectants.

Inoculation
1. For each patient isolation, appropriately label a 25 cm² flask containing 5×10^6 PHA-stimulated peripheral blood lymphocytes in 10 ml of stimulation medium (Appendix C).

2. Allow the lymphocytes to settle and, without disturbing the cells, remove 5-7 ml of the medium.
3. Inoculate the flask with 0.5 ml (1 x 10^7 cells) of the patient lymphocyte suspension.
4. Incubate the flasks in an upright position for 1 hour at 35-37°C.
5. Add propagation medium (Appendix C) to the 10 ml mark.
6. Incubate the cultures for 24 hours at 35-37°C.
7. Remove half the medium and replace it with an equal volume of propagation medium.
8. Incubate the cultures as before for 1 week.
9. One week after initial culture, and every 3-4 days thereafter, resuspend the cells by gentle agitation and remove 5 ml of cell suspension from the flasks for antigen detection.
10. Replace the culture medium with 5 ml of fresh propagation medium.
11. After one week, add 3 x 10^6 fresh PHA-stimulated lymphocytes to the flask.
12. Maintain the cultures for 14 days before discarding them as negative. If the culture becomes contaminated, filter the supernatant fluids through a 0.2 μm filter and add the filtrate to fresh PHA-stimulated lymphocytes.

Culture Confirmation

Positive cultures are characterized by the presence of large balloon-like cells in the culture. Because balloon-like cells can also occur in the absence of infection, immunofluorescent staining is necessary to confirm HHV-6 infection. In addition, negative (uninoculated) control cultures should be stained to assure that the HHV-6 did not originate from the donor PBLs.

Immunofluorescence Assay (IFA)

Indirect immunofluorescence assays can be used to determine if HHV-6 antigens are present in the culture. In this procedure, the cells are washed, placed on a slide, and fixed with acetone. Monoclonal antibodies or human antisera containing IgG to HHV-6 are reacted with the fixed smears. After washing away any unbound antibody, fluorescein-labelled antimouse (or antihuman IgG) antibodies are added together with an Evans blue counterstain. The smears are washed, dried and examined under a fluorescence microscope.

Slide Preparation
1. Centrifuge the cultures at 600-800 x g to pellet the cells.
2. Remove the culture fluids and store at -70°C for future reference.
3. Resuspend the cells in 1 ml of PBS containing 2% FCS.
4. Place 5-10 μl of the suspension on both wells of an acetone-cleaned, two-well slide. Prepare several slides and store the remainder for future reference (see below).
5. Allow the smears to air dry.
6. Fix the slides for 10 minutes in acetone.
7. Slides should be stained immediately or they can be stored at -20°C for up to 3 years. Slides should be stored under desiccated conditions to minimize antigen degradation and background staining.

IFA Procedure
1. Remove positive and negative control slides from the freezer and allow them to come to room temperature.
2. Place one drop (10-30 μl) of the antibody preparation on one of the patient smears and on each of the control smears. Place a drop of PBS on the other (reagent control) smears.
3. Incubate the slides for 30 minutes at 35-37°C in a covered, humidified chamber. Do not allow the antibody to dry on the slide as this could cause nonspecific antibody binding.
4. Remove the excess antibody with a gentle stream of PBS. Do not direct the stream directly at the cell smear as this could dislodge the cells.
5. Soak the slide in PBS for 5 minutes, shake off the excess fluid.
6. Add one drop (15-30 μl) of fluorescein-labelled antimouse antibody to all the smears. If human serum is used, use a fluorescein-labelled goat human IgG. Both antisera should contain 0.01% Evans blue as a counterstain.

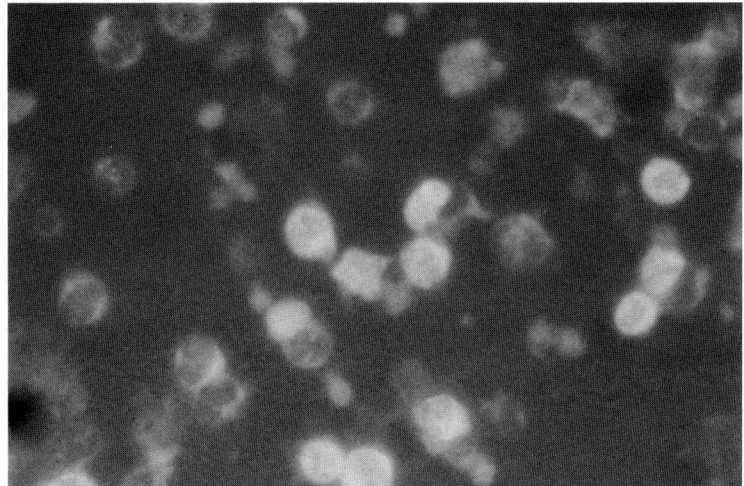

FIG. 1. Fluorescent antibody staining of an HHV-infected cell smear.

7. Incubate the slide for 30 minutes at 33-35°C in a covered, humidified chamber.
8. Remove the excess antibody with a gentle stream of PBS as before.
9. Soak the slide in PBS for 5 minutes, shake off the excess fluid, and allow the slide to air dry.
10. Carefully add a small drop of FA mounting fluid to the center of each well.
11. Place a number 1 coverslip on the mounting fluid and carefully remove all the air bubbles.
12. Immediately examine the slide using a fluorescence microscope. For optimum clarity, use 200-300X magnification for screening and 400X for confirmation of staining pattern.

Interpretation of Results

Positive Test. The presence of HHV-6 is indicated by intense apple-green fluorescence in the nucleus and cytoplasm of infected cells. Uninfected cells will stain red due to the counterstain (Fig. 1). Many cultures will contain cell debris and cell clumps that may exhibit a dull fluorescence which could be misinterpreted as a positive reaction.
REPORT: HHV-6 isolated.

Negative Test. If all control reactions function properly, the absence of specific cytoplasmic fluorescence indicates the absence of detectable virus. If the cultures are repeatedly negative after 14 days, the culture should be reported as negative.
REPORT: HHV-6 not isolated.

QC Procedures.

Use of appropriate controls are necessary because staining patterns and intensities can vary depending upon the type of antibody (monoclonal/polyclonal) used. Therefore, appropriate control slides or cultures must be stained concurrently with the patient specimens. Positive control smears must exhibit typical apple-green fluorescence. If no fluorescence is observed in the positive control, the test is invalid and must be repeated. Negative controls must not stain with the antibody. If specific staining occurs in the negative controls, the test is invalid and must be repeated. Extra control slides prepared from positive and negative cultures (above) can be used as controls.

REFERENCES

1. Salahuddin SZ, Ablashi DV, Markham PD, et al. Isolation of a new virus, HBLV, in patients with lymphoproliferative disorders. *Science* 1986;234:596-601.

2. Becker WB, Engelbrecht S, Becker ML, Piek C,

Robson BA, Wood L, Jacobs P. Isolation of a new human herpesvirus producing a lytic infection of helper (CD4) T-lymphocyte cultures - Another cause of acquired immunodeficiency? *S Afr Med J* 1988;74;610-614.
3. Black J, Sanderlin K, Goldsmith C, Gary H, Lopez C, Pellet P. Growth properties of human herpesvirus-6 strain Z29. *J Virol Meth* 1989;26:133-146.
4. Downing RG, Sewankambo N, Serwadda D, Honess R, Crawford D, Jarrett R, Griffin BE. Isolation of human lymphotropic herpesviruses from Uganda. *Lancet* 1987;2:390.
5. Asano Y, Yoshikawa T, Suga S, Yazaki T, Hata T, Nagai T, Kajita Y, Ozaki T, Yoshida S. Viremia and neutralizing antibody response in infants with exanthem subitum. *J Pediatr* 1989;114:535-539.
6. Stewart JA, Sanderlin KC. Human herpesvirus 6. In: Lennette EH, ed. *Laboratory diagnosis of viral infections*, second edition. New York: Marcel Dekker, Inc. 1992;463-475.
7. Knowles WA, Gardner SD. High prevalence of antibody to human herpesvirus-6 and seroconversion with rash in two infants [letter]. *Lancet* 1988;2:912-913.
8. Gopal M, Thomson B, Fox J, Tedder R, Honess R. Detection by PCR of HHV-6 and EBV DNA in blood and oropharynx of healthy adults and HIV-seropositives [letter]. *Lancet* 1990;335:1598-1599.

CHAPTER 17
Influenza Virus

INTRODUCTION

Influenza viruses are negative-stranded RNA viruses belonging to the family Orthomyxoviridae. Influenza A and influenza B viruses are thought to comprise a single genus while influenza C virus will probably be placed in a separate genus. Influenza virions can be either filamentous or roughly spherical. Filaments may be 400 nm long while spherical virions have an average diameter of 100 nm. These viruses contain an envelope that is derived from the host cell and modified by the insertion of virus-coded proteins. All influenza virus envelopes contain virus-coded matrix and hemagglutinin proteins. Influenza A and B also possess neuraminidase activity. Influenza A and B viruses possess a genome consisting of 8 separate RNA segments that code for at least 8 proteins. The influenza C virus genome is thought to contain 7 RNA segments that code for at least 7 proteins. All influenza viruses are ether, pH, and heat sensitive.

A unique feature of influenza virus is the frequency with which antigenic variation occurs. Relatively minor antigenic changes that occur nearly every year are referred to as antigenic drift. Major antigenic changes are called antigenic shifts. Viruses that undergo antigenic shift have little immunologic similarity between new and old hemagglutinin and/or neuraminidase molecules. Antigenic shift of this type generally heralds pandemic influenza. Although antigenic variation is a frequent event with influenza A, antigenic variation occurs less frequently with influenza B virus and has not been associated with influenza C virus. Nonetheless, antibodies used to detect influenza viruses should be checked regularly against the prevalent virus strains to assure that the current strains can be detected by existing reagents.

Influenza moves rapidly throughout the population each winter, causing more than 20,000 excess deaths, principally in the elderly, the immunocompromised, and in patients with chronic lung or kidney conditions (1). In these populations, influenza can also cause polyneuritis, encephalopathy and inflammation of cardiac and skeletal muscles. Reye's syndrome may also occur after influenza infections (2). Influenza generally spreads through a population in an epidemic fashion and at least 10 global pandemics have occurred during the past 200 years. Major epidemics of influenza A occur at 2-3 year intervals in the United States while influenza B epidemics occur every 4-6 years. Influenza C does not occur in epidemics (3).

Influenza virus is efficiently transmitted through inhalation of virus-laden aerosol droplets formed when an infected individual coughs or sneezes. Once transmitted, influenza viruses cause acute respiratory tract disease characterized by an abrupt onset of fever, chills, headache, and myalgia. Coryza, sore throat and cough are typical of the disease and these symptoms are often severe and prolonged. In healthy adults, the disease is usually self-limiting and typically resolves within a week.

The incubation period for influenza is 18 to 70 hours. Influenza viruses infect the columnar epithelium of the upper and lower respiratory tract and progeny virus can be detected 24 hours before the onset of illness. Virus titers rise rapidly thereafter, peaking 24-48 hours after the onset of symptoms. Subsequently, virus titers rapidly decline and viable virus is not detectable after 5-10 days except in young children who may shed large amounts of virus for a prolonged period of time.

Laboratory diagnosis of influenza can be made

AT A GLANCE...

INFLUENZA VIRUS

Virus Detection Methods
Tube cultures in RMK, MRC-5, RD, or MDCK cells.
Centrifugation-enhanced (shell vial) cultures using RMK, MRC-5, RD, or MDCK cells.
Direct fluorescent antibody (DFA) staining of respiratory smears.
Enzyme immunoassays (EIA) for influenza A.

Specimen Source
The specimen of choice is a nasal washing or an NP and a throat swab where both swabs are placed in the same viral transport tube. Influenza can be isolated from nasal washes, NP swab specimens, nasal aspirates, throat swab specimens, sputum, and from broncheoalveolar lavage specimens. The best time for specimen collection is one day before the onset of symptoms to three days after the onset of symptoms. Although influenza usually cannot be isolated 5-10 days after the onset of symptoms, DFA and EIA methods may detect virus during this time.

Time to Result
Standard Culture: Positive culture - 2 days to 2 weeks. Negative culture - 2 weeks.
Shell Vial Method: Positive culture - 1-2 days. Negative culture - 2 days.
DFA Method: 15-60 minutes.
EIA Methods: 20 minutes to 3.5 hours.

Influenza Serology

IgM   First detectable 5-14 days after onset of symptoms. Peak titers are observed 2-4 weeks after infection. IgM antibodies often drop to undetectable levels after 4-6 weeks.

IgG   Strain-specific IgG antibodies are first detectable 10-14 days after onset of symptoms. Antibody levels peak after 4-6 weeks and remain elevated for years.

IgA   First detectable 5 days after the onset of symptoms with peak levels at 14 days postinfection. Influenza serum IgA levels can remain elevated for 2-5 months.

Epidemiology
Most cases of influenza occur in the winter and early spring (January through April) with peak incidence in January and February. Influenza A causes major epidemics at 2-3 year intervals. Influenza B epidemics occur every 4-6 years. Influenza C does not produce epidemics.

Transmission/Incubation Period
Influenza is very contagious and the disease is transmitted through inhalation of virus-laden aerosols or by autoinoculation after handling fomites contaminated with nasal or throat secretions. The incubation period is 18-70 hours. Patients are infectious one day before the onset of symptoms and for 3-4 days thereafter. Infants may shed virus for longer periods of time.

Inactivators and Disinfectants
Influenza is quickly inactivated by heat, alcohols, and organic solvents. Effective disinfectants include 70% alcohol, 10% household bleach solutions, quaternary ammonium compounds, phenols, iodophor compounds, and glutaraldehyde compounds.

by direct staining of respiratory epithelial cells; inoculation of respiratory specimens into eggs; inoculation of RMK cells in conventional culture tubes or shell vial cultures; or by EIA testing. Serologic confirmation of influenza infections are rarely required because antibody levels do not rise within a clinically relevant timeframe. Complement-fixing (CF) antibody tests and hemagglutination-inhibition (HAI) tests are used when serological confirmation is requested. Both tests require paired (acute and convalescent) sera with the convalescent serum obtained 10-20 days after the acute-phase serum specimen. Fourfold or greater rises in antibody titer are considered diagnostic of infection. Although not a recommended procedure, a single high titer convalescent specimen is suggestive of a recent infection.

Increasing influenza antibody levels can be detected by virus neutralization (Nt), HAI, CF, EIA, and indirect fluorescent antibody methods two weeks after the onset of influenza. Antibody levels usually peak 4-6 weeks after infection. During reinfection, the antibody response may be more rapid. The complement fixing antibody response is directed primarily against the viral ribonucleoprotein. Therefore, CF antibody responses are genus-specific. CF antibodies are predominantly of the IgM class. CF antibodies are first detectable 5 days after the onset of disease and peak levels are achieved 2-4 weeks thereafter. The CF antibody response declines rapidly and is usually undetectable after 3 months.

The HAI and Nt antibody responses are primarily composed of IgG antibodies to the viral hemagglutinin (HA) antigen. HAI and Nt antibodies persist for months to years and then gradually decline. In contrast with the CF test, HAI and Nt antibody responses are subtype specific and these methods are used to determine the antigenic relatedness among hemagglutinins of influenza viruses. Although there is no exact correlation between HAI and Nt titers, serum HAI titers of $\geq$ 1:40 and Nt titers of $\geq$ 1:8 are associated with protection against influenza in most subjects. Similarly, patients with nasal neutralizing antibody titers of $\geq$ 1:4 are protected against influenza.

Secretory IgA develops in the respiratory tract approximately five days after infection and near-peak titers are achieved after 14 days. Neutralizing secretory IgA are detectable in the saliva, nasal secretions, sputum, and tracheal washings. Serum IgA appears at the same time as the secretory IgA although serum IgA titers usually decline to undetectable levels within a few months.

VIRUS ISOLATION

Introduction

In most clinical laboratories, tube cultures of primary rhesus monkey kidney (RMK) cells have replaced embryonated chicken eggs as the gold standard for influenza A and B isolations. Although a new cell line (HMV-II) has been described for the isolation of influenza C (6), most influenza C isolations are still done in embryonated chicken eggs. Influenza A and B can also be isolated in MRC-5, RD, A549, HL and MDCK tube or shell vial cultures. Some studies indicate that propagation of the virus in the presence of a proteolytic enzyme such as trypsin may increase viral isolation rates (7).

In RMK and MDCK cells, CPE is typically observed 5-10 days postinoculation. Influenza B produces CPE earlier than influenza A under most circumstances. However, influenza CPE may not be obvious or infected cells may appear toxic. Hemagglutination (HA) or hemadsorption (HAd) testing is done to circumvent CPE interpretation problems. Hemadsorption procedures can detect the presence of viral hemagglutinin before CPE becomes apparent. In RMK cells, about two-thirds of all positive cultures are HAd-positive within 3 days of inoculation while the remainder are HAd-positive after 5-7 days (8). Cultures are usually hemagglutination-positive a day after they are HAd positive. Final identification of the hemagglutinating or hemadsorbing virus is usually accomplished by staining the cells with fluorescein-labelled monoclonal antibodies. Several studies have shown that centrifugation-enhanced (shell vial) methods produce antibody-positive cultures earlier than tube cultures (9-14). While shell vial methods have been shown to be rapid and specific, they are less sensitive than conventional tube cultures. Unless otherwise specified, specimens submitted for influenza isolation should be inoculated into at least two RMK tube cultures and incubated at 33-35°C in a roller drum at 10-15 rph. Two RMK, MDCK, or MRC-5 shell vials may also be inoculated if more rapid virus detection is required.

Specimen Collection

Influenza replicates in the ciliated columnar epithelial cells of the respiratory tract. The best time for specimen collection in this area is 1 day before the onset of symptoms to 3-4 days after the onset of symptoms. Influenza viruses are generally not detectable 5-10 days after onset of symptoms except in young children who may shed large amounts of virus for prolonged periods.

Influenza A and influenza B can be readily isolated from nasopharyngeal (NP) and throat swab specimens, nasal and throat washes, or from sputum. If the patient is producing sputum, sputum may be the best specimen. Alternatively, a nasal wash or an NP plus a throat swab specimen in the same viral transport tube are preferred specimens.

Although nasal washings produce a larger quantity of virus than swabs, they are not used in all clinical settings because they cause more discomfort. In addition, nasal washes could pose an aspiration risk in children (15). In most situations, swab specimens are the easiest specimens to obtain. Specimens should be sent to the virology laboratory on wet ice. **DO NOT FREEZE SPECIMENS AT -20°C.** If specimens cannot be inoculated onto cell cultures within 72 hours, they should be frozen at -70°C.

Specimen Preparation
Swabs
1. Vortex the specimen with several glass beads for 20-30 seconds to release any bound cells or virus.
2. Remove the swabs from the transport medium and firmly roll them against the inside of the tube to remove as much fluid as possible.

Wash and Aspirate Specimens
1. Vortex the specimen with several glass beads for 20-30 seconds to release any cell-associated virus.
2. Centrifuge suspension at 1000 x g for 5 minutes to remove mucus and cell debris.
3. Carefully remove the mucus and use the remaining specimen to inoculate cell cultures.

Standard Tube Culture
Procedure
1. Appropriately label 2 tubes containing freshly confluent RMK cells.
2. Remove the medium from the cells and wash the monolayers twice with HBSS or serum-free MEM containing antibiotics. This wash step is important because some serum components may interfere with influenza virus replication.
3. Add 2 ml of serum-free MEM (with antibiotics) to each tube.
4. Add 0.2-0.5 ml of specimen to each tube.
5. Place the tubes in a roller drum and incubate at 33-35°C and 10-15 rph.
6. Examine the cells at least every other day for CPE.
7. Test the tubes for hemagglutination or hemadsorption 2-3 days after inoculation or when CPE is observed. Test negative cultures again on day 10 or day 14.
8. Perform culture confirmation testing (see below) on any monolayer that exhibits CPE, HA, and/or HAd activity.

Interpretation of Results

Influenza CPE is characterized by areas of large, irregularly shaped, granular or vacuolated cells that are spread randomly over the cell sheet. As the CPE progresses, the cell monolayer will degenerate leaving a significant amount of cell debris in the culture medium. No report should be made based upon CPE alone.

Hemadsorption (HAd) Procedure

During replication, influenza, mumps, and parainfluenza viruses produce hemagglutinin molecules that are inserted into the plasma membrane of the host cell. Because these hemagglutinins bind to red blood cells, virus-infected cells can be readily identified because red cells adsorb to them. Hemadsorption assays are rapid, convenient, and relatively inexpensive. In addition, the proportion of cells with attached erythrocytes at 4°C compared with the number at 22°C can provide some indication as to whether the isolate is parainfluenza, mumps, or influenza virus. In parainfluenza 1, 2, and 3 infections, RBCs often elute from the infected cells when the culture is warmed to 22°C. Influenza and mumps virus infections will generally have the same amount of hemadsorption at 4°C and at 22°C.

Hemadsorption assays can be performed with guinea pig, fowl, and human O erythrocytes. However, most laboratories use guinea pig cells because they are smaller than the other RBCs and more of them can attach to an infected cell. Thus, guinea pig RBCs produce a more uniform HAd pattern and reactions are easier to interpret than fowl or human erythrocyte HAd tests.

Because nearly all of the influenza viruses isolated during the past few years have produced CPE in RMK cells, many laboratories now wait until CPE forms before performing the HAd assay. Other laboratories test the first tube 2-3 days after inoculation and the second tube tested on day 9-10. Negative cultures need not be discarded after HAd testing. Once the RBCs have been removed, the monolayers are washed twice with serum-free MEM, 2 ml of fresh MEM is added, and the cultures can be incubated for an additional 5 days.

Procedure

1. Prepare a 0.5% suspension of guinea pig RBCs in sterile saline or phosphate buffered saline (Appendix C).
2. Add 0.4 ml of the RBC suspension to each culture tube.
3. Incubate the tubes on their side (cell monolayer downward) for 20-60 minutes at 4°C in a slanted (4°) test tube rack.
4. Examine the cell monolayer under the microscope (100-200X) for RBC adsorption and record the results (Table 1).
5. Incubate the tubes for one hour at room temperature. Examine the tubes for hemadsorption as before and record the results.

Table 1. Recording HAd Reactions.

Percent of Cells With HAd	Score
75-100	4+
50-75	3+
25-50	2+
< 25	1+
Questionable	+/-

Interpretation of Results

Positive Test. Specific hemadsorption is characterized by chains, rosettes, or clumps of red cells that adhere to the surface of the cell sheet (Fig. 1). The appearance of this HAd pattern indicates a positive Had test. However, results should be interpreted carefully because monkey kidney cells are sometimes infected with hemadsorbing simian viruses. Comparing the level of HAd to an uninoculated control culture is important to eliminate false positive results. Influenza and mumps viruses will generally have the same amount of hemadsorption after incubation at 4°C and after warming to 22°C. Positive cultures should be scraped from their tubes and stained with specific monoclonal antibody reagents.

REPORT: No report should be made based upon the results of a hemadsorption test alone. All HAd positive cultures should be confirmed with type-specific monoclonal antibody staining.

Negative Test. A negative HAd test is characterized by RBCs that float freely in the medium and the absence of RBC chains, rosettes, or clumps adhering to the cell sheet. Because endogenous simian viruses can produce a background level of HAd, cultures that exhibit some HAd activity may be considered negative if the HAd levels are no higher than the background (uninoculated culture)

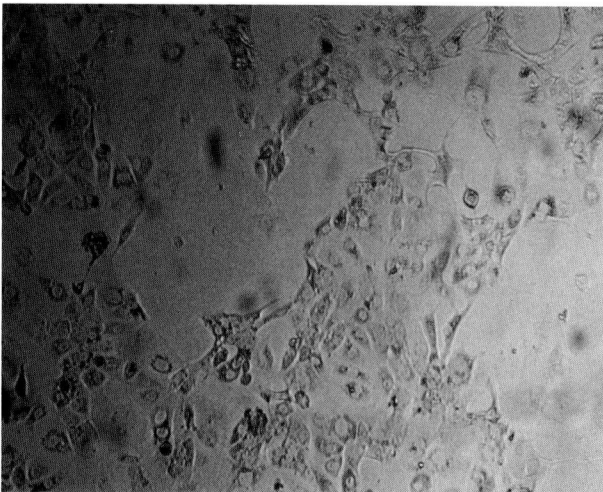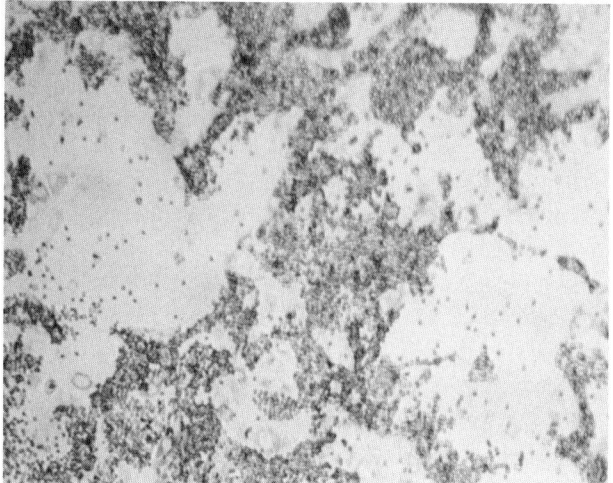

FIG. 1. Influenza-infected RMK cells before (left) and after (right) allowing guinea pig red blood cells to adsorb.

levels.

REPORT: Influenza not isolated.

QC Procedures.

Two influenza-infected RMK (positive control) tubes should be inoculated with each batch of influenza isolations. Test the first tube for HAd activity on day 2-3 and the second tube on day 10. Both tubes should exhibit specific HAd activity as described above. When inoculating the positive control cultures, it may be advantageous to dilute the inoculum for the day 10 tube 1:10 to preclude monolayer destruction. Two uninoculated RMK tubes serve as negative controls. Because monkey kidney cells are sometimes infected with SV5 or other hemadsorbing simian viruses, the negative control tubes must come from the same lot of cells as the patient isolation tubes. A negative control tube should be tested whenever patient isolation tubes are tested. Positive HAd tests will have more HAd activity than the negative control tube. If uninoculated control tubes have more than 25% of the cell sheet exhibiting HAd (> 1+ HAd), the cell lot should be discarded.

Hemagglutination (HA) Procedure

Some laboratories prefer to use hemagglutination (HA) rather than HAd for screening respiratory cultures (8). HA testing does not disturb the cell sheet and, if necessary, a single culture tube could be tested every other day for HA activity. The HA procedure is especially convenient when handling a large number of cultures because the cultures are handled only once per test. In addition, cultures can be tested repeatedly without washing RBCs from the cultures and hemadsorbing simian viruses do not produce as much background.

In this procedure, cultures are screened for HA activity at least twice (day 5 and day 10 postinoculation) or when CPE is observed. If the laboratory prefers, HAd assays can be performed on all cultures before a negative result is reported.

1. Appropriately label a 96-well U-bottom microtiter plate with culture identification and control information.
2. Prepare a 0.5% suspension of guinea pig RBC suspension (Appendix C).
3. Place 100 µl of saline into each well.
4. Add 25 µl of cell culture supernatant into each well.
5. Place 50 µl of the 0.5% guinea pig RBC suspension into each well.
6. Cover and gently tap the plate to mix the contents. Incubate for one hour at 4°C.
7. Examine the plate using a magnifying mirror

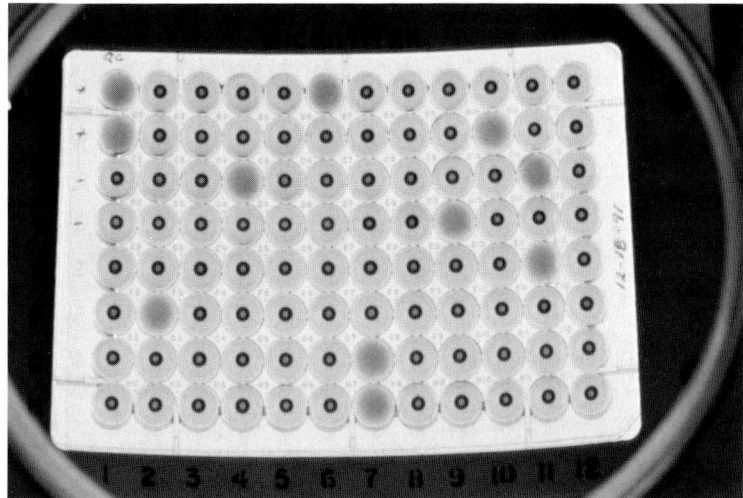

FIG. 2. Influenza hemagglutination. Wells A1 and B1 contain positive (influenza A) controls while wells C1 and D1 contain culture medium (negative controls). This plate contains culture fluids from nine positive cultures. The apparent curvature of the plate is an artifact resulting from the curved mirror used to photograph the plate.

and record the results.
8. Warm the plate to 22-25°C, examine the plate, and record the results as before.

Interpretation of Results
Positive Test. The presence of hemagglutinating virus prevents RBCs from forming a "button" in the bottom of the microtiter well (Fig. 2). Because this is only a screening procedure for hemagglutinin activity, positive cell cultures should be scraped and stained with type specific monoclonal antibodies for final identification of influenza A and influenza B.
REPORT: No report should be made on the results of a hemagglutination test alone. See cell culture confirmation section.

Negative Test. The presence of a red cell button on the bottom of the well indicates that hemagglutinin activity was not detected (Fig 2). Since this test is only a screening procedure, a final hemadsorption test should be run before the cell culture is considered free of hemagglutinin activity.
REPORT: Influenza not isolated.

Inconclusive Test. The lack of a button in the negative (no virus) control well could indicate the presence of simian virus contamination in primary monkey kidney cells which may lead to erroneous interpretation. Samples can be diluted 1:5 in PBS and tested again. Simian virus contaminants will often dilute out faster than the influenza virus.

QC Procedures.
Positive (with influenza virus) and negative (no virus) controls must be tested from each lot of cell cultures. The procedure for preparing positive controls follows in the next section on cell culture isolation and confirmation.

Culture/HA/Had Confirmation
1. Remove all but a few drops of the cell culture medium.
2. Scrape the cells from the tube surface.
3. Resuspend the cells in the remaining fluid.
4. Using a pipette, place a small drop of the cell suspension in the wells of an acetone-cleaned slide. The remainder of the cell suspension should be retained for inoculation onto fresh RMK cells if the staining is negative.
NOTE: Two well slides may be used for influenza A and B staining. However, in the

case of an inconclusive HA/HAd positive specimen, an 8-well slide may be used so that simultaneous staining can be done for influenza A, influenza B, adenovirus, RSV, and parainfluenza 1, 2, and 3.

5. The slide should be processed as directed by the manufacturer of the fluorescein-labeled antibody to influenza virus or as described below.
6. Allow the slide to air dry, then fix it in acetone for 10-15 min. Fixed slides may be stored for up to a week at 2-8°C or for one year at -20°C under desiccated conditions.
7. Place enough of the appropriate fluorescein-labelled monoclonal antibody on each well to cover the cell smear (15-30 µl).
8. Incubate the slide for 15-30 minutes at 33-35°C in a covered, humidified chamber. Do not allow the antibody to dry on the slide as this could cause nonspecific antibody binding.
9. Remove the excess antibody with a gentle stream of PBS. Do not direct the stream directly at the cell smear as this could dislodge the cells.
10. Soak the slide in PBS for 5 minutes, shake off the excess fluid, and allow the slide dry.
11. Carefully add a small drop of FA mounting fluid to the center of each well.
12. Place a number 1 coverslip on the mounting fluid and carefully remove all the air bubbles.
13. Immediately examine the slide using a fluorescence microscope. For optimum clarity, use 200-300X magnification for screening and 400X for confirmation of cell morphology.

Interpretation of Results

Positive Test. The presence of influenza virus is indicated by apple-green fluorescence in the nucleus and cytoplasm of the infected cells. Only individuals experienced in reading FA reactions should examine the cell smears because cell debris and cell clumps may exhibit a dull fluorescence which can be misinterpreted as a positive reaction.
REPORT: Influenza (A or B) virus isolated.

Negative Test. The absence of influenza virus is indicated by the lack of specific fluorescence in the infected cells. If the cultures were CPE-, HA-, and/or HAd-positive and FA negative, they should be stained with antibodies to other respiratory viruses and mumps virus.
REPORT: Influenza virus not isolated.

QC Procedures.

Subpassages of recent influenza isolates or ATCC cultures should be inoculated into cell culture tubes with each batch of influenza isolations. Uninfected cell cultures serve as negative controls. Infected and uninfected cell monolayers should be scraped and stained as described above. Positive controls must hemadsorb and exhibit a typical pattern of fluorescence. Negative controls must not exhibit specific fluorescent staining. Antibodies used to detect influenza A and influenza B should be checked regularly against prevalent influenza strains to assure that the current influenza strains can be detected by existing reagents.

Preparation of Positive Control Cultures

1. Add 20 µl of a recent influenza isolate (or an ATCC culture) to 2 ml of sterile GLB. Hold the diluted virus on ice until it is used to inoculate the monolayer.
2. Remove the growth medium from one 75 cm^2 flask of newly confluent MDCK cells.
3. Rinse the monolayer twice with 10 ml of serum-free MEM.
4. Add 2 ml of the diluted virus to the monolayer and allow the virus to adsorb for 1-2 hours at 33-35°C. Note: The flask should be rocked every 15 minutes to assure even virus distribution and to prevent monolayer desiccation.
5. Add 20 ml of serum-free MEM containing L-glutamine, antibiotics, and 2 µg/ml trypsin.
6. Incubate the flask at 33-35°C until the CPE involves 80-100% of the monolayer and/or HA testing is positive to a titer of ≥ 128.
7. Scrape the cells from the flask into the cell culture medium.
8. Transfer the medium to a 50 ml polypropylene centrifuge tube.

9. Add GLB to the 50 ml mark.
10. Quickly freeze the medium at -70°C.
11. Thaw the medium quickly in a 37°C water bath.
12. Centrifuge the tube at 600-800 x g for 15 minutes to pellet the cell debris.
13. Dispense 1.0 ml of the diluted supernatant fluids into each of 50 freezing vials and freeze the vials at or below -70°C. Store the vials in liquid nitrogen for up to 5 years or at -70°C for up to 2 years.

Centrifugation-Enhanced (Shell Vial) Method

Rapid detection of influenza is essential for the timely institution of appropriate infection control measures and for the initiation of antiviral chemotherapy. In order to address the need for rapid influenza detection, centrifugation-enhanced (shell vial) methodologies originally described for cytomegalovirus (14) have been adapted for the detection of influenza A and influenza B (9-13). Shell vial methods can detect 50% of all isolates 24 hours after inoculation and 70-80% of all isolates after 48 hours. Although shell vial methods are more rapid than conventional tube cultures, these methods are inherently less sensitive. Therefore, each set of shell vials should be backed up by at least one RMK tube culture. In some laboratories, the combination of shell vials plus one RMK tube have proven to be as cost effective as inoculating two RMK tubes and performing serial hemadsorptions.

Because shell vials are usually stained before CPE develops, a variety of cell lines, including A549, HL, RD, and MDCK cells may be used. Unless otherwise indicated, four MDCK, RD, or MRC-5 shell vials and one RMK tube culture should be inoculated for each specimen. The RMK tube culture serves as a backup for specimens containing small quantities of viable influenza virus and for the isolation of other viruses that may be in the specimen.

Procedure
1. Appropriately label 4 shell vials and 1 RMK tube for each specimen.
2. Remove the medium from the shell vials and wash the monolayers twice with serum-free MEM containing antibiotics.
3. Inoculate each vial with 0.2-0.5 ml of the specimen.
4. Centrifuge the vials at 700 x g for 1 hour at room temperature.
5. Add 1 ml of serum-free MEM to each vial. For MDCK cells, the overlay medium should consist of serum-free MEM, supplemented with 2 μg/ml of trypsin.
6. Incubate the vials at 33-35°C.
7. Two vials are fixed and stained (one for influenza A and the other for influenza B) after 24 hours. The second set of vials are stained at 40-48 hours postinoculation.

Staining of Coverslips

Identification of influenza in shell vial systems is accomplished by staining the coverslips with monoclonal antibodies to influenza. Most specimens will produce 1-20 small foci after 48 hours. Staining should be accomplished as directed by the manufacturer of the monoclonal antibodies or as described below.

1. Carefully remove the culture fluid from the vials and wash the coverslips twice by adding 1-2 ml of PBS and soaking for 5 minutes.
2. Remove the PBS and add 2-3 ml of acetone to each vial. Fix the coverslips for 10 minutes at room temperature.
3. Remove the acetone and allow the coverslips to air dry.
4. Rinse the coverslips briefly with PBS. (The moisture trapped between the coverslip and the vial will keep the stain from wicking under the coverslip.)
5. Add enough monoclonal antibody to the vial to cover the coverslip (100-150 μl) and incubate for 30 minutes at 33-35°C in a humidified chamber to prevent the antibody from drying on the coverslip.
6. Wash the coverslips twice as previously described (step 1).
7. Place a small amount of mounting fluid on a slide. Remove the coverslips from the vial

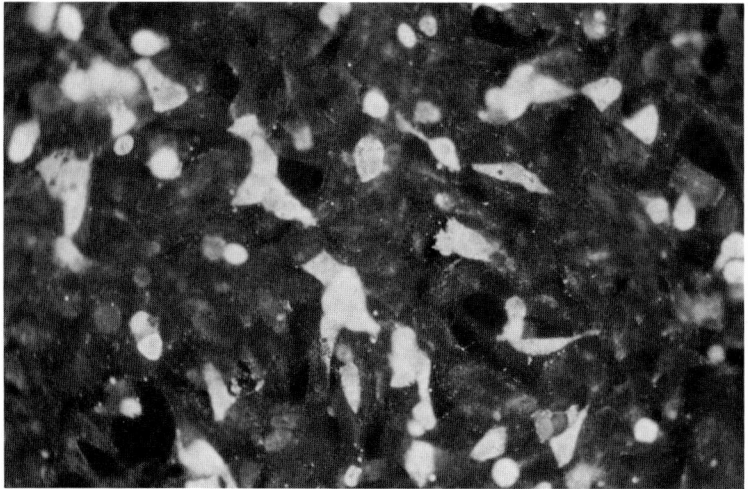

FIG. 3. RMK shell vial coverslip culture stained with monoclonal antibodies to influenza A virus.

and lay them, cell side down, on the mounting fluid.
8. Immediately examine the coverslips at 100-300X using a fluorescence microscope.

Interpretation of Results

Coverslips that have individual or small groups of cells that exhibit brilliant apple-green cytoplasmic or nuclear fluorescence (Fig. 3) are positive for influenza. Coverslips without the specific fluorescence described above are considered negative. Some specimens may trap the monoclonal antibodies between the cell monolayer and inoculum cells. Careful examination of the cell morphologies can distinguish this type of reaction from specific, influenza reactions. In addition, nonspecific staining can occur if the antibody reagents are allowed to dry on the coverslip. In this case, the entire cell sheet may stain apple-green.

Positive Test. The presence of influenza virus is indicated by brilliant, apple-green cytoplasmic and/or nuclear fluorescence (Fig. 3).
REPORT: Influenza (A or B) virus isolated.

Negative Test. The absence of influenza is indicated by the lack of specific fluorescence described above. The entire cell sheet should be stained red by the counterstain.
REPORT: Influenza virus not isolated.

QC Procedures.

A 1:10 dilution of recent influenza isolates or ATCC cultures (see Preparation of Positive Control Cultures, above) should be inoculated with each batch of influenza shell vial cultures. Uninfected shell vial cultures serve as negative controls. Infected and uninfected shell vial cultures should be processed and stained as described for the patient cultures. Positive controls must exhibit the typical fluorescence patterns. Negative controls must not exhibit specific fluorescent staining.

DIRECT SMEARS

Although centrifugation-enhanced (shell vial) methods can reduce the influenza isolation times to 16-48 hours, this time interval is often too lengthy when antiviral therapy and chemoprophylaxis are being contemplated. In a effort to reduce the influenza detection time to a clinically useful interval, a number of investigators have shown that direct staining of acetone-fixed nasopharyngeal smears provides a rapid and relatively sensitive method for the detection of influenza virus (16-18).

In this procedure, the nasopharynx is vigorously swabbed and two smears are made on an acetone-cleaned microscope slide. The slide is fixed in acetone and one smear is stained with monoclonal antibodies to influenza A and the second, with monoclonal antibodies to influenza B.

Although rapid and specific, direct smears are less sensitive than either conventional tube cultures or shell vial cultures (13) and direct smears should be backed up by culture methods to detect other viruses that may be in the specimen. Direct smears require the presence of intact, influenza-infected columnar epithelial cells before a positive result can be reported. Inadequate specimen collection and smear preparation are the principal reasons for the failure of this method to detect influenza viruses in direct smears. In addition, excess mucus can interfere with antibody binding. Mucus can cause false positive reactions by nonspecifically trapping the fluorescent antibody reagents and false negative reactions by clumping the cells, and thereby, masking the influenza antigens. However, skilled technicians can usually distinguish these types of reactions from influenza-specific reactions.

Despite these drawbacks, evaluation of direct smears for influenza virus is an effective method for the rapid detection of influenza infections. Direct smears can often detect influenza infections even after the patient stops shedding infectious virus. In addition, direct smears may be especially valuable in remote laboratories where cold-chain specimen transport is not easily accomplished. For these reasons, many laboratories employ a combination of direct smear evaluations and cell culture isolation for the detection of influenza.

Specimen Collection

The success of the direct fluorescent antibody (DFA) procedure depends upon the preparation of a well made cell smear. Smears that are too thick or "lumpy" can cause nonspecific trapping of the influenza antibody reagents. Smears that are grossly contaminated with red blood cells can cause interpretation difficulties due to red cell autofluorescence. As mentioned previously, mucus can cause false positive and false negative reactions. However, the principal reasons direct smears fail to detect influenza infections are (a) inadequate specimen collection and (b) the presence of too few cells on the slide. Specimen adequacy and its definition are significant sources of laboratory-to-laboratory variation and an area of increasing regulatory concern. Many laboratories require at least 50 cells per smear while others require 1-2 cells/high power field. Whatever the number, too many cells can cause nonspecific trapping of antibody and with too few cells the laboratory may not be able to find an infected cell.

Specimen Processing

Nasal Aspirates, Nasal Washes, and Throat Washes
1. Centrifuge the specimen at 1000-1500 x g for 5 minutes to pellet the cells.
2. Remove all but 100-200 μl of the fluid.
3. Suspend the cell pellet in the remaining fluid.
4. Spread one drop of the cell suspension onto each well of an acetone cleaned two-well slide.
5. Allow the smears to air dry at room temperature. Slides may be stored for up to 48 hours at 2-8°C.

Swabs in Viral Transport Medium
1. Vortex the specimen vigorously to release as many cells from the swabs as possible.
2. Remove the swabs from the viral transport medium and firmly roll them against the inside of the tube to remove as much fluid as possible.
3. Centrifuge the specimen at 1000-1500 x g for 5 minutes to pellet the cells.
4. Remove all but 100-200 μl of the viral transport medium.
5. Suspend the cell pellet in the remaining fluid.
6. Spread one drop of the cell suspension onto each well of an acetone cleaned two-well slide.
7. Allow the smears to air dry at room temperature. Slides may be stored for up to 48 hours at 2-8°C.

Test Procedure

For optimum performance, slides should be fixed and stained within 1-2 hours of specimen collection.

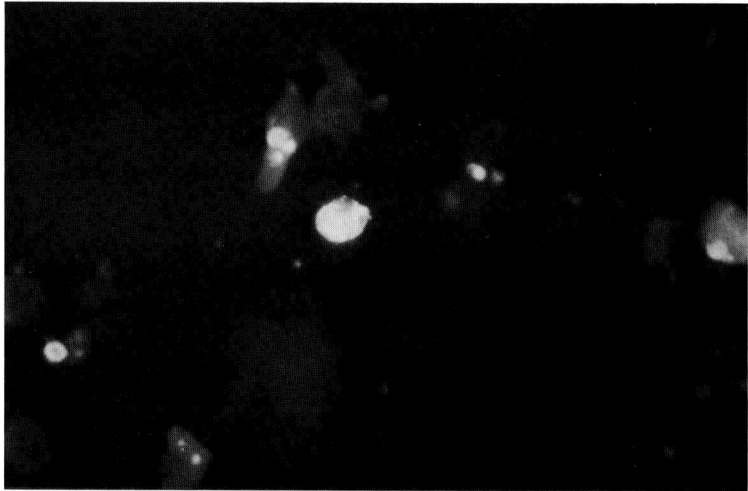

FIG. 4. Direct fluorescent antibody staining of a nasopharyngeal smear using monoclonal antibodies to influenza A virus.

Alternatively, unfixed slides may be stored for up to 48 hours at 2-8°C. Fixed slides may be stored at 2-8°C for 1 week or frozen at -20°C for up to 1 year. Storing slides under desiccated conditions will decrease the background staining and minimize antigen degradation.

1. Upon arrival in the laboratory, examine the slide wells at 100-300X magnification. Ideally, a minimum of 50 cells should be visible on each well before the specimen is considered adequate for further processing.
2. The slide should be processed as directed by the manufacturer of the fluorescein-labeled antibody to influenza virus or as described below.
3. Fix the cells by immersing the slides in cold acetone for 10 minutes. Allow the slides to air dry.
4. Place enough of the fluorescein-labeled monoclonal antibody on each well to cover the cell smear (15-30 μl). One well should be stained with antibodies to influenza A and the other with antibodies to influenza B.
5. Incubate the slide for 30 minutes at 35-37°C in a covered, humidified chamber. Do not allow the antibody to dry. Drying can cause nonspecific antibody binding.
6. Remove the excess antibody with a gentle stream of PBS. Do not direct the stream directly at the cell smear as this could dislodge the cells.
7. Soak the slide in PBS for 5 minutes, shake off the excess fluid, and allow the slide to dry.
8. Carefully add a small drop of FA mounting fluid to the center of each well.
9. Place a number 1 coverslip on the mounting fluid and carefully remove all the air bubbles.
10. Immediately examine the slide using a fluorescence microscope. For optimum clarity, use 200-300X magnification for screening and 400X for confirmation of cell morphology.

Interpretation of Results

Positive Test. The presence of influenza virus is indicated by intense apple-green fluorescence in the nucleus and cytoplasm of the columnar epithelial cells (Fig. 4). Only intact cells should be examined because cell debris and clumps of normal cells may exhibit a dull, yellow-green fluorescence which could be misinterpreted as a positive reaction.
REPORT: Influenza (A or B) virus detected.

Negative Test. The absence of influenza virus is indicated by the lack of intense apple-green

fluorescence in the cells.
REPORT: Influenza virus not detected.

Inconclusive Test. Specimens with fewer than 25 cells on each well (or 1-2 cells/HPF) may give inconclusive results.
REPORT: Unacceptable specimen - too few cells.

QC Procedures.

Positive and negative controls should be stained at least once each day to assure that the antibody reagent is performing properly. Positive cells should stain intensely as described above. Negative cells should stain red and should not exhibit specific apple-green fluorescence.

Because the same antibody reagent is used for confirmation of tube cultures, shell vial cultures, and direct smears, culture confirmation testing can be used to demonstrate that the reagent is working properly. In laboratories that do not perform influenza isolations or laboratories where no isolations are ongoing, prepared slides should be stained whenever direct smears are stained. Control slides can be purchased commercially from a number of vendors or they can be prepared as described below.

Preparation of Control Slides
1. Add 20 µl of the positive control culture (described above) to 2 ml of sterile GLB. Hold the diluted virus on ice until used to inoculate the monolayer.
2. Remove the growth medium from one 75 cm² flask of newly confluent MDCK cells.
3. Rinse the monolayer twice with 10 ml of serum-free MEM.
4. Add 2 ml of the diluted virus to the monolayer and allow the virus to adsorb for 1-2 hours at 33-35°C. Note: The flask should be rocked every 15 minutes to assure even distribution of the virus and to prevent monolayer desiccation.
5. Add 20 ml of serum-free MEM containing L-glutamine, antibiotics, and 2 µg/ml trypsin.
6. Incubate the flask at 33-35°C until the CPE involves 50-60% of the monolayer.
7. Trypsinize the cells and remove them from the flask.
8. Centrifuge the cells at 250 x g for 5 minutes.
9. Resuspend the cell pellet in 5 ml of PBS. Perform a cell count and adjust the cell concentration to 2-5 x 10^6 cells/ml.
10. Dispense 3-10 µl of the cell suspension onto each well and allow the suspension to air dry.
11. Fix the slides in acetone at room temperature for 10 minutes and allow the slides to air dry.
12. Store the slides at -20°C or below for up to one year. Slides should be stored under desiccated conditions to minimize antigen degradation and background fluorescence.

NOTE: Slides prepared in this manner will contain both positive and negative cells. Therefore, a single slide can be used for QC purposes. When slides are made from flasks with 100% CPE, two slides must be used - one containing a positive cell smear and a second slide containing an uninfected cell smear.

REFERENCES

1. Centers for Disease Control. Public health burden of vaccine-preventable diseases among adults: Standards for adult immunization practice. *MMWR* 1990;39:725-729.
2. Kendal A, Harmon MW. *Orthomyxoviridae*: The influenza viruses. In: Lennette EH, Halonen P, Murphy FA. *Laboratory diagnosis of infectious diseases - principles and practice*. Volume II. New York: Springer-Verlag, 1988;602-625.
3. Jordan WS, Denny FW, Badger GF, et al. A study of illness in a group of Cleveland families. XVII. The occurrence of Asian influenza. *Am J Hyg* 1958;68:190-196.
4. Moriuchi H, Oshima T, Nishimura H, Nakamura K, Katsushima N, Numazaki Y. Human malignant melanoma cell line (HMV-II) for isolation of influenza C and parainfluenza viruses. *J Clin Microbiol* 1990;28:1147-1150.
5. Murphy BR, Webster RG. Orthomyxoviruses. In: Fields BN, Knipe DM, Chanock RM, Hirsch MS, eds. *Virology*, Second Edition. New York: Raven Press, 1990;2011-2054.
6. Mills RD, Cain KJ, Woods GL. Detection of influenza virus by centrifugal inoculation of

MDCK cells and staining with monoclonal antibodies. *J Clin Microbiol* 1989;27:2505-2508.
7. Seno M, Kanamoto Y, Tako S, Takei N, Fukuda S, Umisa H. Enhancing effect of centrifugation on isolation of influenza virus from clinical specimens. *J Clin Microbiol* 1990;28:1669-1670.
8. Johnston SLG, Wellens K, Siegel C. Comparison of hemagglutination and hemadsorption tests for influenza detection. *Diagn Microbiol Infect Dis* 1992;15:363-365.
9. Waris M., Ziegler T, Kivivirta M, Ruuskanen O. Rapid detection of respiratory syncytial virus and influenza A virus in cell cultures with immunoperoxidase staining with monoclonal antibodies. *J Clin Microbiol* 1990;28:1159-1162.
10. Bartholoma NY, Forbes BA. Successful use of shell vial centrifugation and 16 to 18-hour immunofluorescent staining for the detection of influenza A and B in clinical specimens. *Am J Clin Pathol* 1989;92:487-490.
11. Stokes CE, Bernstein JM, Kyger SA, Hayden FG. Rapid diagnosis of influenza A and B by 24-h fluorescent focus assays. *J Clin Microbiol* 1988;26:1263-1266.
12. Espy MJ, Smith TF, Harmon MW, Kendal AP. Rapid detection of influenza by shell vial assay with monoclonal antibodies. *J Clin Microbiol* 1986;24:677-679.
13. Johnston SLG, Siegel CS. A comparison of direct immunofluorescence, shell vial culture, and conventional culture for the rapid detection of influenza A and B. *Diagn Microbiol Infect Dis* 1991;14:131-134.
14. Gleaves CA, Smith TF, Shuster EA, Pearson GR. Rapid detection of cytomegalovirus using low-speed centrifugation and monoclonal antibody to an early antigen. *J Clin Microbiol* 1984;19:917-919.
15. Frayha H, Castriciano S, Mahony J, Chernesky M. Nasopharyngeal swabs and nasopharyngeal aspirates equally effective for the diagnosis of viral respiratory disease in hospitalized children. *J Clin Microbiol* 1989;27:1387-1389.
16. McQuillin J, Madeley CR, Kendal AP. Monoclonal antibodies for the rapid diagnosis of influenza A and B virus infections by immunofluorescence. *Lancet* 1985;ii:911-914.
17. Shalit I, McKee PA, Beauchamp H, Wayner JL. Comparison of polyclonal antiserum versus monoclonal antibodies for the rapid diagnosis of influenza A virus infections by immunofluorescence in clinical specimens. *J Clin Microbiol* 1985;22:877-879.
18. Minnich L, Ray CG. Comparison of immunofluorescent staining of clinical specimens for respiratory virus antigens with conventional techniques. *J Clin Microbiol* 1980;12:391-394.

CHAPTER 18

Measles Virus

INTRODUCTION

Measles virus is a *Morbillivirus* and a member of the Paramyxoviridae. On electron microscopy, measles virions are pleomorphic spheres with a diameter of 120-250 nm. Virions consist of a 17 nm nucleocapsid, a single-stranded (negative sense) RNA genome, and an envelope that contains short surface projections. Although morphologically indistinguishable from other paramyxoviruses, the measles virus surface projections do not have neuraminidase activity. The measles virus genome codes for six polypeptides including an RNA-dependent RNA polymerase. Measles is thermolabile and is quickly inactivated by heat, organic solvents, alcohols, ultraviolet and visible light.

Measles is endemic in every country in the world and prior to widespread vaccine usage, measles was one of the most common exanthems. During the prevaccination era, urban areas of the Unites states would typically experience epidemics at 2-5 year intervals, usually in the late winter and early spring. The mass immunization program of the early 1960's caused a decrease in the number of measles cases seen in the United States each year. Since 1989, there has been a marked increase in the number of reported measles cases (1-4). While the majority of these cases occurred in unvaccinated preschool children, 45% were in patients with a history of measles vaccination (1,3-4). These vaccine failures were most common in persons 12-19 years of age.

Measles is an acute, generalized infection that begins as an infection of the respiratory epithelium. Local replication in the respiratory mucosa produces a primary and secondary viremia which in turn causes involvement of the entire respiratory mucosa. This involvement accounts for the cough, coryza, and fever that are classic signs of measles. Koplik spots, which are pathognomonic of measles, usually appear a few days after the generalized respiratory tract involvement. Koplik spots are bright red spots with a blue-white center that often appear on the oral mucosa opposite the second molars. However in severe measles infections, Koplik spots may involve the entire mucus membrane of the mouth.

Measles is highly communicable when the virus is replicating in the respiratory tract (5). The measles rash usually occurs 4-5 days after the onset of the respiratory symptoms. The rash characteristically is an erythematous macular or maculopapular exanthem that usually begins on head and face then proceeds down the body to involve the extremities and finally the palms and soles. The rash generally lasts 5 days and starts to clear on the skin that was first involved. The entire illness from respiratory symptoms to resolution of the fever and rash lasts 7-10 days in uncomplicated cases. Secondary cases within a household are often more severe than primary cases (6-8). This phenomenon has been attributed to intense exposure of siblings to a larger virus inoculum.

Although most measles cases are uneventful, encephalitis has been reported as a rare complication. Immunocompromised patients can develop giant cell pneumonia without evidence of rash (9). In these cases, the clinical diagnosis of measles may be difficult or impossible to make. In addition, bacterial or fungal superinfection of the respiratory tract may occur secondarily to the local tissue damage caused by the virus. Pneumonia accounts for about 60 percent of deaths in infants dying of measles whereas death due to acute encephalitis is more often observed in children aged 10-14 years (10).

Presumptive diagnosis of measles is typically

AT A GLANCE...

MEASLES VIRUS

Virus Detection Methods
Tube cultures of HEK, human amnion cells; or RMK, or AgMK cells.
Centrifugation enhanced (shell vial) cultures using RMK or A549 cells.
Direct fluorescent antibody staining of buccal, nasopharyngeal, or urinary sediment smears.

Specimen Source
Best isolation rates are achieved from nasopharyngeal and throat swabs collected 1-2 days after the onset of symptoms. Virus can be isolated from conjunctival swabs 1-2 days after the onset of illness; from blood 1-4 days after onset; and from urine sediment for 1 week. Isolation of measles from spinal fluids is rare.

Time to Result
Standard Culture: Positive Culture - 5 to 10 days. Negative Culture - 4 weeks.
Shell Vial Method: Positive Culture - 2-5 days. Negative Culture - 5 days.
Indirect Fluorescent Antibody Stain Method: 1-2 hours.

Measles Serology

IgM Usually present at the onset of rash. Antibody levels peak 2-4 weeks later and decline below detectable levels after 30-60 days. IgM testing of CSF may be useful in diagnosing measles encephalitis.

IgG First appears 1-2 days after the onset of rash and peak levels are achieved 2-3 weeks later. Diagnosis of active infection can be achieved by demonstrating seroconversion or a 4-fold rise in antibody titer in acute and convalescent sera. After infection, measles IgG persists for the lifetime of the patient.

Epidemiology
Measles occurs worldwide in temperate, tropical, and arctic climates. Prior to widespread measles vaccination, epidemics occurred at 2-5 year intervals in late winter and early spring. Since 1989, there have been a marked increase in the number of measles cases reported in the United States with 40% of cases having a history of previous vaccination.

Transmission/Incubation Period
Measles is extremely contagious and is transmitted via respiratory droplets. Measles may be transmitted by direct contact with oropharyngeal secretions or rarely, by articles freshly soiled with nasal or throat secretions. Fever usually occurs 9-12 days after virus exposure and the rash develops 10-14 days after exposure. Patients are infectious from 1-2 days before the appearance of the prodromal respiratory symptoms to the fourth day after the appearance of rash. Communicability is minimal after the second day of rash.

Inactivators and Disinfectants
Measles is thermolabile and is quickly inactivated by heat and organic solvents. Effective disinfectants include 70% ethanol, 10% household bleach solutions, quaternary ammonium compounds, phenols, iodophor compounds, and glutaraldehyde solutions.

based upon clinical presentation, the presence of Koplik's spots, and the distinctive exanthem. However, the reduced incidence of measles in the United States and other industrialized countries has produced a generation of young clinicians who are unfamiliar with this distinctive symptomatology (11). Therefore, viral serologies, virus isolations, and fluorescent antibody testing of buccal smears may be necessary for confirmation of the clinical diagnosis (5,9,11,13).

Serological confirmation of measles can be accomplished by demonstrating seroconversion, measles-specific IgM, or a four-fold rise in measles antibody titers. Hemagglutination inhibition, virus neutralization, complement fixation, indirect immunofluorescence, dot immunobinding, and enzyme immunoassays have been used for measles antibody determinations. In unusual cases such as suspected SSPE where lymphocytes may carry virus; in immunocompromised patients developing pneumonia without a rash; in unexplained encephalitis; and autopsy specimens, virus isolations may provide more definitive information than serologic testing.

VIRUS ISOLATION

Introduction

Measles can be isolated in primary human embryonic kidney (HEK), human amnion cells, primary rhesus or African green monkey kidney cells. Other cell lines such as Vero or BSC-1 may be used for virus isolation but these continuous cell lines are usually less susceptible to infection than primary cultures. Measles virus grows slowly in cell culture and CPE is usually not apparent until 5-10 days after inoculation. During primary isolation, measles virus produces giant cells characterized by a ring of nuclei surrounding a central cytoplasmic area. Intracytoplasmic and intranuclear inclusions may be observed in these cells. Measles virus CPE is nearly identical to the cytopathology produced by SV-5, SV-40, foamy viruses, and herpes-like viruses that latently infect some primary monkey kidney cultures. Therefore, fluorescent antibody staining is used to confirm the presence of measles virus in cell cultures.

With prolonged passage, some measles virus strains lose their ability to produce syncytia. Some high passage measles virus strains produce a stellate or spindle cell CPE that is characterized by long narrow cells with branched cytoplasm. Wild measles viruses rarely produce this type of CPE.

Specimen Collection and Storage

Measles is highly contagious and care should be taken during specimen collection and handling to minimize contamination of laboratory personnel and the environment. The best isolation rates are achieved from nasopharyngeal and throat swabs (both swabs placed in a single viral transport tube) collected 1-2 days after the onset of symptoms (11). Virus can be isolated from conjunctival swabs during first 1-2 days after the onset of illness; from blood (buffy coat) 1-4 days after onset; and from urine sediment for up to 1 week. Isolation of measles from spinal fluids is rare.

Measles is extremely heat labile and specimens should be inoculated onto cell cultures as soon as possible after collection. Specimens should be transported to the laboratory on wet ice. If temporary storage is unavoidable, specimens may be stored at 2-8°C for up to 48 hours. DO NOT FREEZE AT -20°C. If specimens cannot be inoculated onto cell cultures within 72 hours, they should be frozen at -70°C.

Specimen Preparation

Urine Sediments
1. Centrifuge 10 ml of urine at 1,500-2,000 x g.
2. Remove all the urine, leaving the cellular pellet.
3. Add 1.0 ml of viral transport medium.
4. Vortex vigorously to suspend the sediment.

Standard Tube Culture

Procedure
1. Appropriately label two primary monkey kidney cell culture tubes.
2. Remove the culture medium from the cells and add 2 ml of refeed medium containing 2% FCS.

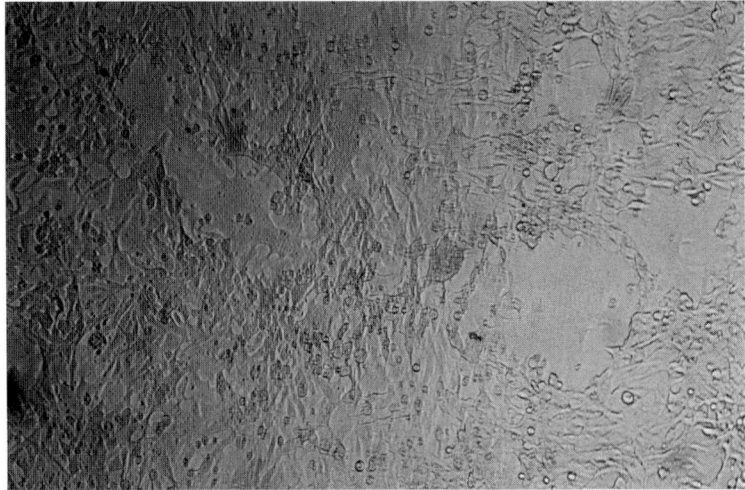

FIG. 1. Measles virus CPE in primary rhesus monkey kidney cells.

3. Add 0.2-0.5 ml of the specimen to each tube.
6. Place the tubes in a roller rack (10-15 rph) and incubate at 35-37°C.
7. Cell cultures should be examined at least every other day for 30 days. Replace the culture medium every 3-5 days or when the medium becomes acidic. Cultures that become toxic should be scraped and inoculated onto new cell cultures. On average, measles virus CPE develops within 5-10 days (range = 2-15 days).
8. If no CPE is present after 14 days, one tube should be scraped and transferred to fresh culture tubes. The remaining cells should be stained for measles virus as described below. Specimens containing low virus concentrations may need to be passaged twice before virus is detectable.

Interpretation of Results

Presumptive identification of measles in primary monkey kidney cultures is based upon the presence of syncytia where the cell nuclei form a ring around the central cytoplasmic mass (Fig. 1). Characteristic intranuclear and intracytoplasmic inclusions may be present and can serve to distinguish measles virus CPE from that of other paramyxoviruses. Care must be taken when examining tube cultures because measles virus CPE is nearly identical to the CPE produced by SV-5, SV-40, and other adventitious agents found in primary monkey kidney cultures. To control for this problem, uninoculated tubes from the same lot of primary monkey kidney cells should be examined and compared with inoculated tubes. In addition, confirmatory testing must be done using specific antibodies to measles virus.

Culture Confirmation Procedure
1. Remove all but a few drops of medium from the cell culture tube.
2. Scrape the cells from the surface of the tube.
3. Suspend the cells in the remaining culture medium.
4. Using a pipette, place a small drop of the cell suspension on an acetone-cleaned slide. The remainder of the cell suspension can be retained for inoculation onto fresh RMK cells if the staining is negative.
5. Allow the cell suspension to dry at room temperature.
6. Slides should be fixed and stained as directed by the manufacturer of the measles virus monoclonal antibodies or as described below.
7. Fix the cells by placing the slide in cold acetone for 10-15 minutes.
8. Allow the slide to air dry. For optimum performance, slides should be stained

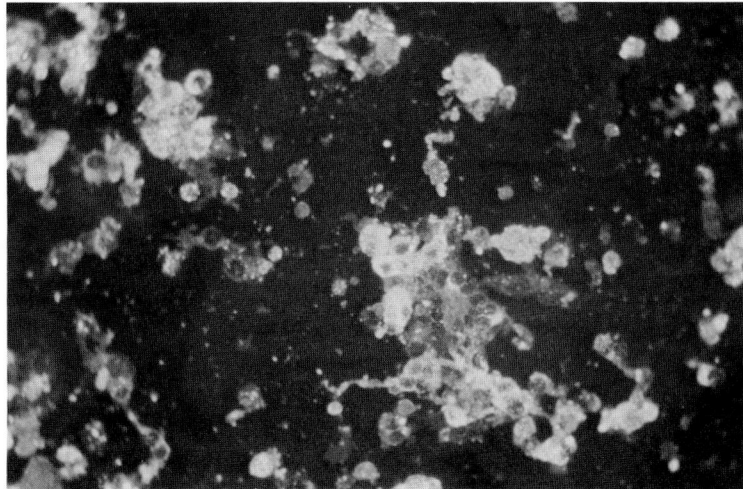

FIG. 2 Fluorescent antibody staining of RMK cells infected with measles virus. Cells were scraped from their tube and stained.

immediately after fixation. Alternatively, fixed slides may be stored for up to 1 week at 2-8°C or one year at -20°C or below. Slides should be stored under desiccated conditions in order to minimize antigen degradation and background staining.

9. Add enough of the monoclonal antibody to cover the cell smear (15-50 µl).
10. Incubate the slide for 30 minutes at 35-37°C in a covered, humidified chamber. Do not allow the antibody to dry as this could cause nonspecific antibody binding.
11. Remove the excess antibody with a gentle stream of PBS. Do not direct the stream directly at the cell smear as this could dislodge the cells.
12. Soak the slide in PBS for 5 minutes, shake off the excess fluid.
13. Add enough of the FITC-labelled antimouse conjugate to cover the cell smear (approximately 15-50 µl).
14. Incubate the slide for 30 minutes at 35-37°C in a covered, humidified chamber. Do not allow the antibody to dry.
15. Remove the excess conjugate with a gentle stream of PBS. Do not direct the stream directly at the cell smear as this could dislodge the cells.
16. Soak the slide in PBS for 5 minutes, shake off the excess fluid, and allow the slide to air dry.
17. Carefully add a small drop of FA mounting fluid to the center of each smear.
18. Place a number 1 coverslip on the mounting fluid and carefully remove all the air bubbles.
19. Immediately examine the slide using a fluorescence microscope. For optimum clarity, use 200-300X magnification for screening and 400X for confirmation of cell morphology.

Interpretation of Results

Positive Test. The presence of measles is indicated by the appearance of intense apple-green fluorescence in the cytoplasm of the infected cells (Fig. 2). Cells stained late in the infection cycle may also display intranuclear staining. Only individuals experienced in reading FA reactions should examine the cell smears because cell debris and cell clumps may exhibit a dull fluorescence which can be misinterpreted as a positive reaction.

REPORT: Measles virus isolated.

Negative Test. The absence of measles virus is indicated by the uniform red coloration of the cells and the absence of intense apple-green fluorescence after 30 days of culture.
REPORT: Measles virus not detected.

QC Procedures.

Subpassages of measles virus isolates or ATCC cultures should be inoculated with each batch of measles isolations. Uninfected cell cultures can serve as negative controls. Some ATCC cultures and high passage clinical isolates produce stellate or spindle-like CPE rather than the characteristic syncytia found in primary isolates. Therefore, QC inocula should be monitored carefully to assure that the proper type of CPE is produced for the particular application.

Infected and uninfected cell monolayers should be scraped and stained as described above. Positive controls must exhibit CPE typically found in primary isolates and the cell smears must contain cells exhibiting intense apple-green cytoplasmic (and nuclear) fluorescence. Negative controls must stain with the red counterstain and should not exhibit any CPE or specific fluorescent staining.

To control for "spontaneous" CPE produced by SV-5, SV-40, or other adventitious agents that can be found in primary monkey kidney cells, the amount of CPE present in the same lot of uninoculated culture tubes should be examined and compared with inoculated tubes.

Preparation of Positive Control Inocula
1. Add 20 μl of a recent measles isolate (or an ATCC culture) to 2 ml of sterile GLB. Hold the diluted virus on ice until it is used to inoculate the monolayer.
2. Remove the growth medium from one 75 cm^2 flask containing newly confluent Vero cell monolayer.
3. Add 2.0 ml of the virus dilution to the flask.
4. Allow the virus to adsorb for 1-2 hours at 35-37°C. The flask should be rocked every 15 minutes to assure even virus distribution and to prevent monolayer desiccation.
5. Add 12 ml of refeed medium containing 2% FCS.
6. Incubate the flask at 35-37°C until the CPE involves 80-100% of the monolayer (usually 5-30 days). Replace the medium every 3 days or whenever it becomes acidic.
7. Scrape the cells from the flask into the culture medium.
8. Transfer the medium to a 50 ml polypropylene centrifuge tube and freeze quickly at -70°C.
9. Thaw the medium in a 37°C water bath.
10. Centrifuge the cells at 600-800 x g for 15 minutes.
11. Transfer the supernatant fluids to a sterile tube.
12. Add sterile MEM containing 20% FCS to the 50 ml mark.
13. Dispense 1 ml of the virus suspension to each of 50 freezing vials and freeze the vials at -70°C or below. Virus inocula can be stored at -70°C for up to 2 years or in liquid nitrogen for up to 7 years.

Centrifugation Enhanced (Shell Vial) Method

In contrast with the 6-28 day isolation times required for traditional tube cultures, centrifugation enhanced (shell vial) methods can reduce the time-to-result to as little as 24 hours. In this procedure, two vials are inoculated. One vial is stained after 24 hours and the other, after 5 days. Up to 80% of positive isolates will be positive by shell vial methods after 36 hours (12). Followup staining of the second vial at 5 days postinoculation can increase the isolation rate to 100%. Standard tube cultures are usually inoculated in parallel with shell vial isolations in order to maximize isolation rates (11,14). Shell vial isolations are usually done in RMK or A549 cells.

Procedure
1. Appropriately label two shell vial cultures for each specimen and aseptically remove the culture medium.
2. Add 0.2-0.5 ml of the specimen to each vial and centrifuge at 700 x g for 60 minutes at room temperature.
3. Add 1 ml of refeed medium containing 2%

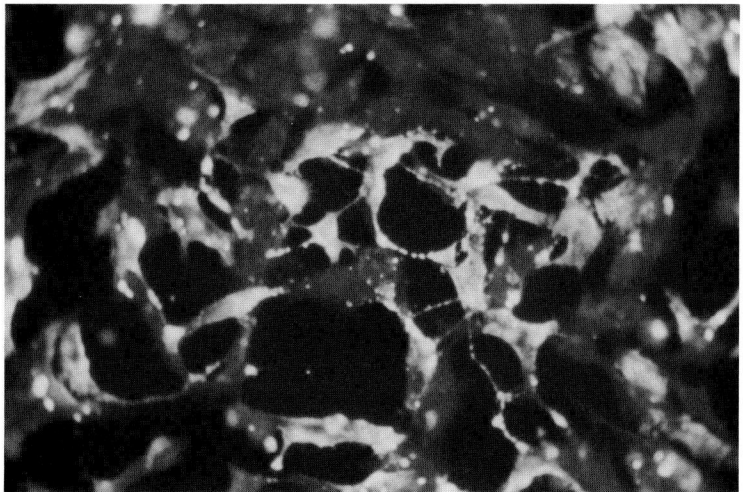

FIG. 3. Fluorescent antibody staining of A549 shell vial cultures infected with measles virus.

FCS and antibiotics to each vial.
4. Incubate the vials at 35-37°C. Stain one vial 36-48 hours after inoculation and the other vial after 5 days.

Staining of Coverslips
1. Coverslip cultures should be fixed and stained as directed by the manufacturer of the monoclonal antibodies or as described below.
2. Carefully remove the culture fluids from the vial and wash the coverslip twice by adding 1-2 ml of PBS and soaking for 5 minutes.
3. Remove the PBS and add 2-3 ml of acetone to each vial. Fix the coverslips for 10 minutes at room temperature.
4. Remove the acetone and allow the coverslip to air dry.
5. Rinse the coverslip briefly with PBS. (The moisture trapped between the coverslip and the shell vial will keep the stain from wicking under the coverslip.)
6. Add enough monoclonal antibody to the vial to cover the coverslip (100-150 µl) and incubate 30 minutes at 35-37°C. The shell vials should be covered or incubated in a humidified container to prevent the antibody from drying.
7. Wash the coverslips twice as previously described (step 1).
8. Add enough FITC-labelled conjugate to the vial to cover the coverslip (100-150 µl) and incubate 30 minutes at 35-37°C.
9. Wash the coverslips twice as previously described (step 1).
10. Place a small amount of mounting fluid on a slide.
11. Remove the coverslips from the vial and lay them cell side down on the mounting fluid.
12. Immediately examine the coverslips at 100-300X using a fluorescence microscope.

Interpretation of Results

Positive Test. The presence of measles virus is indicated by brilliant apple-green fluorescence in the cytoplasm of infected cells (Fig. 3). Some of the cells may appear as multinucleated syncytia. (see Color Plates 1C and 1D following Chapter 7)
REPORT: Measles virus isolated.

Negative Test. The absence of measles virus isolation is indicated by the lack of specific fluorescence described above. The entire cell sheet should be stained red by the counterstain.
REPORT: Measles virus not isolated.

QC Procedures.

Subpassages of measles clinical isolates or ATCC cultures should be inoculated with each batch of measles isolations. Uninfected cell cultures can serve as negative controls. Infected and uninfected cell monolayers should be scraped and stained as described above. Positive controls must exhibit the typical focal fluorescence patterns. Negative controls must not exhibit specific fluorescent staining. Positive control inocula can be prepared as in the previous section.

DIRECT SMEARS

Introduction

Although virus isolation is the gold standard for measles virus detection, the instability of the virus and the long isolation times (5-30 days) limit the diagnostic utility of most culture systems. Recently, direct detection of measles virus giant cells or measles antigens in cells from the pharynx, nasal and buccal mucosa, conjunctiva, or urinary tract have been shown to be more sensitive than virus isolation (12,13). During the prodrome and the early period of the exanthem, more than 90% of patients are positive by this method (5). Fluorescent antibody methods are sensitive and specific and laboratory results can be reported within 2 hours of their receipt in the laboratory. However, this method can only detect measles-specific antigens and these tests will produce false negative results when testing specimens containing other viruses. For this reason, many laboratories employ a combination of direct smear and culture methods for the diagnosis of measles.

Specimen Collection

The success of the fluorescent antibody procedure depends upon the submission of a well made cell smear. Smears that are grossly contaminated with red blood cells can also cause interpretation difficulties due to red cell autofluorescence. Specimen adequacy and its definition are significant sources of laboratory-to-laboratory variation and an area of increasing regulatory concern. Many laboratories require at least 50 cells per smear while other laboratories require 1-2 cells/high power (400X) field. Whatever the number, too many cells can cause nonspecific trapping of the conjugate and with too few cells, the laboratory may not be able to find an infected cell.

Specimen Processing

Swabs in Viral Transport Medium
1. Vortex the specimen vigorously to release as many cells from the swab as possible.
2. Remove the swab from the viral transport medium and firmly roll it against the inside of the tube to remove as much fluid as possible.
3. Centrifuge the specimen at 1000-1500 x g to pellet the cells.
4. Remove all but 100-200 μl of the viral transport medium.
5. Suspend the cell pellet in the remaining fluid.
6. Spread one drop of the cell suspension onto each well of an acetone cleaned two-well slide.
7. Allow the smears to air dry at room temperature. Slides may be stored for up to 48 hours at 2-8°C.

Urine
1. Centrifuge 10 ml of urine at 1,000-1,500 x g to pellet any cells.
2. Remove all but 100-200 μl of the fluid.
3. Suspend the cell pellet in the remaining fluid.
4. Spread one drop of the cell suspension onto each well of an acetone cleaned, two-well slide.
5. Allow the smears to air dry at room temperature. Slides may be stored for up to 48 hours at 2-8°C.

Test Procedure
1. Examine the smears at 100-300X magnification. Ideally, a minimum of 50 cells should be visible on each slide before the specimen is considered adequate for further processing.
2. The slide should be processed as directed by the manufacturer of the labelled antibody or as described below.
3. Fix the cells by immersing the slides in cold acetone for 10 minutes. Allow the slides to air dry.
4. Add enough monoclonal antibody to cover the cell smear (15-30 μl).

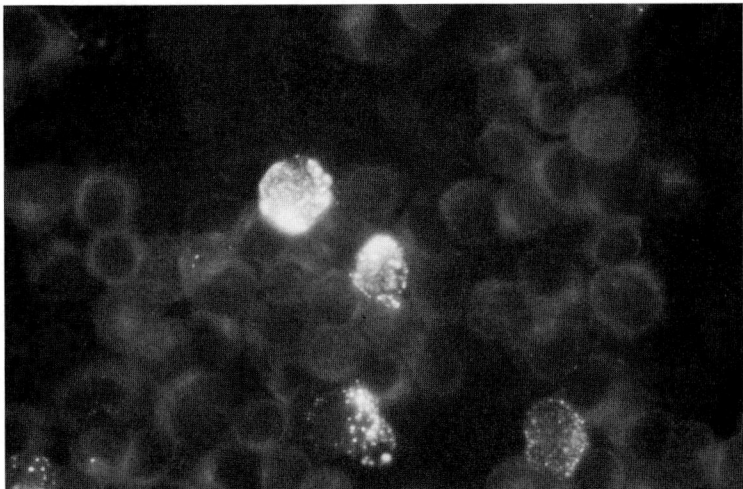

FIG. 4. Direct fluorescent antibody staining of a measles virus control slide.

5. Incubate the slide for 30 minutes at 35-37°C in a covered, humidified chamber. Do not allow the antibody to dry. Drying can cause nonspecific antibody binding.
6. Remove the excess antibody with a gentle stream of PBS. Do not direct the stream directly at the cell smear as this could dislodge the cells.
7. Soak the slide in PBS for 5 minutes, and shake off the excess fluid.
8. Add enough FITC-labelled conjugate to cover the cell smear (15-30 μl).
9. Incubate the slide for 30 minutes at 35-37°C in a covered, humidified chamber.
10. Remove the excess conjugate with a gentle stream of PBS. Do not direct the stream directly at the cell smear as this could dislodge the cells.
11. Soak the slide in PBS for 5 minutes, shake off the excess fluid, and allow the slide to air dry.
12. Carefully add a small drop of FA mounting fluid to the center of the smear.
13. Place a number 1 coverslip on the mounting fluid and carefully remove all the air bubbles.
14. Immediately examine the slide using a fluorescence microscope. For optimum clarity, use 200-300X magnification for screening and 400X for confirmation.

Interpretation of Results

Positive Test. The presence of measles is indicated by intense apple-green fluorescence in the cytoplasm of the cells (Fig. 4). Only intact cells should be examined because cell debris and clumps of normal cells may exhibit a dull fluorescence which could be misinterpreted as a specific, positive reaction.
REPORT: Measles virus detected.

Negative Test. The absence of measles is indicated by the lack of intense apple-green fluorescence in the exfoliated cells.
REPORT: Measles virus not detected.

Inconclusive Test. Specimens with fewer than 50 cells on each well (and no positive cells) may give erroneous results.
REPORT: Unacceptable specimen - Too few cells.

QC Procedures.

Positive and negative controls should be stained concurrently with the patient slides to assure the antibody reagent is performing properly. Positive cells should stain intensely as described above. Negative cells should be stained with the counterstain and should not exhibit specific apple-green fluorescence. Because the same antibody reagent is used for confirmation of tube cultures, shell vial cultures, and direct smears, these procedures can be used to demonstrate that the reagent is working properly. In laboratories that do not perform measles isolations or laboratories where no isolations are ongoing, prepared slides should be stained whenever direct smears are stained (Fig. 4). Control slides can be purchased commercially from a number of vendors or they can be prepared in the laboratory as described below.

Preparation of Control Slides
1. Prepare a 1:1000 dilution of a vigorous measles isolate or an ATCC culture in GLB. Hold the virus dilution on ice until used to inoculate the cell monolayer.
2. Remove the medium from one 75 cm^2 flask of freshly confluent Vero cells.
3. Add 2.0 ml of the virus dilution to the flask.
4. Allow the virus to adsorb for 1-2 hours at 35-37°C. Rock the flask every 15 minutes to assure even virus distribution and to prevent monolayer desiccation.
5. Add 12 ml of refeed medium containing 2% FCS.
6. Incubate the flask at 33-35°C until the CPE involves 40-60% of the monolayer (3-7 days). Replace the culture medium if it becomes acidic.
7. Trypsinize the cells and remove them from the flask. Centrifuge the cells at 250 x g for 5 minutes.
8. Resuspend the cell pellet in 5 ml of PBS containing 2% FCS. Perform a cell count and adjust the cell concentration to 2-5 x 10^6 cells/ml.
9. Dispense 3-10 µl of the cell suspension onto each slide and allow the suspension to air dry.
10. Fix the slides in cold acetone for 10 minutes and allow the slides to air dry.
11. Store the slides at -20°C or below for up to one year. Slides should be stored under desiccated conditions to minimize antigen degradation and background fluorescence.

NOTE: Slides prepared in this manner will contain both positive and negative cells (Fig. 4). Therefore, a single slide can be used for QC purposes. When slides are made from flasks with 100% CPE, two slides must be used - one containing a positive cell smear and a second slide containing an uninfected cell smear.

REFERENCES

1. Centers for Disease Control. Measles-United States, 1988. *MMWR* 1989;38:601-605.
2. Farizo KM, Stehr-Green PA, Simpson DM, Markowitz LE. Pediatric emergency room visits: a risk factor for acquiring measles. *Pediatrics* 1991;87:74-79.
3. Friedman S. Measles in New York City. *JAMA* 1991;266:1220.
4. The National Vaccine Advisory Committee. The measles epidemic, the problems, barriers, and recommendations. *JAMA* 1991;266:1547-1552.
5. Norrby E, Oxman MN. Measles Virus. In: Fields, BN, Knipe DM, Chanock RM, Hirsch MS, Melnick JL, Monath TP, Roizman B, eds. *Virology*, Second Edition. New York: Raven Press, 1990;1013-1044.
6. Aaby P, Bukh J, Hoff G, et al. High measles mortality in infancy related to intensity of exposure. *J Pediatr* 1986;109:40-44.
7. Hull HF. Increased measles mortality in households with multiple cases in The Gambia. *Rev Infect Dis* 1988;10:463-467.
8. Koster FT. Mortality among primary and secondary cases of measles in Bangladesh. *Rev Infect Dis* 1988;10:471-473.
9. Siegel C, Johnston SLG, Adair S. Isolation of measles virus in primary rhesus monkey cells from a child with acute interstitial pneumonia who cytologically had giant-cell pneumonia without a rash. *Am J Clin Pathol* 1990;94:464-469.
10. Gershon AA. Measles virus (rubeola). In: Mandell GL, Douglas RG Jr, Bennett JE, eds. *Principles and practice of infectious diseases*, Third Edition. New York: Churchill Livingstone, 1989;1279-1289.
11. Schiff GM. Measles (Rubeola). In: Lennette EH, ed. *Laboratory Diagnosis of Viral Infections*, Second Edition. New York: Marcel Dekker, Inc. 1992;535-547.
12. Minnich LL, Goodenough F, Ray CG. Use of immunofluorescence to identify measles virus infections. *J Clin Microbiol* 1991;29:1148-1150.

13. Smaron MF, Saxon E, Wood L, McCarthy C, Morello JA. Diagnosis of measles by fluorescent antibody and culture of nasopharyngeal secretions. *J Virol Meth* 1991;33:223-229.

14. Gleaves CA, Wilson DJ, Wold AD, Smith TF. Detection and serotyping of herpes simplex virus in MRC-5 cells by use of centrifugation and monoclonal antibodies 16 h post-inoculation. *J Clin Microbiol* 1985;21:29-32.

CHAPTER 19

Mumps Virus

INTRODUCTION

Mumps virus belongs to the family *Paramyxoviridae*, and the genus *Paramyxovirus*. Mumps virus possesses a negative-stranded, 15 kilobase RNA genome that codes for five major structural proteins. In electron micrographs, mumps virus has an irregular spherical shape with an average diameter of 200 nm (range 90-300 nm). The virus possesses a lipid-containing envelope that is studded with glycoprotein spikes. These spikes possess neuraminidase, hemagglutinin, and cell fusion activities. Mumps virus is relatively labile and is quickly inactivated by 0.1% formalin, ultraviolet light, organic solvents, detergents, and heating to 56°C. There is only one known antigenic type of mumps virus.

Mumps is endemic throughout the world and before widespread vaccinations, the "mumps season" extended from January through May with major epidemics occurring every 2-5 years. During the prevaccine era, more than 50% of mumps cases occurred in 5 to 9 year olds and 90% of disease occurred in children under 14 years of age (1). Vaccination has caused a shift in disease demographics and today, more than 50% of mumps cases occur in teenagers and young adults.

Two types of mumps vaccines have been used. A killed virus vaccine was administered from 1950 through 1978 and was eventually replaced by a live attenuated vaccine. In the United States, mumps vaccination programs were responsible for a 95% decrease in the annual incidence of mumps (1). However, there was a 2.5-fold increase in the incidence of mumps virus infections in 1986 and 1987 (1-4). This resurgence was subsequently linked to underimmunization and use of the killed vaccine (1-4).

Mumps virus is moderately infectious and transmission occurs through direct contact with infected individuals and through inhalation of infectious aerosols. Mumps virus can also be spread indirectly by autoinoculation of the nose or mouth after handling contaminated fomites, soiled handkerchiefs, or tissues. The incubation period is 16-18 days with a range of 2-4 weeks.

Once transmitted, mumps virus can produce symptomatic and asymptomatic infections. Typical mumps is an acute, self-limiting disease characterized by bilateral or unilateral parotitis (5). However, mumps virus infections can sometimes cause meningoencephalitis, orchitis, ovaritis, pancreatitis, thyroiditis, or infections of the eye or inner ear.

Diagnosis of mumps is usually based upon clinical history and the presence of parotid swelling, tenderness, and mild to moderate constitutional symptoms. Laboratory confirmation of mumps virus infection is usually not required (6). However, laboratory testing is often indicated when the differential diagnosis of mumps includes other agents such as parainfluenza type 3, coxsackie viruses, and influenza A virus, that also causes acute parotitis (6). Laboratory testing is also indicated when parotitis is absent or recurrent, in aseptic meningitis, in cases where extrasalivary gland involvement is prominent, and when documentation of specific viral etiology is desired (6).

Laboratory confirmation of mumps virus infection can be accomplished by virus isolation, by the presence of mumps virus IgM in a single serum, or by documenting either seroconversion or a four-fold rise in mumps virus antibody levels. A number of methods can be used to detect mumps virus antibodies including serum neutralization, hemagglutination inhibition, complement fixation, EIA, and IFA. EIA and IFA methods are preferred

AT A GLANCE...

MUMPS VIRUS

Virus Detection Methods
Tube cultures of primary RMK or CMK, HeLa, HEp-2, HEK, or HFF cells.
Centrifugation enhanced (shell vial) cultures using RMK, HeLa, or HEp-2 cells.

Specimen Source
Saliva collected 2-3 days before, to 4-5 days after, the onset of parotitis is the specimen of choice. Mumps virus is frequently isolated from CSF during the first 3 days of mumps virus meningitis. Mumps virus also can be isolated from urine specimens during the first 5 days of illness. Viremia is rarely detectable.

Time to Result
Standard Culture: Positive culture - 6-8 days. Negative culture - 2 weeks.
Shell Vial culture: Positive culture - 48 hours. Negative culture - 5 days.

Mumps Virus Serology

IgM   First detectable 1-5 days after the onset of symptoms. IgM levels reach peak titers 1-2 weeks later and IgM antibody levels usually persist for 3-6 months.

IgG   First detectable 7-10 days after the onset of symptoms. Once present, IgG levels remain detectable for the lifetime of the patient. Eighty to ninety percent of adults have antibody to mumps virus.

Epidemiology
Mumps is endemic throughout the world. Prior to licensing of the live attenuated mumps vaccine, epidemics occurred every 2-5 years with the peak incidence occurring from January through May. In the prevaccine era, more than 50% of cases occurred in the 5- to 9-year group and 90% of disease occurred in children under 14 years of age. Today, more than 50% of mumps cases occur in teenagers and young adults. There is only one serotype of mumps virus.

Transmission/Incubation Period
Mumps is transmitted by direct contact and through inhalation of infectious aerosols. In addition, the virus can be spread indirectly by autoinoculation of the nose or mouth after handling infected fomites, soiled handkerchiefs, or tissues. The incubation period is 16-18 days with a range of 2-4 weeks. Patients are infectious from 9 days prior, to 8 days after, the development of parotitis. Urine from mumps patients should be considered infectious for two weeks after onset of symptoms.

Inactivators and Disinfectants
Mumps virus is relatively unstable and is quickly inactivated heat, alcohols, detergents, and organic solvents. Effective disinfectants include 0.1% formalin solutions, 70% alcohol, 10% household bleach solutions, quaternary ammonium compounds, phenols, iodophor compounds, and glutaraldehyde compounds.

methods because they are more sensitive and they allow for the detection of mumps virus IgG and IgM. Mumps virus IgG tests are commonly used for immune status determinations. However, IgG test results must be carefully interpreted because up to 10% of serum IgG reactions may be caused by other paramyxovirus infections (7). In contrast with the IgG response, IgM responses do not appear to cross-react with the other paramyxoviruses and the presence of mumps virus IgM in a single serum specimen is considered diagnostic of recent infection.

Mumps virus IgG can be detected in the CSF of 80-90% of patients with mumps virus meningitis (8). Antibody levels are usually present at the onset of symptoms and IgG titers usually peak after the onset of meningitis. In most cases, mumps virus IgM is also present in the CSF. However, some patients with moderate to high serum IgM titers do not have mumps virus IgM in CSF (8).

VIRUS ISOLATION

Introduction

Mumps virus can be propagated in primary RMK or CMK, HEp-2, HeLa, HFF, Vero, and HEK cells. However, primary RMK or CMK cells are used most often because they are more sensitive than other cell types. Like influenza and parainfluenza viruses, mumps virus is sensitive to serum components. Therefore, the cultures should be washed before inoculation and mumps virus isolations should be performed using serum-free medium.

Although some mumps virus isolates produce CPE in RMK cells 6-8 days after inoculation, CPE production is not predictable. In addition, mumps virus CPE (i.e., syncytia, vacuolation and cell degeneration) is difficult to distinguish from simian virus CPE that is sometimes present in primary monkey kidney cells. Therefore, cell monolayers should be tested for the ability to hemadsorb guinea pig RBCs. Cultures should be tested at least once within the first week of incubation (5-7 days after inoculation) and again on day 14. Positive HAd tests should be confirmed by staining with monoclonal antibody reagents.

Specimen Collection and Storage

Saliva collected 2-3 days before, to 4-5 days after, the onset of parotitis is the specimen of choice. Mumps virus has been isolated from 50-70% of CSF specimens collected during the first 3 days of mumps virus meningitis (6). Mumps virus can also be isolated from urine specimens during the first 5 days of illness. However, viremia is rarely detectable.

Saliva or swabbings of Stenson's duct should be placed in viral transport medium immediately after collection in order to retard the growth of oral flora. All specimens, including CSF and first voided urine samples, should be transported to the laboratory on wet ice. Specimens may be stored at 2-8°C for up to 48 hours before inoculation onto cell cultures. If inoculation is delayed by more than 48 hours, specimens should be rapidly frozen in dry ice and stored at or below -70°C. However, freezing will decrease the infectivity of the samples. **DO NOT FREEZE AT -20°C.** Storage at this temperature will rapidly inactivate mumps virus.

Specimen Processing

Saliva

1. Dilute specimen with an equal volume of viral transport medium containing twice the normal concentration of antibiotics.
2. Place the diluted specimen in a tightly stoppered sterile container and vortex vigorously with several glass beads.
3. Centrifuge the specimen for 15 minutes 1000-3000 x g.
4. Transfer the supernatant fluids to another sterile container and incubate for 1 hour at room temperature to reduce the bacterial load.

Urine

1. Mix the specimen to suspend any sediments. Place 10-20 ml of the specimen in a centrifuge tube.
2. Centrifuge the specimen at 1000-3000 x g for 15 minutes at 2-8°C to pellet the cellular fraction.
3. Suspend the pellet in 1.0 ml of viral transport

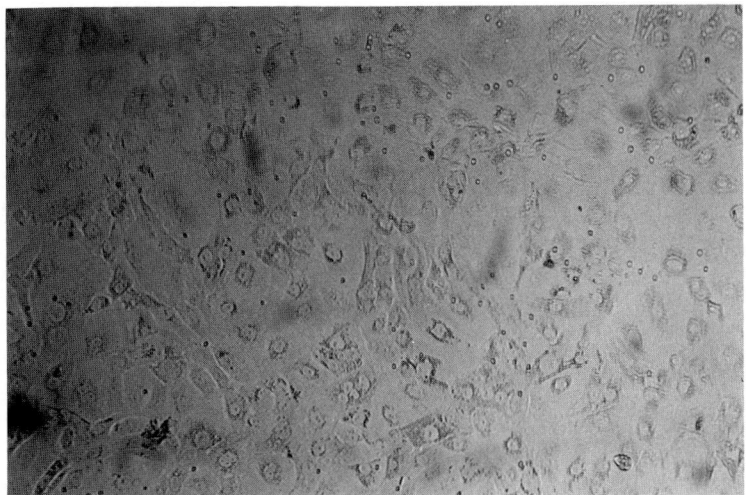

FIG. 1. Mumps virus CPE in RMK cells.

medium.
4. Use this specimen to inoculate cell cultures.

Standard Tube Culture
Procedure
1. Appropriately label 2 tubes containing freshly confluent RMK cells.
2. Remove the medium from the cells and wash the monolayers twice with HBSS or serum-free MEM containing antibiotics. This wash step is important because some serum components may interfere with mumps virus replication.
3. Add 2 ml of serum-free MEM with antibiotics to each tube.
4. Add 0.2 to 0.5 ml of specimen to each tube.
5. Place the tubes in a roller drum and incubate at 35-37°C and 10-15 rph.
6. Observe the cells at least every other day for CPE (Fig. 1).
7. Test each tube for HAd activity twice during the first 7 days of culture and again on day 14.
8. Tubes displaying positive CPE or hemagglutinin activity should be scraped and stained with antibody reagents.

Interpretation of Results

Mumps virus CPE is usually visible 6-8 days after inoculation. Mumps virus CPE is characterized by areas of syncytia formation, vacuolation and cell degeneration.

Report: No report should be made based upon CPE alone.

Hemadsorption Procedure

During replication, mumps, influenza, and parainfluenza viruses produce hemagglutinin molecules that are inserted into the plasma membrane of the host cell. Because these hemagglutinins bind to red blood cells, virus-infected cells can be readily identified because red cells adsorb to them (Fig. 2). Hemadsorption assays are rapid and relatively inexpensive. In addition, the proportion of cells with attached erythrocytes at 4°C compared with the number at 22°C can provide some indication as to whether the isolate is mumps, parainfluenza, or influenza virus. In parainfluenza 1, 2, and 3 infections, RBCs often elute from the infected cells when the culture is warmed to 22°C. Influenza and mumps virus infections will generally have the same amount of hemadsorption at 4°C and at 22°C.

Hemadsorption assays can be performed with guinea pig, fowl, and human O erythrocytes. However, most laboratories use guinea pig cells because they are smaller than the other RBCs and

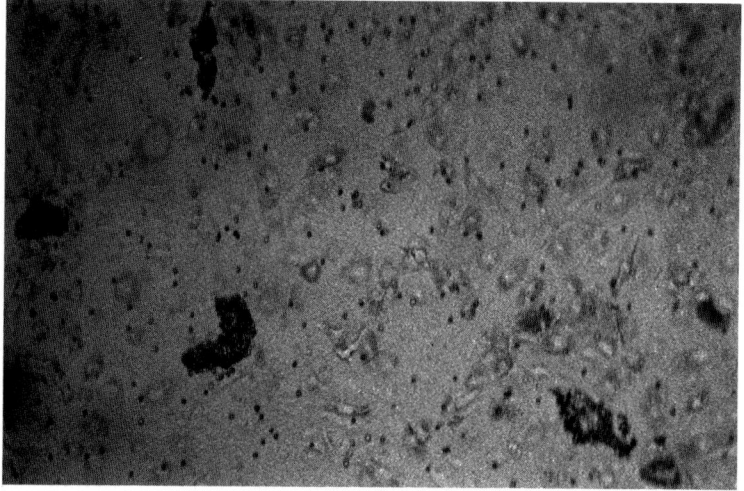

FIG. 2 Hemadsorption testing of mumps virus-infected RMK cells. The denser areas indicate areas where guinea pig red blood cells adsorb to infected cells.

more of them can attach to an infected cell thereby creating a more uniform and easier-to-interpret HAd pattern.

Procedure
1. Prepare a 0.5% suspension of guinea pig RBCs in sterile saline or phosphate buffered saline (Appendix C).
2. Add 0.4 ml of the RBC suspension to each culture tube.
3. Incubate the tubes on their side (cell monolayer downward) for 20-60 minutes at 4°C in a slanted (4°) test tube rack.
4. Examine the cell monolayer under the microscope (100-200X) for the presence of RBC adsorption and record the results (Table 1).

Table 1. Recording HAd Reactions.

Percent of Cells With HAd	Score
75-100	4+
50-75	3+
25-50	2+
< 25	1+
Questionable	+/-

5. Incubate the tubes for one hour at room temperature. Examine the tubes for hemadsorption as before and record the results.

Interpretation of Results

Positive Test. Specific hemadsorption is characterized by chains, rosettes, or clumps of red cells that adhere to the surface of the cell sheet (Fig. 2). The appearance of this HAd pattern indicates a positive Had test. However, results should be interpreted carefully because monkey kidney cells are sometimes infected with hemadsorbing simian viruses. Comparing the level of HAd to an uninoculated control culture is important to eliminate false positive results. Influenza and mumps viruses will generally have the same amount of hemadsorption after incubation at 4°C and after warming to 22°C. Positive cultures should be removed from the tube and stained with specific monoclonal antibody reagents.

REPORT: No report should be made based upon the results of a hemadsorption test alone. All HAd positive cultures should be confirmed with type-specific monoclonal antibody staining.

Negative Test. A negative HAd test is characterized by RBCs that float freely in the medium and the <u>absence</u> of RBC chains, rosettes, or clumps

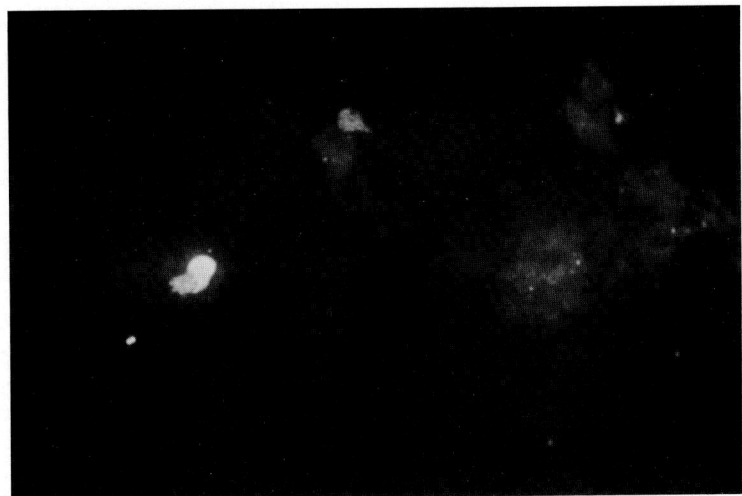

FIG. 3. Fluorescent antibody staining of mumps virus-infected RMK cells. Cells were scraped from their tube and stained with monoclonal antibodies.

adhering to the cell sheet. Because endogenous simian viruses can produce a background level of HAd, cultures that exhibit some HAd activity may be considered negative if the HAd levels are no higher than the background (uninoculated culture) levels. A negative isolation report should be generated when the cultures are HAd negative after 14 days of culture.

REPORT: Mumps virus not isolated.

QC Procedures.

Two RMK tubes should be inoculated with mumps virus (positive controls) for each batch of mumps virus isolations. Test the first tube for HAd activity on day 5-7 and the second tube on day 14. Both tubes should exhibit specific HAd activity as described above. When inoculating the positive control cultures, it may be advantageous to dilute the inoculum for the day 14 tube 1:10 to prevent monolayer destruction. Two uninoculated RMK tubes serve as negative controls. Because monkey kidney cells are sometimes infected with SV5 or other hemadsorbing simian viruses, the negative control tubes must be from the same lot of cells as the patient isolation tubes. A negative control tube should be tested whenever patient isolation tubes are tested. Positive HAd tests will have more HAd activity than the negative control tube. If uninoculated control tubes have more than 25% of the cell sheet exhibiting HAd (> 1+ HAd), the cell lot should be discarded.

Culture/HAd Confirmation

1. Remove all but a few drops of the cell culture medium.
2. Scrape the cells from the tube surface.
3. Suspend the cells in the remaining fluid.
4. Using a pipette, place a small drop of the cell suspension onto the wells of an acetone-cleaned multiwell slide. The remainder of the cell suspension should be retained for inoculation onto fresh RMK cells if staining is negative.
5. The slide should be processed as directed by the manufacturer of the fluorescein-labeled antibodies or as described below.
6. Allow the slide to air dry, then fix it in acetone for 10-15 min. Fixed slides may be stored for up to a week at 2-8°C or for one year at -20°C under desiccated conditions.
7. Place enough of the appropriate fluorescein-labeled monoclonal antibody on each well to cover the cell smear (15-30 µl).
8. Incubate the slide for 30 minutes at 35-37°C in a covered, humidified chamber. Do not allow the antibody to dry on the slide as this

could cause nonspecific antibody binding.
9. Remove the excess antibody with a gentle stream of PBS. Do not direct the stream directly at the cell smears as this could dislodge the cells.
10. Soak the slide in PBS for 5 minutes, shake off the excess fluid, and allow the slide to air dry.
11. Carefully add a small drop of FA mounting fluid to the center of each well.
12. Place a number 1 coverslip on the mounting fluid and carefully remove all the air bubbles.
13. Immediately examine the slide using a fluorescence microscope. For optimum clarity, use 200-300X magnification for screening and 400X for confirmation of cell morphology.

Interpretation of Results
Positive Test. The presence of mumps virus is indicated by apple-green fluorescence in the cytoplasm infected cells (Fig. 3). Only individuals experienced in reading FA reactions should examine the cell smears because cell debris and cell clumps may exhibit a dull fluorescence which can be misinterpreted as a positive reaction.
REPORT: Mumps virus isolated.

Negative Test. The absence of mumps virus is indicated by the lack of specific fluorescence in the infected cells. If the cultures were CPE negative; HA or HAd-positive; and FA negative, they should be stained with antibodies to parainfluenza and influenza viruses.
REPORT: Mumps virus not isolated.

QC Procedures.
Subpassages of recent mumps virus isolates or ATCC cultures should be inoculated into cell culture tubes with each batch of mumps isolations. Uninfected cell cultures serve as negative controls. Infected and uninfected cell monolayers should be scraped and stained as described above. Positive controls must hemadsorb and exhibit a typical CPE pattern and appropriate fluorescent antibody staining. Negative controls must not hemadsorb more than cell culture control and they should not exhibit specific fluorescent staining.

Preparation of Positive Control Inocula
1. Add 20 μl of a recent mumps virus isolate (or an ATCC culture) to 2 ml of sterile GLB. Hold the diluted virus on ice until it is used to inoculate the monolayer.
2. Remove the growth medium from one 75 cm^2 flask of 80% confluent HEp-2 cells.
3. Rinse the monolayer twice with 12 ml of serum-free MEM.
4. Add 2 ml of the diluted virus to the monolayer and allow the virus to adsorb for 1-2 hours at 35-37°C. Note: The flask should be rocked every 15 minutes to assure even virus distribution and to prevent monolayer desiccation.
5. Add 20 ml of serum-free MEM containing antibiotics.
6. Incubate the flask at 35-37°C until the CPE involves 80-100% of the monolayer.
7. Scrape the cells from the flask into the cell culture medium.
8. Transfer the medium to a 50 ml polypropylene centrifuge tube.
9. Add GLB to the 50 ml mark.
10. Quickly freeze the medium at -70°C.
11. Thaw the medium quickly in a 37°C water bath.
12. Centrifuge the tube at 600-800 x g for 15 minutes to pellet the cell debris.
13. Dispense 1.0 ml of the diluted supernatant fluids into each of 50 freezing vials and freeze the vials at or below -70°C. Control cultures can be stored at -70°C for up to 2 years or in liquid nitrogen for 5 years.

Centrifugation-Enhanced (Shell Vial) Method
In contrast with a typical 5-6 day tube culture isolation time, shell vial cultures can identify mumps virus as early as 48 hours after inoculation. A variety of cell lines including RMK, HeLa, and Vero cells can be used for shell vial isolations. However, RMK cells are preferred because they are more sensitive than other cell lines. Because shell vial cultures are less sensitive than tube cultures for mumps virus isolation, an RMK tube culture should be inoculated for each shell vial specimen. The

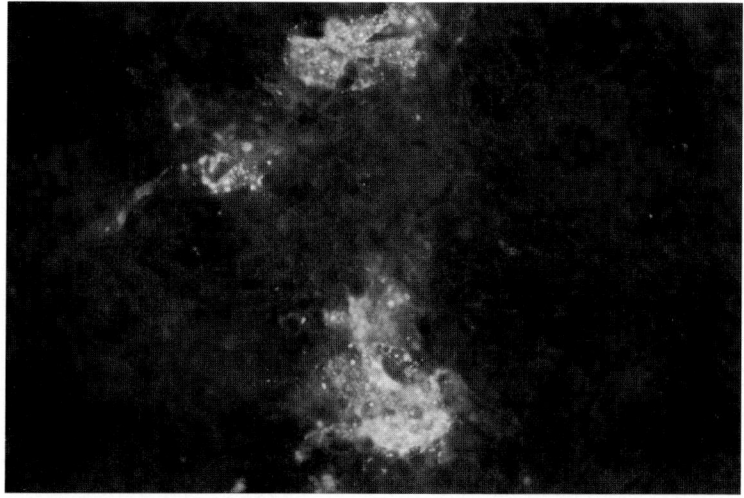

FIG. 4. Fluorescent antibody staining of a mumps virus-infected Vero coverslip culture.

RMK tube serves as a backup for specimens containing low concentrations of virus and for the isolation of other viruses from the specimen.

Procedure
1. Appropriately label 2 shell vials for each specimen.
2. Remove the medium from the shell vials and wash the monolayers twice with serum-free MEM containing antibiotics.
3. Inoculate each vial with 0.2-0.5 ml of the specimen.
4. Centrifuge the vials at 700 x g for 1 hour at room temperature.
5. Allow the virus to adsorb for one hour at 35-37°C.
6. Add 1 ml of serum-free MEM to each vial.
7. Incubate the vials at 35-37°C.
8. Fix and stain one vial after 48 hours and the other vial after 5 days after inoculation.

Staining of Coverslips

Identification of mumps virus is accomplished by staining the coverslips with monoclonal antibodies. Most specimens will produce 1-50 small fluorescent foci after 48 hours of culture. Fixation and staining should be done as directed by the manufacturer of the monoclonal antibodies or as described below.

1. Carefully remove the culture fluid from the vials and wash the coverslips twice by adding 1-2 ml of PBS and soaking for 5 minutes.
2. Remove the PBS and add 2-3 ml of acetone to each vial. Fix the coverslips for 10 minutes at room temperature.
3. Remove the acetone and allow the coverslips to air dry.
4. Rinse the coverslips briefly with PBS. The moisture trapped between the coverslip and the vial will keep the stain from wicking under the coverslip.
5. Add enough monoclonal antibody to the vial to cover the coverslip (100-150 µl) and incubate for 30 minutes at 35-37°C in a humidified chamber to prevent the antibody from drying on the coverslip.
6. Wash the coverslips twice as previously described (step 1).
7. Place a small amount of mounting fluid on a slide. Remove the coverslips from the vial and lay them, cell side down, on the mounting fluid.
8. Immediately examine the coverslips at 100-300X using a fluorescence microscope.

Interpretation of Results

Coverslips that have individual cells or small groups of cells that exhibit brilliant apple-green cytoplasmic fluorescence (Fig. 4) are positive for mumps virus. Coverslips without the specific fluorescence described above are considered negative. Some specimens may trap the monoclonal antibodies between the cell monolayer and inoculum cells. Careful examination of the cell morphologies can distinguish this type of reaction from mumps reactions. In addition, nonspecific staining can occur when the antibody reagents are allowed to dry on the coverslip. In this case, the entire cell sheet may stain apple-green.

Positive Test. The presence of mumps virus is indicated by brilliant, apple-green fluorescence in the cytoplasm of infected cells (Fig. 4).
REPORT: Mumps virus isolated.

Negative Test. The absence of mumps is indicated by the lack of specific fluorescence described above. The entire cell sheet should be stained red by the counterstain.
REPORT: Mumps virus not isolated.

QC Procedures.

A 1:10 dilution of recent mumps virus isolates or ATCC cultures (see Preparation of Positive Control Inocula, above) should be inoculated with each batch of mumps shell vial cultures. Uninfected shell vial cultures serve as negative controls. Infected and uninfected shell vial cultures should be processed and stained as described for the patient cultures. Positive controls must exhibit the typical fluorescence patterns. Negative controls must not exhibit specific fluorescent staining.

REFERENCES

1. Cochi SL, Preblud SR, Orenstein WA. Perspectives on the relative resurgence of mumps in the United States. *Am J Dis Child* 1988;142:499-507.
2. Centers for Disease Control. Summary of notifiable diseases in the United States. *MMWR* 1987;35:1-48.
3. Sosin DM, Cochi SL, Gunn RA, Jennings CE, Preblud SR. Changing epidemiology of mumps and the impact on university campuses. *Pediatr* 1989;84:770-784.
4. Wharton M, Cochi SL, Hutcheson RH, Bistowish JM, Schaffner W. A large outbreak of mumps in the post-vaccine era. *J Infect Dis* 1988;158:1258-1260.
5. Brunell PA. Mumps. In: Feigin RD, Cherry JD, eds. *Pediatric Infectious Diseases*, Second Edition. Philadelphia: W.B. Saunders, 1987.
6. Baum SG, Litman N. Mumps virus In: Mandell GL, Douglas RG, Jr., Bennett JE, eds. *Principles and practice of infectious disease*, Third Edition. New York: Churchill Livingstone, Inc., 1990;1260-1265.
7. Mufson MA. Parainfluenza viruses, mumps virus, Newcastle disease virus. In: Schmidt NJ, Emmons RW, eds. *Diagnostic procedures for viral, rickettsial and chlamydial infections*, Sixth edition. 1989;669-692.
8. Forsberg P, Fryden A, Link H, Orvell C. Viral IgM and IgG antibody synthesis within the central nervous system in mumps meningitis. *Acta Neurol Scan* 1986;73:372-380.

CHAPTER 20

Parainfluenza Virus

INTRODUCTION

Parainfluenza viruses belong to the genus *Paramyxovirus* of the family Paramyxoviridae. Parainfluenza viruses possess a negative-stranded, 15 kilobase RNA genome that codes for 6 major structural proteins. In electron micrographs, parainfluenza is a pleomorphic enveloped virus with a helical, 18 nm nucleocapsid and a 120-300 nm envelope containing glycoprotein spikes. These spikes possess hemagglutinin, neuraminidase, and cell fusion activities. Parainfluenza viruses are relatively labile and they are quickly inactivated at low pH, by heat, detergents, formaldehyde, alcohols, and organic solvents.

Four human parainfluenza virus serotypes have been identified (types 1, 2, 3, and 4) and type 4 virus has been divided into two subtypes, 4A and 4B. Although parainfluenza viruses share antigenic determinants with mumps virus, Newcastle disease virus, and each other, each parainfluenza type and subtype is serologically distinct. Types 1-3 can be readily isolated by cell culture while type 4 is more difficult to grow.

Parainfluenza viruses are important causes of respiratory disease in infants and young children. These viruses cause a wide spectrum of disease ranging from an afebrile cold to croup, bronchiolitis, and pneumonia (1-4). Parainfluenza type 1 causes bronchitis in 25% of infected children while croup is the dominating symptom in type 2 infections. Croup (laryngotracheobronchitis) occurs frequently following primary infection with parainfluenza types 1 and 2. Parainfluenza type 3 has a predilection for the epithelial cells lining the small air passages of the lung. Parainfluenza type 3 is second only to respiratory syncytial virus as the cause of pneumonia and bronchiolitis in infants less than 6 months of age (5). Although type 4 infections usually cause mild symptoms, they can cause serious illness in immunosuppressed individuals. In older children and adults, parainfluenza infections are typically mild and are often self-diagnosed as common colds. Local and systemic antibodies are usually produced during parainfluenza infections. However, immunity is of short duration and reinfections occur frequently (6).

Parainfluenza viruses are transmitted by direct contact with infected individuals and by respiratory droplets generated when these individuals cough or sneeze (1,5,7). During reinfection, children and adults can shed virus asymptomatically for up to 10 days. These individuals may provide an important vehicle for disseminating the virus in pediatric wards and newborn nurseries. The incubation period varies from 3-10 days in primary infections. Following transmission parainfluenza viruses initially replicate in the mucous membranes of the nose and throat. Aspiration of virus-laden respiratory secretions is responsible for spreading the virus to the lower respiratory tract.

Laboratory diagnosis of parainfluenza can be made by direct staining of respiratory epithelial cells, cell culture isolation, or by serological methods. Complement fixation, hemagglutination inhibition, EIA, indirect fluorescent antibody, and virus neutralization assays have been used for the detection of parainfluenza antibodies. Serological detection of an acute parainfluenza infection can be based upon the presence of high levels of virus-specific IgM or a four-fold rise in antibody titer in acute and convalescent sera. However, antibody responses may not be type-specific and infection with one virus frequently causes a heterotypic antibody response to another parainfluenza virus or mumps virus, especially during reinfections. Therefore, the serotype responsible for infection cannot be determined by serologic testing.

> AT A GLANCE...
>
> # PARAINFLUENZA VIRUSES
>
> **Virus Detection Methods**
> Tube cultures of primary rhesus monkey kidney cells.
> Centrifugation-enhanced (shell vial) cultures using RMK cells.
> Direct fluorescent antibody staining of respiratory cells.
>
> **Specimen Source**
> Throat swabs, nasopharyngeal swabs, washes or aspirates are the specimens of choice. Specimens should be collected 1 day before, to 3 days after, the appearance of symptoms.
>
> **Time to Result**
> Standard Culture: Positive culture - 2 days to 2 weeks. Negative culture - 2 weeks.
> Shell Vial Culture: Positive or negative culture - 48 hours.
> Direct Fluorescent Antibody Stain: 45 minutes.
>
> **Epidemiology**
> Parainfluenza viruses are found throughout the world. There are four serotypes of human parainfluenza virus. In the United States, type 1 and type 2 viruses produce disease outbreaks the Fall of odd-numbered years while type 3 viruses produce outbreaks in the Spring of even-numbered years. The epidemiology of type 4 viruses is not well characterized because the disease caused by viruses is so mild that it rarely requires medical attention.
>
> **Transmission/Incubation Period**
> Parainfluenza viruses are transmitted by direct contact with infected individuals and by respiratory droplets generated when an infected person coughs or sneezes. Patients are infectious during the first 3-5 days of illness. The incubation period is 3-10 days.
>
> **Inactivators and Disinfectants**
> Parainfluenza is inactivated by elevated temperatures, organic solvents, alcohols, detergents, and low pH conditions. Effective disinfectants include 70% alcohol, 10% household bleach solutions, quaternary ammonium compounds, phenols, iodophor compounds, and glutaraldehyde solutions.

Nonetheless, serological testing may provide a sensitive indicator of infection, especially in adults who may shed virus for only a short time (8).

VIRUS ISOLATION

Introduction

Parainfluenza viruses 1, 2, and 3 can be propagated in rhesus monkey kidney (RMK), HEp-2, HEK, HFF, A549, HL, Vero, and LLC-MK2 cells. However, RMK cells are usually used for primary isolations because they are more sensitive than other cell types (1,9,10). Like influenza, parainfluenza replication can be suppressed by inhibitors present in some serum preparations. Therefore, cell cultures should be washed with serum free medium or HBSS before specimen inoculation and virus isolation should be done in serum-free medium.

When present, parainfluenza CPE is usually detectable 5-10 days after inoculation. However, parainfluenza produces little if any visible CPE and the first indication that the virus is present in the culture is usually a positive hemadsorption (HAd) test. Final confirmation and serotyping of parainfluenza viruses are accomplished by scraping the cell monolayer and staining the cells with type-specific monoclonal antibodies. Because CPE is often inapparent, hemagglutination (Chapter 17) and hemadsorption testing provide useful screening methods for the detection of parainfluenza 1, 2, or 3. Cultures should be tested for HAd activity at least once during the first five days of culture and again, before the culture is considered negative. Some laboratories perform an HAd test 24 hours after inoculation and every second day thereafter for 14 days or until CPE develops.

Many laboratories routinely inoculate two RMK tubes for parainfluenza virus isolations. If more rapid virus isolation is required, two RMK shell vials can also be inoculated. However, shell vial cultures are not as sensitive as tube cultures and tube cultures should be used to back up all shell vial cultures.

Specimen Collection

Parainfluenza replicates in the ciliated columnar epithelial cells of the respiratory tract. The best time for specimen collection is from 1 day before the onset of symptoms to 3-4 days afterwards. Parainfluenza viruses are generally not detectable 5-10 days after onset of symptoms except in young children who may shed large amounts of virus for prolonged periods.

Parainfluenza 1, 2, and 3 can be isolated from nasopharyngeal (NP) and throat swab specimens, or from nasal and throat washes. The preferred specimens are nasal washes or an NP plus a throat swab. Although nasal washes produce a larger quantity of virus than swabs, they are not used in all clinical settings because they cause more discomfort (11) and, in most situations, swab specimens are the easiest specimens to obtain. If swab specimens are used, NP and throat swabs should be obtained and placed in the same viral transport tube. Specimens should be sent to the virology laboratory on wet ice. **DO NOT FREEZE SPECIMENS AT -20°C.** If specimens cannot be inoculated onto cell cultures within 72 hours, they should be frozen at -70°C.

Specimen Preparation

Swabs
1. Vortex the specimen with several glass beads for 20-30 seconds to release any bound cells or virus.
2. Remove the swabs from the transport medium and firmly roll them against the inside of the tube to remove as much fluid as possible.

Wash and Aspirate Specimens
1. Vortex the specimen with several glass beads for 20-30 seconds to release any cell-associated virus.
2. Centrifuge suspension at 200 x g for 5 minutes to remove any mucus or cell debris.
3. Carefully remove the mucus and use the supernatant fluids to inoculate cell cultures.

Standard Tube Culture

Procedure
1. Appropriately label 2 tubes containing freshly confluent RMK cells.
2. Remove the medium from the cells and wash the monolayers twice with HBSS or serum-free MEM containing antibiotics. This wash step is important because some serum components may interfere with parainfluenza virus replication.
3. Add 2 ml of serum-free MEM (with antibiotics) to each tube.
4. Add 0.2-0.5 ml of specimen to each tube.
5. Place the tubes in a roller drum and incubate at 33-35°C and 10-15 rph.
6. Examine the cells at least every other day for CPE.
7. Test the tubes for hemagglutination (Chapter 17) or hemadsorption 2-4 days after inoculation or when CPE is observed. Test negative cultures again on day 10 or day 14.
8. Perform culture confirmation testing (see below) on any monolayer that exhibits CPE, HA, and/or HAd activity.

Interpretation of Results

When present, CPE is usually visible 5-10 days

after inoculation. Parainfluenza type 1 CPE consists of small rounded cells that may not be readily apparent. Parainfluenza type 2 is the only serotype that produces a discernable CPE. Type 2 virus CPE is characterized by the presence of dark, granular, and irregular syncytia which retract from the cell monolayer producing a "Swiss cheese" effect. Type 3 virus causes poorly recognizable CPE consisting of elongated, fusiform cells that eventually retract and pull away from the cell monolayer. Type 4 virus CPE resembles type 1 CPE and is usually not apparent. Hemadsorption testing is the preferred method for achieving a preliminary identification.

Report: No report should be made based upon CPE alone.

Hemadsorption (HAd) Procedure

During replication, influenza, mumps, and parainfluenza viruses produce hemagglutinin molecules that are inserted into the plasma membrane of the host cell. Because these hemagglutinins bind to red blood cells, virus-infected cells can be identified because red cells adsorb to them. Hemadsorption assays are rapid, convenient, and relatively inexpensive. In addition, the proportion of cells with attached erythrocytes at 4°C compared with the number at 22°C can provide some indication as to whether the isolate is parainfluenza, mumps, or influenza virus. In parainfluenza 1, 2, and 3 infections, RBCs often elute from the infected cells when the culture is warmed to 22°C. Influenza and mumps virus infections will generally have the same amount of hemadsorption at 4°C and at 22°C.

Hemadsorption assays can be performed with guinea pig, fowl, and human O erythrocytes. However, most laboratories use guinea pig cells because they are smaller than other RBCs and more of them can attach to an infected cell. Thus, guinea pig RBCs produce a more uniform HAd pattern and reactions are easier to interpret than fowl or human erythrocyte HAd tests.

Procedure

1. Prepare a 0.5% suspension of guinea pig RBCs in sterile saline or phosphate buffered saline (Appendix C).
2. Add 0.4 ml of the RBC suspension to each culture tube.
3. Incubate the tubes on their side (cell monolayer downward) for 20-60 minutes at 4°C in a slanted (4°) test tube rack.
4. Examine the cell monolayer under the microscope at 100-200X for the presence of RBC adsorption and record the results (Table 1).

TABLE 1. *Recording HAd Reactions*

Percent of Cells With HAd	Score
75-100	4+
50-75	3+
25-50	2+
< 25	1+
Questionable	+/-

5. Incubate the tubes for one hour at room temperature. Examine the tubes for hemadsorption as before and record the results.

Interpretation of Results

Positive Test. Specific hemadsorption is characterized by chains, rosettes, or clumps of red cells that adhere to the cell sheet (Fig. 1). The appearance of this HAd pattern indicates a positive test. The presence of less hemadsorption after warming the tube to room temperature suggests that the isolate is a parainfluenza virus. However, results should be interpreted carefully because monkey kidney cells are sometimes infected with hemadsorbing simian viruses. Comparing the level of HAd to an uninoculated control culture is important to eliminate false positive results. HAd-positive cells should be removed from the tube and stained with specific monoclonal antibody reagents.

REPORT: No report should be made based upon the results of a hemadsorption test alone. All HAd positive cultures should be confirmed with

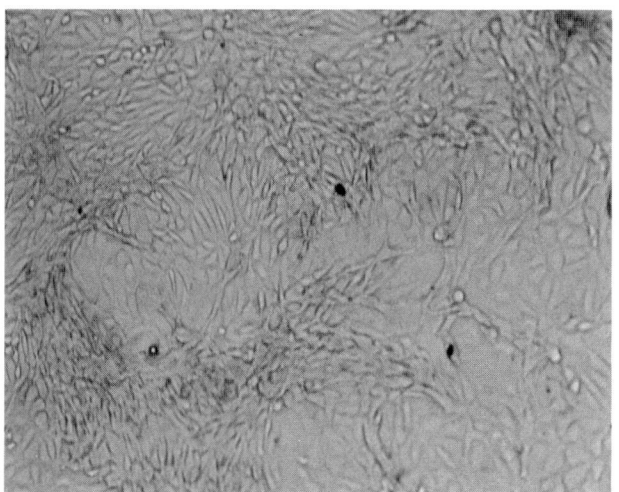 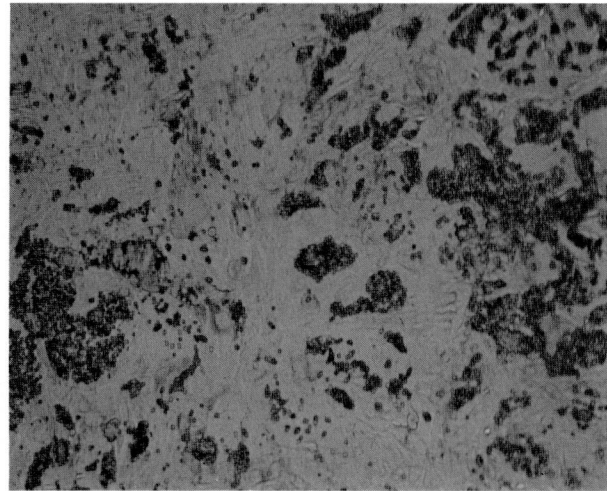

FIG. 1. Hemadsorbing pattern of uninoculated (left) and parainfluenza-infected RMK cells.

type-specific monoclonal antibody staining.

Negative Test. A negative HAd test is characterized by red cells that float freely in the medium and the <u>absence</u> of RBC chains, rosettes, or clumps adhering to the cell sheet (Fig. 1). Because endogenous simian viruses can produce a background level of HAd, cultures that exhibit some HAd activity may be considered negative if the HAd levels are no higher than the background (uninoculated culture) levels. A negative isolation report should be generated if the cultures are HAd negative after 14 days of culture.
REPORT: Parainfluenza not isolated.

QC Procedures.

Two parainfluenza-infected RMK (positive control) tubes should be inoculated with each batch of parainfluenza isolations. The first tube should be tested for HAd activity on day 2-4 and the second tube on day 10-14. Both tubes should exhibit specific HAd activity as described above. When inoculating the positive control cultures, it may be advantageous to dilute the inoculum for the day 10 tube 1:10 to prevent monolayer destruction. Two uninoculated RMK tubes serve as negative controls. Because monkey kidney cells are sometimes infected with SV5 or other hemadsorbing simian viruses, negative control tubes must come from the same lot of cells as the patient isolation tubes. A negative control tube should be tested whenever patient isolation tubes are tested. Positive HAd tests will have more HAd activity than the negative control tube. If uninoculated control tubes have more than 25% of the cell sheet exhibiting HAd (> 1+ HAd), the cell lot should be discarded.

Culture/HA/Had Confirmation
1. Remove all but a few drops of the cell culture medium.
2. Scrape the cells from the tube surface.
3. Suspend the cells in the remaining fluid.
4. Using a pipette, place a small drop of the cell suspension onto each of three wells of an acetone-cleaned multiwell slide. The remainder of the cell suspension should be retained for inoculation onto fresh RMK cells if staining is negative.
5. The slide should be processed as directed by the manufacturer of the fluorescein-labeled antibodies or as described below.
6. Allow the slide to air dry, then fix it in acetone for 10-15 min. Fixed slides may be stored for up to a week at 2-8°C or for one year at -20°C under desiccated conditions.
7. Place enough of the appropriate fluorescein-labeled monoclonal antibody on each well to

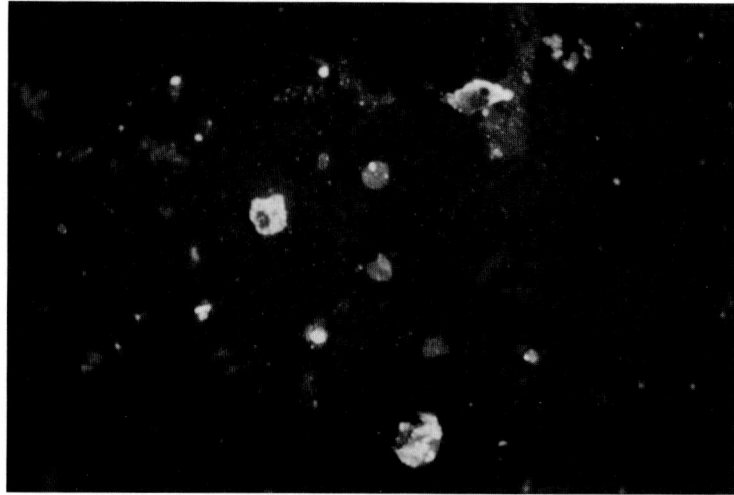

FIG. 2 Fluorescent antibody staining of parainfluenza 1-infected RMK cells. Cells were scraped from their tube and stained with monoclonal antibodies.

cover the cell smear (15-30 µl).
8. Incubate the slide for 30 minutes at 35-37°C in a covered, humidified chamber. Do not allow the antibody to dry on the slide as this could cause nonspecific antibody binding.
9. Remove the excess antibody with a gentle stream of PBS. Do not direct the stream directly at the cell smears as this could dislodge the cells.
10. Soak the slide in PBS for 5 minutes, shake off the excess fluid, and allow the slide to air dry.
11. Carefully add a small drop of FA mounting fluid to the center of each well.
12. Place a number 1 coverslip on the mounting fluid and carefully remove all the air bubbles.
13. Immediately examine the slide using a fluorescence microscope. For optimum clarity, use 200-300X magnification for screening and 400X for confirmation of cell morphology.

Interpretation of Results
Positive Test. The presence of parainfluenza virus is indicated by apple-green fluorescence in the cytoplasm infected cells (Fig. 2). Only individuals experienced in reading FA reactions should examine the cell smears because cell debris and cell clumps may exhibit a dull fluorescence which can be misinterpreted as a specific, positive reaction.
REPORT: Parainfluenza (1, 2, or 3) virus isolated.

Negative Test. The absence of parainfluenza virus is indicated by the lack of specific fluorescence in the infected cells. If the cultures were CPE negative; HA or HAd-positive; and FA negative, they should be stained with antibodies to other respiratory viruses and mumps virus.
REPORT: Parainfluenza virus not isolated.

QC Procedures.
Subpassages of recent parainfluenza isolates or ATCC cultures should be inoculated into cell culture tubes with each batch of parainfluenza isolations. Uninfected cell cultures serve as negative controls. Infected and uninfected cell monolayers should be scraped and stained as described above. Positive controls must hemadsorb and exhibit a typical pattern of fluorescence. Negative controls must not hemadsorb more than cell culture control and they should not exhibit specific fluorescent staining.

Preparation of Positive Control Cultures
1. Add 20 µl of a recent parainfluenza isolate (or an ATCC culture) to 2 ml of sterile GLB.

Hold the diluted virus on ice until it is used to inoculate the monolayer.
2. Remove the growth medium from one 75 cm² flask of newly confluent RMK cells.
3. Rinse the monolayer twice with 10 ml of serum-free MEM.
4. Add 2 ml of the diluted virus to the monolayer and allow the virus to adsorb for 1-2 hours at 33-35°C. Note: The flask should be rocked every 15 minutes to assure even virus distribution and to prevent monolayer desiccation.
5. Add 20 ml of serum-free MEM containing antibiotics.
6. Incubate the flask at 33-35°C until the 50-60% of the monolayer is HAd positive or until the supernatant fluids have a titer of 1:64 or higher.
7. Scrape the cells from the flask into the cell culture medium.
8. Transfer the medium to a 50 ml polypropylene centrifuge tube.
9. Add GLB to the 50 ml mark.
10. Quickly freeze the medium at -70°C.
11. Thaw the medium quickly in a 37°C water bath.
12. Centrifuge the tube at 600-800 x g for 15 minutes to pellet the cell debris.
13. Dispense 1.0 ml of the diluted supernatant fluids into each of 50 freezing vials and freeze the vials at or below -70°C. Control cultures can be stored at -70°C for up to 2 years or in liquid nitrogen for 5 years.

Centrifugation-Enhanced (Shell Vial) Method

Shell vial cultures can be used to identify parainfluenza isolates as early as 36-48 hours after inoculation. Although other cell lines can be used, RMK shell vials are preferred because of their increased sensitivity. Shell vial cultures are less sensitive than tube cultures for parainfluenza isolation. Therefore, an RMK tube culture should be inoculated with each shell vial specimen. The RMK tube serves as a backup for specimens that contain small quantities of virus and for the isolation of other viruses.

Procedure
1. Appropriately label 2 shell vials and 1 RMK tube for each specimen.
2. Remove the medium from the shell vials and wash the monolayers twice with serum-free MEM containing antibiotics.
3. Inoculate each vial with 0.2-0.5 ml of the specimen.
4. Centrifuge the vials at 700 x g for 1 hour at room temperature.
5. Allow the virus to adsorb for one hour at 33-35°C
6. Add 1 ml of serum-free MEM to each vial.
7. Incubate the vials at 33-35°C.
8. Fix and stain one vial after 48 hours and the other vial 5 days after inoculation.

Staining of Coverslips

Identification of parainfluenza is accomplished by staining the coverslips with polyclonal antibodies to parainfluenza or a pool of parainfluenza monoclonal antibodies. Most specimens produce 1-20 small foci after 48 hours of culture. Fixation and staining should be done as directed by the manufacturer of the monoclonal antibodies or as described below.
1. Carefully remove the culture fluid from the vials and wash the coverslips twice by adding 1-2 ml of PBS and soaking for 5 minutes.
2. Remove the PBS and add 2-3 ml of acetone to each vial. Fix the coverslips for 10 minutes at room temperature.
3. Remove the acetone and allow the coverslips to air dry.
4. Rinse the coverslips briefly with PBS. The moisture trapped between the coverslip and the vial will keep the stain from wicking under the coverslip.
5. Add enough monoclonal antibody to the vial to cover the coverslip (100-150 µl) and incubate for 30 minutes at 35-37°C in a humidified chamber to prevent the antibody from drying on the coverslip.
6. Wash the coverslips twice as previously described (step 1).
7. Place a small amount of mounting fluid on a slide. Remove the coverslips from the vial

and lay them, cell side down, on the mounting fluid.
8. Immediately examine the coverslips at 100-300X using a fluorescence microscope.

Interpretation of Results

Coverslips that have individual cells or small groups of cells that exhibit brilliant apple-green cytoplasmic fluorescence are positive for parainfluenza. Coverslips without the specific fluorescence described above are considered negative. Some specimens may trap the monoclonal antibodies between the cell monolayer and inoculum cells. Careful examination of the cell morphologies can distinguish this type of reaction from specific, parainfluenza reactions. In addition, nonspecific staining can occur when the antibody reagents are allowed to dry on the coverslip. In this case, the entire cell sheet may stain apple-green.

Positive Test. The presence of parainfluenza virus is indicated by brilliant, apple-green cytoplasmic and/or nuclear fluorescence in infected cells.
REPORT: Parainfluenza (1, 2, or 3) virus isolated.

Negative Test. The absence of parainfluenza is indicated by the lack of specific fluorescence described above. The entire cell sheet should be stained red by the counterstain.
REPORT: Parainfluenza virus not isolated.

QC Procedures.

A 1:10 dilution of recent parainfluenza isolates or ATCC cultures (see Preparation of Positive Control Cultures, above) should be inoculated with each batch of parainfluenza shell vial cultures. Uninfected shell vial cultures serve as negative controls. Infected and uninfected shell vial cultures should be processed and stained as described for the patient cultures. Positive controls must exhibit the typical fluorescence patterns. Negative controls must not exhibit specific fluorescent staining.

DIRECT SMEARS

Although rapid and specific, direct smears are less sensitive than either conventional tube cultures or shell vial cultures (9). Therefore, direct smears should be backed up by tube cultures to detect other viruses that may be in the specimen. Direct smears require the presence of intact, parainfluenza-infected respiratory epithelial cells before a positive result can be reported. Inadequate specimen collection and smear preparation are the principal reasons for the failure of this method to detect parainfluenza viruses. In addition, excess mucus can interfere with antibody binding. Mucus can cause false positive reactions by nonspecifically trapping the fluorescent antibody reagents and false negative reactions by clumping the cells and thereby, masking the parainfluenza antigens. However, skilled technicians can usually distinguish these reactions from specific parainfluenza reactions.

Despite these drawbacks, evaluation of direct smears for parainfluenza virus is an effective method for the rapid detection of parainfluenza infections. Direct smears may be especially valuable in remote laboratories where cold-chain specimen transport is not easily accomplished.

Specimen Collection

The success of the direct fluorescent antibody (DFA) procedure depends upon the preparation of a well made cell smear. Smears that are too thick or "lumpy" can cause nonspecific trapping of the antibody reagents. Smears that are grossly contaminated with red blood cells can cause interpretation difficulties due to red cell autofluorescence. However, the principal reasons direct smears fail to detect parainfluenza infections are (a) inadequate specimen collection and (b) the presence of too few cells on the slide. Many laboratories require at least 50 cells per smear while others require 1-2 cells/high power field. Whatever the number, too many cells can cause nonspecific trapping of antibody and with too few cells, the laboratory may not be able to find an infected cell.

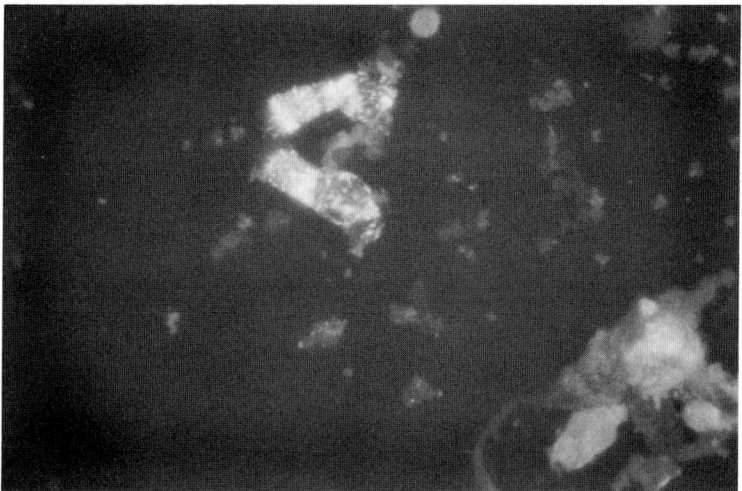

FIG. 3. Positive nasopharyngeal smear stained with monoclonal antibodies to parainfluenza 1.

Specimen Processing

Nasal Aspirates, Nasal Washes, and Throat Washes
1. Centrifuge the specimen at 1000-1500 x g for 5 minutes to pellet the cells.
2. Remove all but 100-200 µl of the fluid.
3. Suspend the cell pellet in the remaining fluid.
4. Spread one drop of the cell suspension onto three wells of an acetone cleaned slide.
5. Allow the smears to air dry at room temperature. Slides may be stored for up to 48 hours at 2-8°C.

Swabs in Viral Transport Medium
1. Vortex the specimen vigorously to release as many cells from the swabs as possible.
2. Remove the swabs from the viral transport medium and firmly roll them against the inside of the tube to remove as much fluid as possible.
3. Centrifuge the specimen at 1000-1500 x g for 5 minutes to pellet the cells.
4. Remove all but 100-200 µl of the viral transport medium.
5. Suspend the cell pellet in the remaining fluid.
6. Spread one drop of the cell suspension onto three wells of an acetone cleaned two-well slide.
7. Allow the smear to air dry at room temperature. Slides may be stored for up to 48 hours at 2-8°C.

Test Procedure

For optimum performance, slides should be fixed and stained within 1-2 hours of specimen collection. Alternatively, unfixed slides may be stored for up to 48 hours at 2-8°C. Fixed slides may be stored at 2-8°C for 1 week or frozen at -20°C for up to 1 year. Storing slides under desiccated conditions will decrease the background staining and minimize antigen degradation.

1. Upon arrival in the laboratory, examine the slide wells at 100-300X magnification. Ideally, a minimum of 50 cells should be visible on each smear before the specimen is considered adequate for further processing.
2. The slide should be processed as directed by the manufacturer of the fluorescein-labeled antibodies to parainfluenza virus or as described below.
3. Fix the cells by immersing the slides in acetone (room temperature) for 10 minutes. Allow the slides to air dry.
4. One or all of the smears can be stained simultaneously. Some laboratories will stain the smears with polyclonal antibodies to

parainfluenza while others use type specific monoclonal antibodies and report the presence of a specific parainfluenza type. The latter procedure is described below (step 5).

5. Place enough of the fluorescein-labeled monoclonal antibody on each well to cover the cell smear (15-30 µl). Each well should be stained with one of the monoclonal antibodies to parainfluenza 1, 2, or 3.
6. Incubate the slide for 30 minutes at 35-37°C in a covered, humidified chamber. Do not allow the antibody to dry. Drying can cause nonspecific antibody binding.
7. Remove the excess antibody with a gentle stream of PBS. Do not direct the stream directly at the cell smear as this could dislodge the cells.
8. Soak the slide in PBS for 5 minutes, shake off the excess fluid, and allow the slide to air dry.
9. Carefully add a small drop of FA mounting fluid to the center of each well.
10. Place a number 1 coverslip on the mounting fluid and carefully remove all the air bubbles.
11. Immediately examine the slide using a fluorescence microscope. For optimum clarity, use 200-300X magnification for screening and 400X for confirmation of cell morphology.

Interpretation of Results
Positive Test. The presence of parainfluenza virus is indicated by intense apple-green fluorescence in the cytoplasm of the exfoliated epithelial cells (Fig. 3). Only intact cells should be examined because cell debris and clumps of normal cells may exhibit a dull, yellow-green fluorescence which could be misinterpreted as a positive reaction.
REPORT: Parainfluenza (1, 2, or 3) virus detected.
Negative Test. The absence of parainfluenza virus is indicated by the lack of specific fluorescence in the cells.
REPORT: Parainfluenza virus not detected.

Inconclusive Test. Specimens with fewer than 50 cells on each well (or 1-2 cells/HPF) may give inconclusive results.

REPORT: Unacceptable specimen - too few cells.

QC Procedures.
Positive and negative controls should be stained at least once each day to assure that the antibody reagent is performing properly. Positive cells should stain intensely as described above. Negative cells should stain red and should not exhibit specific apple-green fluorescence. Because the same antibody reagent is used for confirmation of tube cultures, shell vial cultures, and direct smears, culture confirmation testing can be used to demonstrate that the reagent is working properly. In laboratories that do not perform parainfluenza isolations or laboratories where no isolations are ongoing, prepared slides should be stained whenever direct smears are stained. Control slides can be purchased commercially from a number of vendors or they can be prepared as described below.

Preparation of Control Slides
1. Add 20 µl of the positive control culture (described above) to 2 ml of sterile GLB. Hold the diluted virus on ice until used to inoculate the monolayer.
2. Remove the culture medium from one 75 cm^2 flask of newly confluent RMK cells.
3. Rinse the monolayer twice with 10 ml of serum-free MEM.
4. Add 2 ml of the diluted virus to the monolayer and allow the virus to adsorb for 1-2 hours at 35-37°C. Note: The flask should be rocked every 15 minutes to assure even distribution of the virus and to prevent monolayer desiccation.
5. Add 20 ml of serum-free MEM containing antibiotics.
6. Incubate the flask at 35-37°C until 50-60% of the monolayer hemadsorbs guinea pig RBCs.
7. Rinse the monolayer twice with PBS to remove the RBCs, trypsinize the cells, and remove them from the flask.
8. Centrifuge the cells at 250 x g for 5 minutes.
9. Resuspend the cell pellet in 5 ml of PBS. Perform a cell count and adjust the cell concentration to 2-5 x 10^6 cells/ml.

10. Dispense 3-10 µl of the cell suspension onto each well and allow the suspension to air dry.
11. Fix the slides in acetone at room temperature for 10 minutes and allow the slides to air dry.
12. Store the slides at -20°C or below for up to one year. Slides should be stored under desiccated conditions to minimize antigen degradation and background fluorescence.

NOTE: Slides prepared in this manner will contain both positive and negative cells. Therefore, a single slide can be used for QC purposes. When slides are made from flasks with 100% CPE, two slides must be used - one containing a positive cell smear and a second slide containing an uninfected cell smear.

REFERENCES

1. Chanock RM. Parainfluenza Viruses. *In* Lennette EH, Schmidt NJ, eds. *Diagnostic procedures for viral, rickettsial and chlamydial infections*, fifth edition. Washington: American Public Health Association 1979;611-632.
2. Heilman CA. Respiratory syncytial and parainfluenza viruses. *J Infect Dis* 1990;161:402-406.
3. McIntosh K. Pathogenesis of severe acute respiratory infections in the developing world: Respiratory syncytial virus and parainfluenza viruses. *Rev Infect Dis* 1991;13(Suppl 6):S492-500.
4. Singh-Naz N, Willy M, Riggs N. Outbreak of parainfluenza virus type 3 in a neonatal nursery. *Pediatr Infect Dis J* 1990;9:31-33.
5. Grandien M. *Paramyxoviridae:* The parainfluenza viruses. In: Balows A, Hausler WJ Jr, Lennette EH, eds. *Laboratory diagnosis of infectious diseases - principles and practice*, Volume II. New York: Springer-Verlag, 1988;484-506.
6. Hall CE, Brandt CD, Frothingham TE, Spigland I, Cooney MK, Fox JP. The virus watch program: a continuing surveillance of viral infections in metropolitan New York families. IX. A comparison of infections with several respiratory pathogens in New York and New Orleans families. *Am J Epidemiol* 1971;94:367-385.
7. Ansari SA, Springthorpe VS, Sattar SA, Rivard S, Rahman M. Potential role of hands in the spread of respiratory viral infections: Studies with human parainfluenza virus 3 and rhinovirus 14. *J Clin Microbiol* 1991;29:2115-2119.
8. Bloom HH, Johnson KM, Jacobson R, et al. Recovery of parainfluenza viruses from adults with upper respiratory illness. *Am J Hyg* 1961;74:50.
9. Ray CG, Minnich LL. Efficiency of immunofluorescence for rapid detection of common respiratory viruses. *J Clin Microbiol* 1987;25:355-357.
10. Stout C, Murphy MD, Laurence S, Julian S. Evaluation of a monoclonal antibody pool for rapid diagnosis of respiratory viral infections. *J Clin Microbiol* 1989;27:448-452.
11. Frayha H, Castriciano S, Mahony J, Chernesky M. Nasopharyngeal swabs and nasopharyngeal aspirates equally effective for the diagnosis of viral respiratory disease in hospitalized children. *J Clin Microbiol* 1989;27:1387-1389.

CHAPTER 21

Human Papillomavirus

INTRODUCTION

Papillomaviruses comprise one genus of the family *Papovaviridae*. Human papilloma virus (HPV) virions are relatively small (55 nm in diameter) and possess icosahedral symmetry. The HPV genome consists of a 7.9 kilobase double-stranded, covalently closed circular DNA molecule that codes for at least 2 major proteins. HPV does not have an envelope and is resistant to many organic solvents and alcohols (1). There are more than 50 known types of human papilloma virus.

Although the distinction is not absolute, human papilloma viruses can be divided into two broad groups - those that cause skin lesions and those that infect mucosal surfaces. Skin lesions include common, plantar, and flat warts. Common warts are usually found in clusters on the hands and fingers while flat skin warts tend to localize on the arms, knees, and face of infected individuals. Deep plantar-type warts are commonly found on the toes and soles of the feet. As a group, skin warts are generally benign, self-limiting epithelial tumors that tend to resolve spontaneously. Resolution of these tumors is presumably the result of immune mechanisms (2).

Mucosal HPV infections can be found in the genital tract, the respiratory tract, the oral cavity, and the conjunctiva. More than a dozen HPVs have been recovered from the genital tract and five virus types, HPV-6, 11, 16, 18, and 31, account for the vast majority of genital tract infections (1). HPV-6 and 11 are responsible for most of the exophytic condylomas (genital warts), nearly all respiratory papillomas, and some infections of the oral cavity and conjunctiva.

Epidemiological and virological evidence have consistently linked the presence of some human papilloma viruses with carcinoma. The link is especially strong between HPV-16 and squamous cell carcinoma of the cervix. HPVs can also be recovered from a large proportion of extraepithelial neoplasias, and from 70-90% of invasive cancers. Although HPV-16 and 18 are found in some benign lesions of the genital tract, they are most commonly found in premalignant and malignant lesions of the genital tract (1). In addition, HPV-16 and 18 are preferentially associated with severe lesions (3).

Other HPV types (HPV-31, 38, 52b, and 58) also have been implicated in cervical intraepithelial neoplasia and cervical carcinoma. In benign lesions, HPV is usually present as an episomal DNA element. However, the development of invasive cancer has been associated with the integration of viral DNA into the cellular genome.

HPV is also associated with epidermodysplasia verruciformis (EV), a rare lifelong disease in which individuals acquire warts in childhood and are unable to resolve them (4). About one third of these individuals subsequently acquire squamous cell carcinomas after years of benign disease. These slow growing tumors frequently arise from macular lesions caused by HPV-5, 8, and 14 infection (4).

HPV is transmitted by direct skin contact or by contact with HPV-contaminated surfaces (e.g., bathroom floors). Transmission usually requires local abrasions of the skin or mucosa although autoinoculation by scratching has also been described (1). Genital HPV infections are transmitted by sexual intercourse. Children born through an HPV-infected birth canal may acquire juvenile-onset laryngeal papillomas. Oral-genital sex may account for some cases of oral cavity warts. Some children are also thought to acquire oral warts through habitual chewing of skin warts (2).

Once transmitted, warts probably arise from

> *AT A GLANCE...*
>
> # HUMAN PAPILLOMAVIRUS
>
> **Virus Detection Methods**
> Hybridization techniques - commercially available from various vendors.
> Polymerase chain reaction
> Immunocytochemical staining
>
> **Specimen Source**
> Swabs, scrapings or biopsies of the lesions.
>
> **HPV Serology**
> Serological tests have no role in the diagnosis of HPV infection because the antibody response may require several months to form and antibody levels are generally low. In addition, the presence of antibody has appears to have little impact on the disease course.
>
> **Epidemiology**
> HPV is widespread with skin warts the most common manifestation of HPV infection. Genital warts are common in areas of early sexual activity and in patients with multiple sexual partners. The virus has been associated with the development of cervical cancer.
>
> **Transmission**
> HPV is transmitted by direct skin contact or by contact with HPV-contaminated surfaces (e.g., bathroom floors). Infections usually depend upon local abrasions of the skin or mucosa although autoinoculation by scratching has also been described. Genital HPV infections are transmitted by sexual intercourse. It is thought that infection of the child during passage through an HPV-infected birth canal may be responsible for juvenile-onset laryngeal papillomas. Oral-genital sex may account for some cases of oral cavity warts. In addition some children who habitually chew skin warts may transmit these warts to the oral cavity.
>
> **Inactivators and Disinfectants**
> The effectiveness of disinfectants on HPV infectivity has not been well studied because no method is available for measuring the infectivity of papillomaviruses. It is thought that the resistance of papilloma viruses to physical and chemical agents is similar to that described for polyomaviruses. HPV is thought to be resistant to organic solvents and to heating to 50°C for 1 hour. Infectivity is probably destroyed by treatment with detergents and at alkaline pH. Effective disinfectants are thought to include 10% household bleach solutions, phenols, and glutaraldehyde solutions.

infection of a single cell or a small group of cells in the basal epithelium (5,6). HPV stimulates the increased reproduction of the basal epithelium thereby causing the formation of a lesion.

The increasing prevalence of genital HPV infections makes HPV one of the fastest growing sexually transmitted diseases. However, HPV characterization has been hampered by the inability to propagate the virus in cell cultures or in animal models. At the present time, virions and viral DNA must be isolated from biopsy material, the yield from which is too low to allow for serological or biochemical characterization.

Most warts can be diagnosed by the clinician and laboratory testing is seldom required for this purpose. However, a clinical diagnosis cannot predict whether a benign papilloma is likely to become a malignant lesion. Electron microscopy and viral serologies also fail to provide this type of information. In fact, serological tests have no role in the diagnosis of HPV infection because the HPV antibody response may take several months to develop and when present, antibody levels are generally low. In addition, the presence of antibody appears to have little impact on the disease course (2).

Although the correlation is not certain, molecular typing of HPV DNA is the only method that can provide some statistical information as to the likelihood of malignant transformation. Southern blot hybridization is currently considered the "gold standard" for HPV detection and typing. *In situ* hybridization, dot blot hybridization, polymerase chain reaction, and self-sustained sequence replication methods also have been used for this purpose (7-12) and several HPV typing and detection kits are commercially available (see Appendix B). Immunocytochemical staining appears to be relatively insensitive compared with hybridization techniques.

REFERENCES

1. Shah KV, Gissmann L. Papovaviruses. In: Schmidt NJ, Emmonds RW. eds. *Diagnostic procedures for viral rickettsial and chlamydial infections.* Sixth edition. Washington DC: American Public Health Association 1989;1067-1102.
2. Pfister H. *Papovaviridae:* The papillomaviruses. In: Lennette EH, Halonen P, Murphy FA, eds. *Laboratory diagnosis of infectious diseases - principles and practice.* Volume II. 1988;301-316.
3. Gissmann L, Schneider A. Human papillomavirus DNA in preneoplastic and neoplastic genital lesions. In: Peto R, zur Hausen H, eds. *Banbury Rpt 21.* New York: Cold Spring Harbor Laboratory 1986;217-224.
4. Orth G, Favre M, Breitburd F, Croissant O, Jablonska S, Obalek S. Epidermodysplasia verruciformis. A model for the role of papillomaviruses in human cancer. *Cold Spring Harbor Conf Cell Proliferation* 1980;7:259.
5. Murray R, Hobbs J, Payne B. Possible clonal origin of common warts (verruga vulgaris). *Nature* 1971;232:1-2.
6. Buscema J, Shah K, Hsu S, Rosenshein N, Woodruff JD. Genetic investigation of the cellular origin of vulvovaginal condylomata acuminata. In: Steinberg B, Brandsma J, Taichman L, eds. *Papillomaviruses, cancer cells*, Volume 5. New York: Cold Spring Harbor Laboratory 1987;245-247.
7. Bartholoma NY, Adelson MD, Forbes BA. Evaluation of two commercially available nucleic acid hybridization assays for the detection and typing of human papillomavirus in clinical specimens. *Am J Clin Pathol* 1991;95:21-29.
8. Caussy D, Orr W, Daya AD, Roth P, Reeves W, Rawls W. Evaluation of methods for detecting human papillomavirus deoxyribonucleotide sequences in clinical specimens. *J Clin Microbiol* 1988;26:236-243.
9. Clavel C, Binninger I, Boutterin M, Polette M, Birembaut P. Comparison of four non-radioactive and ^{35}S-based methods for the detection of human papillomavirus DNA by in situ hybridization. *J Virol Methods* 1991;33:253-266.
10. Meyer MP, Markaw CA, Matuscak RR, Saker A, McIntyre-Seltman K, Amortegui AJ. Detection of human papillomavirus DNA in genital lesions by using a modified commercially available in situ hybridization assay. *J Clin Microbiol* 1991;29:1308-1311.
11. Ranki M, Leinonen AW, Jalava T, Nieminen P, Soares VRX, Paavonen J, Kallio A. Use of AffiProbe HPV test kit for detection of human papillomavirus DNA in genital scrapes. *J Clin Microbiol* 1990;28:2076-2081.
12. Schiffman MH, Bauer HM, Lorincz AT, Manos MM, Byrne JC, Glass AG, Cadell DM, and Howley PM. Comparison of southern blot hybridization and polymerase chain reaction methods for the detection of human papillomavirus DNA. *J Clin Microbiol* 1991;29:573-577.

CHAPTER 22

Polyomavirus

INTRODUCTION

Polyomaviruses comprise one genus of the family *Papovaviridae*. There are two known human polyomaviruses - JC virus and BK virus and although some human sera contain antibodies to a lymphotropic polyomavirus of African green monkeys, this virus has not yet been isolated from humans (1). Polyomaviruses possess a relatively small (39-44 in diameter) icosahedral capsid (2). The polyomavirus genome consists of a 5 kilobase double-stranded, covalently closed circular DNA molecule that codes for at least 3 viral capsid proteins and three T-antigens. Polyomaviruses possess hemagglutinin activity and can agglutinate human O red blood cells *in vitro*. Polyomaviruses do not possess an envelope and they are resistant to organic solvents, alcohols, elevated temperatures (50°C for 1 hour) and to sodium deoxycholate.

BKV and JCV are distributed throughout the world and serologic studies indicate that infections usually occur early in life. BK virus antibodies are seen in 50% of children by age 4 and the seroprevalence approaches 100% prevalence by 10-11 years of age. JC virus antibodies are present in 50% of 14 year old children and about 75% of adults. The mechanism of virus transmission is not known. However, the high incidence of virus infection in children suggests transmission by a respiratory route. Although JC and BK virus are excreted in the urine of infected individuals, it is not known if urine serves as a mechanism for disseminating infection. Once transmitted, it is thought that JC and BK viruses replicate in the respiratory tract and a subsequent viremia is responsible for transporting the virus to the target organs (1). JC and BK viruses establish latent infections that can reactivate during times of immune suppression.

Although primary infections with JC and BK virus are thought to be asymptomatic, BK virus seroconversion has been associated with mild respiratory illness in children (1). JC and BK viruses can be isolated from the urine of infected patients during mild or severe immunosuppression. Virus reactivation in the urinary tract has been observed in kidney or bone marrow transplantation, primary immunodeficiency diseases, immunosuppressive chemotherapy, pregnancy, measles, chronic disease, and old age. Unlike other viruses, serious disease seems to follow virus reactivation rather than primary infection. BK virus reactivation has been associated with hemorrhagic cystitis in bone marrow transplant patients and ureteral stenosis in kidney transplant patients (3-6). In contrast, JC virus infects the central nervous system and produces progressive multifocal leukoencephalopathy (PML).

PML is an extremely rare demyelinating disease that results from virus-mediated destruction of oligodendrocytes. Clinically, PML is an asymmetric multifocal brain disease that occurs without signs of increased intracranial pressure. The majority of PML patients are between 40 and 70 years of age and up to 16% of PML patients have AIDS as the underlying disorder (7). PML has an insidious onset and patients usually present with speech and vision impairment followed by general mental deterioration. The disease course is rapid and progressive, causing paralysis, cortical blindness and sensory abnormalities. Death usually occurs within 3-6 months of onset.

Diagnosis of PML and polyoma-associated hemorrhagic cystitis are usually based upon clinical symptoms and a history of immune suppression. Although virus isolation can be used to confirm the clinical diagnosis, isolations are rarely requested because they can take up to 2 months and

> *AT A GLANCE...*
>
> # POLYOMAVIRUSES
>
> **Virus Detection Methods**
> JC and BK viruses can be isolated in HEK and primary human fetal glial cells, WI-38, and HEL cells.
>
> **Specimen Source**
> Urine is the specimens of choice for the detection of active infection.
> Brain tissue is required for PML.
>
> **Time to Result**
> Tube Cultures: Positive culture - 3-7 weeks. Negative Culture - 7 weeks.
> However, 1-2 secondary passages are often required.
> Shell Vial Culture: Positive culture - 48 hours. Negative Culture - 72 hours.
>
> **Virus Serology**
> Serologic methods are of limited value in routine diagnosis.
>
> **Epidemiology**
> BKV and JCV are distributed throughout the world and serologic studies indicate that infections occur early in life. BK virus antibodies are seen in 50% of children by age 4 and the seroprevalence approaches 100% prevalence by 10-11 years of age. JC virus antibodies are present in 50% of children by the age of 14 years, and about 75% of the adults are seropositive.
>
> **Transmission/Incubation Period**
> The exact manner of virus transmission is not known. However, it is thought that infection is spread through the respiratory route. JC and BK virus are excreted in the urine of infected individuals but is not known if urine serves as a mechanism for disseminating infection.
>
> **Inactivators and Disinfectants**
> Polyomaviruses are relatively stable viruses and they are resistant to organic solvents, to elevated temperatures (50°C for 1 hour) and to sodium deoxycholate. Infectivity is destroyed by treatment with detergents and at alkaline pH. Effective disinfectants include 10% household bleach solutions, phenols, and glutaraldehyde solutions.

confirmatory reagents are not widely available. However, centrifugation enhanced (shell vial) methods may make virus isolations more practical in the future. PCR methods have also been used to detect JC and BK viruses in urine (6-10) and these methods may find increased usage because they are faster and at least as sensitive as culture. Hemagglutination inhibition assays have been developed for JC and BK virus antibodies but the high seroprevalence in the population severely limits the diagnostic value of serological methods.

VIRUS ISOLATION

Introduction
BK virus and JC are fastidious viruses and are difficult to grow in culture. BK virus replicates slowly in several human and monkey cell lines including HEK, WI-38, Vero, CV-1 and HFF cells. JC virus has a more restricted host range than BK virus. While JC virus can replicate in HEK cells, primary fetal human glial cells (rich in spongioblasts) or origin-defective SV40 transformed human

fetal glial cells are the most sensitive cultures. JC and BK viruses can produce CPE as early as 2-4 weeks after inoculation. However, blind passages may be required before CPE is visible. Polyomavirus CPE consists of cytoplasmic vacuolation, enlarged nuclei with inclusions, cell rounding, and eventually, cell destruction. These viruses replicate in the nucleus of infected cells and intranuclear inclusions can be observed after staining.

Specimen Collection and Storage

Although JC virus has been found in brain tissue from PML patients, urine remains the specimen of choice for JC and BK viruses. Although polyomaviruses are relatively hardy, specimens should be inoculated into cell culture as soon as possible after collection. If immediate inoculation is not possible, specimens may be stored at 2-8°C for up to 72 hours. If longer delays are anticipated, specimens should be frozen at or below -70°C. **DO NOT FREEZE SPECIMENS AT -20°C.**

Specimen Preparation
1. Place 15 ml of urine in a sterile centrifuge tube.
2. Centrifuge the tube at 1500 x g for 20 minutes at room temperature.
3. Discard the supernatant.
4. Suspend the sedimented cells in 2 ml of sterile water.
5. Rapidly freeze and thaw the suspension three times to release any intracellular virus particles.

Standard Tube Culture
Procedure
1. Appropriately label 2 tubes containing freshly confluent HEK or primary human fetal glial cells.
2. Remove the medium from the cells.
3. Add 2 ml of maintenance medium containing 2% FCS and antibiotics to each tube.
4. Add 0.2 to 0.5 ml of specimen to each tube.
5. Place the tubes in a roller drum and incubate at 35-37°C and 10-15 rph.
6. Examine the cells every 2-3 days for CPE.
7. Test the tubes for human O red blood cell agglutination 10-14 days after inoculation or when CPE is observed. Test negative cultures again on day 21 and on day 30.
8. Perform culture confirmation testing (see below) on any monolayer that exhibits CPE, and/or HA activity.

Interpretation of Results
Polyomavirus CPE is characterized by areas of large, irregularly shaped, vacuolated cells that are spread randomly over the cell sheet. As the CPE progresses, the cell monolayer degenerates leaving a significant amount of cellular debris in the culture medium. No report should be made based upon CPE alone.

REPORT: No report should be made based upon CPE alone.

Culture/HAd Confirmation
1. Remove all but a few drops of the cell culture medium.
2. Scrape the cells from the tube surface.
3. Resuspend the cells in the remaining fluid.
4. Using a pipette, place a small drop of the cell suspension in the wells of an acetone-cleaned slide. The remainder of the cell suspension should be retained for inoculation onto fresh cells if the staining is negative.
5. The slide should be processed as directed by the manufacturer of the fluorescein-labeled antibody to the virus or as described below.
6. Allow the slide to air dry, then fix it in acetone for 10-15 min. Fixed slides may be stored for up to a week at 2-8°C or for one year under desiccated conditions.
7. Place enough of the appropriate fluorescein-labelled monoclonal antibody on each well to cover the cell smear (15-30 μl).
8. Incubate the slide for 15-30 minutes at 35-37°C in a covered, humidified chamber. Do not allow the antibody to dry on the slide as this could cause nonspecific antibody binding.
9. Remove the excess antibody with a gentle stream of PBS. Do not direct the stream directly at the cell smear as this could

dislodge the cells.
10. Soak the slide in PBS for 5 minutes, shake off the excess fluid, and allow the slide to air dry.
11. Carefully add a small drop of FA mounting fluid to the center of each well.
12. Place a number 1 coverslip on the mounting fluid and carefully remove all the air bubbles.
13. Immediately examine the slide using a fluorescence microscope. For optimum clarity, use 200-300X magnification for screening and 400X for confirmation of cell morphology.

Interpretation of Results
Positive Test. The presence of JC or BK virus is indicated by apple-green fluorescence in the nucleus of the infected cells. Only individuals experienced in reading FA reactions should examine the cell smears because cell debris and cell clumps may exhibit a dull fluorescence which can be misinterpreted as a positive reaction.
REPORT: JC or BK virus isolated.

Negative Test. The absence of virus is indicated by the lack of specific fluorescence in the infected cells.
REPORT: JC or BK virus not isolated.

QC Procedures.

Subpassages of recent isolates or ATCC cultures should be inoculated into cell culture tubes with each batch of isolations. Uninfected cell cultures serve as negative controls. Infected and uninfected cell monolayers should be scraped and stained as described above. Positive controls must agglutinate human O red blood cells and exhibit a typical pattern of fluorescence. Negative controls must not exhibit specific fluorescent staining.

Preparation of Positive Control Cultures
1. Add 20 μl of a recent isolate (or an ATCC culture) to 2 ml of sterile GLB. Hold the diluted virus on ice until it is used to inoculate the monolayer.
2. Remove the growth medium from one 75 cm^2 flask of newly confluent HEK cells.
3. Add 2 ml of the diluted virus to the monolayer and allow the virus to adsorb for 1-2 hours at 35-37°C. Note: The flask should be rocked every 15 minutes to assure even virus distribution and to prevent monolayer desiccation.
6. Add 20 ml of maintenance medium containing 2% FCS and antibiotics.
7. Incubate the flask at 35-37°C until the CPE involves 80-100% of the monolayer and/or HA testing is positive to a titer of $\geq$1:128.
8. Scrape the cells from the flask into the cell culture medium.
9. Transfer the medium to a 50 ml polypropylene centrifuge tube.
10. Add GLB to the 50 ml mark.
11. Quickly freeze the medium at -70°C.
12. Thaw the medium quickly in a 37°C water bath.
13. Centrifuge the tube at 600-800 x g for 15 minutes to pellet the cell debris.
14. Dispense 1.0 ml of the diluted supernatant fluids into each of 50 freezing vials and freeze the vials at or below -70°C. Store the vials in liquid nitrogen for up to 10 years or at -70°C for up to 5 years.

Centrifugation Culture (Shell Vial) Method

In contrast with 2-7 week tube culture isolation times, centrifugation-enhanced (shell vial) methods can detect the presence of polyomaviruses as early as 16 hours after inoculation (11). However, staining after 48 hours provides a more sensitive assay. This procedure utilizes a monoclonal antibody to SV40 T antigen that cross-reacts with BK and JC virus T antigens (PAb416 from Oncogene Science, Inc., Mineola, NY). This cross-reactivity allows for the detection of JC and BK viruses before CPE or HA is evident. However, JC and BK infections cannot be differentiated by this procedure.

Although more rapid than conventional tube cultures, shell vial methods are less sensitive than conventional culture. Therefore, shell vial cultures should be backed up by at least one tube culture.

Procedure

1. Appropriately label 2 HEK vials and 1 cell HEK culture tube for inoculation.
2. Inoculate each vial with 0.2-0.5 ml of the specimen.
3. Centrifuge the vials at 700 x g for 45 minutes at room temperature.
4. Add 1 ml of MEM containing 5% FCS to each vial.
5. Incubate the vials at 35-37°C.
6. Fix and stain one vial after 48 hours and the other after 72 hours.

Staining of Coverslips

1. Carefully remove the culture fluid from the vials and wash the coverslips twice by adding 1-2 ml of PBS and soaking for 5 minutes.
2. Remove the PBS and add 2-3 ml of cold acetone to each vial. Fix the coverslips for 10 minutes at room temperature.
3. Remove the acetone and allow the coverslips to air dry.
4. Rinse the coverslips briefly with PBS. (The moisture trapped between the coverslip and the vial will keep the stain from wicking under the coverslip.)
5. Add enough monoclonal antibody to the vial to cover the coverslip (100-150 μl) and incubate for 30 minutes at 35-37°C in a humidified chamber to prevent the antibody from drying on the coverslip.
6. Wash the coverslips twice as previously described (step 1).
7. Place a small drop of mounting fluid on a slide. Remove the coverslips from the vial and lay them, cell side down, on the mounting fluid.
8. Immediately examine the coverslips at 100-300X using a fluorescence microscope.

Interpretation of Results

Coverslips that have individual or small groups of cells that exhibit brilliant apple-green nuclear fluorescence are positive for polyomavirus. Coverslips without the specific fluorescence described above are considered negative. Some specimens may trap the monoclonal antibodies between the cell monolayer and inoculum cells. Careful examination of the cell morphologies can distinguish this type of reaction from specific reactions. In addition, nonspecific staining can occur if the antibody reagents are allowed to dry on the coverslip. In this case, the entire cell sheet may stain apple-green.

Positive Test. The presence of polyomavirus is indicated by brilliant, apple-green nuclear fluorescence in infected cells.
REPORT: Polyomavirus virus isolated.

Negative Test. The absence of polyomavirus is indicated by the lack of specific fluorescence described above. The entire cell sheet should be stained red by the counterstain.
REPORT: Polyomavirus not isolated.

QC Procedures.

A 1:10 dilution (about 100 HA units) of a recent polyomavirus isolate or an ATCC culture (see Preparation of Positive Control Cultures, above) should be inoculated with each batch of shell vial cultures. Uninfected shell vial cultures serve as negative controls. Infected and uninfected shell vial cultures should be processed and stained as described for the patient cultures. Positive controls must exhibit the typical fluorescence patterns. Negative controls must not exhibit specific fluorescent staining.

REFERENCES

1. Shah KV. Polyomaviruses. In: Fields BN, Knipe DM, Chanock RM, Hirsch MS, Melnick JL, Monath TP, Roizman B, eds. *Virology*, Second Edition. New York: Raven Press, 1900:1609-1623.
2. Griffith JP, Griffith DL, Rayment I, Murakami WT, Caspar DLD. Inside polyomavirus at 25-Å resolution. *Nature* 1992;355:652-654.
3. Major EO, Amemiya K, Tornatore CS, Houff SA, Berger JR. Pathogenesis and molecular biology of progressive multifocal leukoencephalopathy, the JC virus-induced demyelinating disease of the human brain. *Clin Microbio Rev* 1992;5:49-73.

4. Weiner LP, Herndon RM, Narayan O, Johnson RT, Shah K, Rubinstein LJ, Preziosi TJ, Conley FK. Isolation of virus related to SV40 from patients with progressive multifocal leukoencephalopathy. *New Eng J Med* 1972;286:385-390.
5. Mahony J, Zapata M, Chernesky M. Characteristics of different solid-phase immunoassay formats for the measurement of BK virus immunoglobulin M in sera of patients on renal dialysis or with kidney allografts. *J Clin Microbiol* 1989;27:1626-1630.
6. Lehrich JR. JC, BK, and other polyomaviruses (Progressive multifocal leukoencephalopathy). In: Mandell GL, Douglas RG Jr, Bennett JE, eds. *Principles and practice of infectious diseases*, Third edition. New York: Churchill Linvingstone, 1990;1200-1203.
7. Walker DL. Progressive multifocal leukoencephalopathy. In: Binken PJ, Bruyn GW, Klawans HL, eds. *Handbook of clinical neurology*. Volume 47. Amsterdam: Elsevier/North-Holland 1985;503-524.
8. Telenti AA, Aksamit J, Proper J, Smith TF. Detection of JC virus DNA by polymerase chain reaction in patients with progressive multifocal leukoencephalopathy. *J Infect Dis* 1990;162:858-861.
9. Marshall WF, Telenti A, Proper J, Aksamit AJ, Smith TF. Survey of urine from transplant recipients for polyomaviruses JC and BK using the polymerase chain reaction. *Molecular and Cellular Probes* 1991;5:125-128.
10. Arthur RR, Dagostin S, Shah KV. Detection of BK virus and JC virus in urine and brain tissue by the polymerase chain reaction. *J Clin Microbiol* 1989;27:1174-1179.
11. Marshall WF, Telenti A, Aksamit AJ, Smith TF. Rapid detection of polyomavirus BK by a shell vial cell culture assay. *J Clin Microbiol* 1990;28:1613-1315.

CHAPTER 23

Human Parvovirus

INTRODUCTION

Human parvovirus (B19) is an autonomously replicating parvovirus and a member of the *Parvoviridae*. B19 is a small (20-25 nm) nonenveloped icosahedral virus that possesses a single-stranded DNA genome of approximately 5.5 kilobases. During virus assembly, an equal number of positive- and negative-sense DNAs are packaged into separate virions. The parvovirus genome codes for two structural proteins and at least one nonstructural protein. One restriction endonuclease study of 17 B19 isolates demonstrated 5 different genotypes among the isolates (1). However, only one serotype has been identified. B19 is very heat-stable and can survive for up to 12 hours at 60°C.

B19 infections occur throughout the world and are relatively common, especially among school-age children. Although B19 infections can occur in any age group, and any time of the year (2), B19 infections are most frequently observed in school outbreaks that begin in late winter and continue until school recesses for the summer. The level of human parvovirus activity in a community can vary from year to year with periods of increased activity being followed by several years of decreased activity (2). Nosocomial transmission of human parvovirus has been reported (3). In most seroepidemiologic studies, 30-60% of adults possess antibodies to B19 virus.

Erythema infectiosum (EI) or fifth disease is the most common manifestation of human parvovirus infection. EI is an acute, self-limiting febrile illness. Patients with EI are typically healthy except for an erythematous rash on the face (slapped-cheek appearance) and an erythematous reticulated rash on the trunk and extremities. As central clearing occurs, the rash may take on a lace-like appearance. The rash usually resolves within a week. However, the rash may recur intermittently for several weeks, especially after exercise, bathing, thermal changes or stress (4). Although most children have few symptoms other than rash, up to 25% of cases may have mild fever, headache, sore throat, and abdominal discomfort (5). Like most childhood exanthems, human parvovirus infections are somewhat more severe in adults who may experience a moderately severe, self-limited arthropathy of the wrists and knees following human parvovirus infection. Approximately 20% of cases may be asymptomatic (6).

Complications of B19 infections can be severe or life threatening, especially in patients with chronic hemolytic anemias (i.e., sickle cell disease, hemoglobin SC disease, hereditary spherocytosis, beta-thalassemia, and autoimmune hemolytic anemia). These patients require increased red blood cell production in order to maintain stable red cell indices. B19's replication in erythroid precursor cells and the subsequent disruption of RBC production may cause a transient aplastic crisis (TAC) in these patients. In immunodeficient patients, B19 infections can cause severe chronic anemia associated with red cell aplasia (2).

During pregnancy, B19 infections can (but usually do not) lead to fetal infection. Fetal infection can, in turn, cause severe anemia, congestive heart failure, generalized edema (fetal hydrops), and fetal death. The fetus is vulnerable to severe anemia because it depends upon a high rate of red cell production for growth and development. In addition, its immune system is immature and the fetus may be unable to mount an adequate antibody response. Fortunately, there is no evidence that B19 is teratogenic (2).

Diagnosis of recent B19 infection can be accomplished demonstrating human parvovirus-specific IgM in the patient serum (7). IgM is

> **AT A GLANCE...**
>
> # HUMAN PARVOVIRUS
>
> **Virus Detection Methods**
> Human Parvovirus cannot be isolated in normal cell culture system. Therefore, EIA and DNA probe techniques must be used to detect the presence of virus.
>
> **Specimen Source**
> Serum and respiratory secretions contain parvovirus DNA 5-10 days after infection (during the prodromal illness) however, parvovirus is usually not detectable at the onset of Erythema infectiosum (EI). In contrast, patients who present with transient aplastic crisis (TAC) are often positive for parvovirus DNA.
>
> **Parvovirus Serology**
>
> IgM Usually detectable at the onset of EI symptoms and 3 days after onset of TAC. IgM antibody titers begin to decline after 1-2 months and become undetectable within 2-3 months.
>
> IgG IgG is usually present by the seventh day of illness and persists for years.
>
> **Epidemiology**
> Human parvovirus occurs most often in the late winter and spring months and epidemics occur at 3-5 year intervals. Erythema infectiosum most commonly occurs in school-aged children aged 5-13 years. Human parvovirus has a worldwide distribution and 60-70% of adults have antibody to the virus. There is only one known serotype.
>
> **Transmission/Incubation Period**
> Human parvovirus is moderately contagious and transmission is presumed to occur through the inhalation of virus-laden droplets. Virus can also be transmitted parenterally by transfusion and vertically from mother to fetus. The incubation period is 4-14 days but can be as long as 20 days. Patients are infectious during the prodromal period (2-10 days after infection), well before Erythema infectiosum develops.
>
> **Inactivators and Disinfectants**
> Human parvovirus B19 is very hardy and can survive for up to 12 hours at 60°C. Parvoviruses are stable over a wide pH range (pH 3-9), and they are resistant to many disinfectants and organic solvents. B19 can be inactivated by betapropiolactone, hydroxylamine, and oxidizing agents. Effective disinfectants include concentrated phenols and 10% household bleach solutions.

usually detectable at the onset of EI symptoms or 3 days after the onset of TAC. IgM antibody titers usually begin to decline after 1-2 months and become undetectable within 3-4 months of infection. Human parvovirus-specific IgG is usually present by the seventh day of illness and antibody titers persist for years. B19 antibody may not be detectable in immunodeficient patients with chronic B19 infection. Documenting infection in these patients may require nucleic acid hybridization tests for B19 DNA or EIA tests for B19 antigen.

Although B19 has been grown in bone marrow explant cultures, the virus cannot be grown in standard cell culture systems or in animals. This inability to efficiently propagate the virus has hampered diagnostic assay development and has limited the availability of B19 antibody testing.

To address this problem, research groups and diagnostics companies have taken several approaches to B19 antigen production including the utilization of synthetic peptides; cloning, expression, and purification of B19 VP1 and VP2 proteins; and production of virion-like B19 particles from cloned DNA inserted into Chinese hamster ovary cells (8). It is anticipated that these approaches could improve the availability of diagnostic tests in the future.

REFERENCES

1. Morinet F, Tratschin JD, Perol Y, et al. Comparison of 17 isolates of the human parvovirus B19 by restriction enzyme analysis. *Arch Virol* 1986;90:165-172.
2. Centers for Disease Control. Risks associated with human parvovirus B19 infection. MMWR 1989;38:81-97.
3. Bell LM, Naides SJ, Stoffman P, Hodinka RL, Plotkin SA. Human parvovirus B19 infection among hospital staff members after contact with infected patients. *N Eng J Med* 1989;321:485-491.
4. Dolin R. Parvoviruses (erythema infectiosum, aplastic crisis). In: Mandell GL, Douglas RG Jr, Bennett JE, eds. *Principles and practice of infectious diseases,* Third Edition. New York: Churchill Livingstone, 1990;1231-1233.
5. Greenwald P, Bashe WJ. An epidemic of erythema infectiosum *Am J Dis Child* 1964;107:30-34.
6. Plummer FA, Hammond GW, Forward K, et al. An erythema infectiosum-like illness caused by human parvovirus infection. *N Eng J Med* 1985;313;74-79.
7. Anderson LJ, Tsou C, Parker RA, et al. Detection of antibodies and antigens of human parvovirus B19 by enzyme-linked immunosorbent assay. *J Clin Microbiol* 1986;24:522-526.
8. Kajigaya S, Shimada T, Fujita S, Young NS. A genetically engineered cell line that produces empty capsids of B19 (human) parvovirus. *Proc Natl Acad Sci USA* 1989;86:7601-7605.

CHAPTER 24

Respiratory Syncytial Virus

INTRODUCTION

Respiratory syncytial virus (RSV) is a negative-stranded RNA virus belonging to the *Pneumovirus* genus and the family Paramyxoviridae. In electron micrographs, RSV virions are pleomorphic with both filamentous and roughly spherical forms. Filamentous forms are 80-500 nm in diameter and up to 2,500 nm in length. Spherical forms possess a 13.5 nm helical nucleocapsid and a 120-300 nm diameter envelope with 12-15 nm glycoprotein spikes. Despite its classification as a paramyxovirus, RSV possesses neither hemagglutinin nor neuraminidase activity. The single-stranded, negative-sense, RNA genome codes for 7-8 structural and 2 nonstructural proteins. Recent information indicates that there are two major RSV strains that circulate simultaneously in the population (1). RSV is very labile and is rapidly inactivated by elevated temperatures, low pH, organic solvents, alcohols, and a number of detergents.

RSV is one of the major causes of acute respiratory illness in infants and children causing up to 40% of pneumonia cases and 90% of bronchiolitis cases during the first months of life (2,3). Nearly half of all infants acquire RSV infections within the first year of life and about 40% of these infections produce lower respiratory tract disease (4). By two years of age, nearly all children have experienced an episode of RSV and 50% of these children may experience two or more episodes. Naturally acquired immunity to RSV is incomplete and reinfections are common in all age groups. However, severe disease rarely occurs after the primary infection.

Adult RSV infections are usually mild or asymptomatic. However, serious upper and lower respiratory tract infections have been described in elderly and immunocompromised patients (5-14). The presence of significant lower respiratory tract disease in the elderly may indicate that RSV has a bimodal distribution - producing serious illness early and late in life (5-7). The incidence of RSV pneumonia in adults is probably under reported because RSV is usually not considered in the differential diagnosis of adults who present with fever and pulmonary infiltrates.

RSV is highly contagious and transmission occurs through contact with respiratory secretions (11,13,14). In respiratory secretions, RSV remains viable for up to 8 hours on countertops and other hard surfaces; for 1.5 hours on latex gloves; and at least 30 minutes on hospital gowns, paper tissues, and skin (12,15,16). Therefore, handling fomites contaminated with respiratory secretions and the subsequent autoinoculation of the nose and eyes provides an important mode of RSV transmission. Virus shedding from asymptomatically infected adults may also provide a significant source of infection, especially in hospitals where RSV is a frequent cause of nosocomial infections in pediatric wards and nurseries (12,15,16). During the winter, the likelihood that an infant or child will acquire an RSV infection increases with the duration of the hospital stay and the number of individuals housed in his/her room (1). RSV infections present a particular hazard for infants with bronchopulmonary dysplasia, congenital heart disease, premature infants, and immunocompromised children. In these children RSV infections are severe and life threatening. Nosocomial transmission can also produce significant disease in nursing home patients and elderly patients on adult medical wards (5-7,10-12).

Following infection, primary RSV replication in the nasopharynx causes fever, cough, malaise, nasal congestion, pharyngitis and/or otitis media. The

AT A GLANCE...

RESPIRATORY SYNCYTIAL VIRUS

Virus Detection Methods
Tube Cultures of HEp-2, A549, HeLa, and primary monkey kidney cells.
Centrifugation-enhanced (shell vial) cultures using HEp-2 cells.
Direct fluorescent antibody staining of respiratory cells.
Commercial EIA tests.

Specimen Sources
Specimens of choice are nasopharyngeal swabs, washes, and aspirates collected 1-3 days after the appearance of symptoms. Sputum is an appropriate specimen for adult RSV.

Time to Results
Standard Culture: Positive culture - 2 days to 2 weeks. Negative culture - 2 weeks.
Shell Vial Culture: Positive culture - 16-48 hours. Negative culture - 48 hours.
Direct Fluorescent Antibody Stain: 30 minutes.

Serologies

- IgM First appears 4-10 days after the onset of illness and antibody levels peak after 10-20 days. IgM is usually not detectable 2-10 weeks after infection. During reinfection, IgM appears earlier, reaches a higher titer, and persists longer than during primary infection.

- IgG First appears 5-10 days after the onset of illness and antibody levels peak after 3-4 weeks. IgG levels decline thereafter but are usually present 1 year after infection. During reinfection, IgG appears earlier, reaches a higher titer, and persists longer than during primary infection.

Epidemiology
RSV has a worldwide distribution and in the Western hemisphere, epidemics occur annually during the winter and early spring (December through April). Although the incidence of RSV may peak within 3-4 weeks, annual RSV epidemics generally last 3-5 months. During the first year of life, 50-68% of infants become infected and nearly all children have been infected by their second birthday. Reinfections occur frequently but disease is usually mild or inapparent. RSV is an important nosocomial agent, causing outbreaks in nurseries throughout the world.

Transmission/Incubation Period
RSV is very contagious and is transmitted through close contact with infected individuals and autoinoculation of the nose and eyes after handling fomites contaminated with respiratory secretions. RSV remains viable for up to 8 hours in respiratory secretions on countertops and other hard surfaces; for 1.5 hours on latex gloves; and at least 30 minutes on hospital gowns, paper tissues, and skin. Once transmitted the incubation period is 2-8 days and patients are infectious during the first 1-2 weeks of the illness.

Inactivators and Disinfectants
RSV is quickly inactivated by heat, alcohols and organic solvents. Effective disinfectants include 70% alcohol, 10% household bleach solutions, quaternary ammonium compounds, phenols, iodophor compounds, and glutaraldehyde compounds.

virus spreads to the lower respiratory tract after 1-3 days. Although the mechanism of spread is not well understood, it is thought to occur through aspiration of RSV-infected secretions or through cell-to-cell transfer. The subsequent viral replication causes necrosis of the broncheolar epithelium. The patient may experience hypoxemia, tachypnea, wheezing and general respiratory distress as the lumen of these small airways becomes obstructed by mucus and sloughed respiratory epithelium.

In 1986, the Food and Drug Administration approved the use of aerosolized ribavirin (1-β-D-ribafuranosyl-1,2,4-triazole-3-carboxamide) for the treatment of RSV infection (17). Ribavirin is a synthetic nucleoside that appears to interfere with the expression of mRNA. For hospitalized infants with RSV bronchiolitis or pneumonia, ribavirin is delivered as a small particle aerosol via an oxygen tent or a ventilator. Ribavirin is usually administered 12 or more hours per day for 3-5 days.

Ribavirin therapy is most effective when given early in the disease (17). Therefore, rapid diagnosis of RSV infections is imperative for efficient patient management. Rapid detection can also influence decisions involving patient cohorting and RSV surveillance of medical personnel during nosocomial RSV outbreaks. A preliminary RSV diagnosis can be based upon clinical presentation and local epidemiology, but a definitive diagnosis requires laboratory testing. Several rapid detection methods are commercially available including direct fluorescent antibody staining of exfoliated respiratory epithelial cells and enzyme immunoassays (3,18-25). Cell culture methods are often used to back up rapid methods because these methods do not have perfect sensitivity or specificity.

Several tests have been developed for the serological diagnosis of RSV infections including complement fixation (CF), indirect immunofluorescence, radioimmunoprecipitation, western blot, virus neutralization, and fusion (syncytia)-inhibition assays. However, serum antibody determinations have little practical value because serum antibody levels have not been predictive of the risk of infection, severity of illness, or recovery in children and adults (4,26,27). In addition, many infected adults and children do not produce a significant rise in serum antibody levels (4). However, these tests do provide the means for studying the prevalence of RSV in the population and the role of antibody in pathogenesis.

VIRUS ISOLATION

Introduction

RSV grows in a number of cell lines including HEp-2, HeLa, HFF, A549, and RMK cells. However, the syncytia that characterizes RSV growth in HEp-2 cells is rarely present in other cell lines. HEp-2 cells are used most often for RSV isolation and the best virus recovery is obtained when using actively growing cultures. Cell overgrowth can also obscure the syncytia. Thus, HEp-2 cells should be seeded lightly (0.5 - 1.5 x 10^5 cells per tube) and tube cultures should be inoculated with patient specimens when 10-20 "islands" of cells are present (Fig. 1).

The use of refeed medium containing 2% FCS is recommended to allow more time before the cultures become confluent and start piling up on one another. Under these conditions, CPE may be visible as early as 2 days after inoculation. However, most isolates require 4-5 days to produce visible CPE. Most laboratories inoculate HEp-2, A549, and RMK tubes for RSV isolations. The reason for using RMK and A549 tubes is to look for dual infections with other respiratory viruses. However, RSV will occasionally grow in these cells and not in HEp-2 cells.

Virus isolation provides the best method for the detection of RSV in clinical specimens. However, traditional cell culture methods have limited clinical utility because physicians may not receive isolation results for two weeks. To decrease isolation times, many laboratories have begun to use the centrifugation-enhanced (shell vial) culture methods originally described by Gleaves, et al. (28) for CMV. Shell vial methods can accommodate the same clinical specimens as traditional tube cultures and can reduce isolation times from 7-10 days to 1-2 days (18, 20-22).

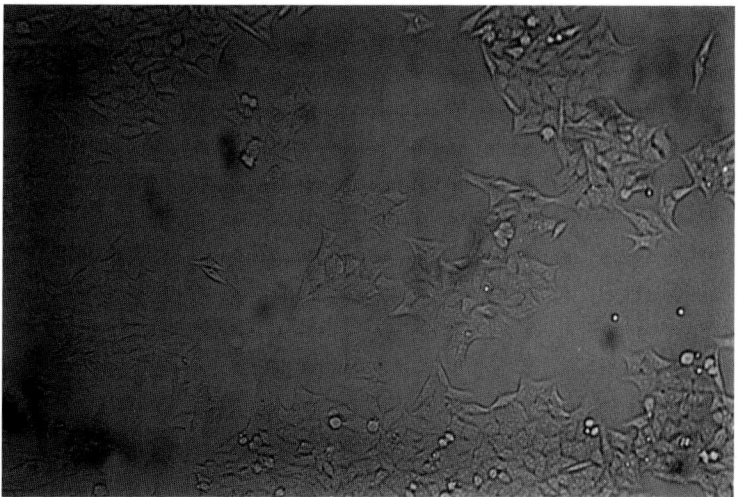

FIG. 1. Proper cell density for isolation of RSV in HEp-2 cells.

Specimen Collection and Storage.

Specimen collection and handling have a greater effect on RSV isolation rates than any other parameter. The ability to isolate RSV from clinical specimens depends upon the type of specimen, time of specimen collection, how the specimen was handled, and the time interval between specimen collection and inoculation into cell culture.

The predictive value of RSV isolation is highest when respiratory specimens are collected 1-3 days after the onset of symptoms. Although RSV has been detected in respiratory specimens for a more than a week after the onset of symptoms, the probability of successful isolation decreases rapidly when specimens are taken later in the disease.

RSV can be isolated from nasopharyngeal (NP) swabs, throat swabs, nasal washes and aspirates, or from adult sputa. In many laboratories, nasal washes or combined NP and throat swabs are the preferred specimens. Although more virus can be recovered from nasal washings, nasal washes are not universally accepted because they cause more discomfort and they can pose an aspiration risk in infants and children (29). For most laboratories, swabs are the easiest specimens to obtain. If swab specimens are used, both NP and throat swabs should be obtained and sent to the laboratory in the same viral transport tube.

RSV is very thermolabile and specimens should be inoculated onto cell cultures as soon as possible after collection, preferably at bedside. If transport is required, specimens should be sent to the virology laboratory on wet ice. RSV loses 90% of its infectivity after 5 minutes at 55°C, 24 hours at 37°C, 48 hours at 25°C, and 4 days at 4°C (1). If necessary, specimens can be stored at 4°C for up to 48 hours. If specimens cannot be inoculated onto cell cultures within that time, they should be quickly frozen at -70°C. **DO NOT FREEZE SPECIMENS AT -20°C.** However, freezing will cause some loss of infectivity.

Specimen Preparation:
Swabs
1. Vortex the viral transport medium, the swabs, and several glass beads for 20-30 seconds to release any bound cells or virus.
2. Remove the swabs from the transport medium and firmly roll them against the inside of the tube to remove as much fluid as possible.

Wash and Aspirate Specimens
1. Nasal washes should be of minimal volume (< 5 ml). Combine the wash with an equal volume of viral transport medium.
2. Nasal aspirates should be used undiluted whenever possible. However, if the specimen volume is not sufficient, the aspirates can be

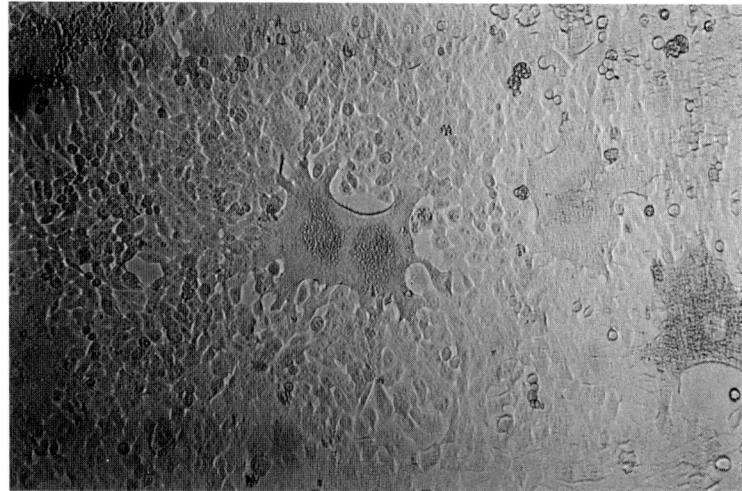

FIG. 2 Syncytia formation in HEp-2 cells four days after infection with respiratory syncytial virus.

diluted in viral transport medium.
3. Vortex the specimens with glass beads to release any cell associated virus.
4. Centrifuge suspension at 600 x g for 5 minutes to remove mucus and cell debris.
5. Carefully remove the mucus and use the remaining specimen to inoculate cell cultures.

Standard Tube Culture
Procedure
1. Appropriately label two HEp-2 culture tubes and one RMK tube.
2. Add 0.2-0.5 ml of specimen to each tube.
3. Incubate the tubes at 35-37°C in a roller drum (10-15 rph) for 10 days.
4. Examine the cultures at least every other day for CPE (Fig. 2).
5. The culture medium should be changed after 5 days or whenever it becomes acidic. Toxic cultures should be transferred to new culture tubes.
6. Perform culture confirmation testing (see below) on any monolayer that exhibits CPE. Some laboratories perform blind staining on one or both HEp-2 monolayers before reporting the culture as negative.

Interpretation of Results
Positive Test. Presumptive identification of RSV is based upon the presence of large irregularly-shaped syncytia that appear as large multinucleated cells with indistinct borders (Fig. 2). Guinea pig red blood cells do not adsorb to RSV-infected cells (Chapter 17). Positive cultures should be scraped and stained with monoclonal antibodies to respiratory syncytial virus.
REPORT: No report should be made based solely upon the presence of CPE.

Negative Test. The lack of characteristic CPE after 10-14 days indicates the absence of viable RSV. Some laboratories scrape and stain all cultures for RSV before reporting a negative isolation result.
REPORT: Respiratory syncytial virus not isolated.

Culture Confirmation Procedure
1. Once CPE develops, remove all but a few drops of culture medium from the tube.
2. Scrape the cells from the surface of the tube and suspend the cells in the remaining fluid.
3. Using a pipette, place a drop of the cell suspension onto an acetone-cleaned slide. Retain the rest of the cell suspension for passage into new culture tubes if necessary.

4. Allow the slide to air dry.
5. The slide should be processed as directed by the manufacturer of the RSV monoclonal antibody or as described below.
6. Fix the slide in cold acetone for 10-15 minutes and allow the slide to air dry. Slides should be stained immediately after fixation. Alternatively, fixed slides may be stored for up to 1 week at 2-8°C or one year at -20°C or below. Slides should be stored under desiccated conditions in order to minimize antigen degradation and background staining.
7. Add enough fluorescein-labelled RSV antibody to cover the cell smear (15-30 µl).
8. Incubate the slide for 30 minutes at 35-37°C in a covered, humidified chamber. Do not allow the antibody to dry as this could cause nonspecific antibody binding.
9. Remove the excess antibody with a gentle stream of PBS. Do not direct the stream directly at the cell smear as this could dislodge the cells.
10. Soak the slide in PBS for 5 minutes, shake off the excess fluid, and allow the slide to air dry.
11. Carefully add a small drop of FA mounting fluid to the center of the smear.
12. Place a number 1 coverslip on the mounting fluid and carefully remove all the air bubbles.
13. Immediately examine the slide using a fluorescence microscope. For optimum clarity, use 200-300X magnification for screening and 400X for confirmation of cell morphology.

Interpretation of Results
Positive Test. The presence of RSV is indicated by the presence of characteristic CPE in the tube cultures and an intense apple-green fluorescence in the cytoplasm of the infected cells (see Color Plate 2D following Chapter 7). Only experienced individuals should interpret FA patterns because cell debris and cell clumps may exhibit a dull fluorescence which can be interpreted as a positive reaction.
REPORT: Respiratory syncytial virus isolated.

Negative Test. The absence of RSV is indicated by the uniform red coloration of the cells and the absence of specific staining.
REPORT: Respiratory syncytial virus not isolated.

QC Procedures.
Subpassages of RSV isolates or ATCC cultures should be inoculated with each batch of RSV isolations. Uninfected cell cultures can serve as negative controls. Infected and uninfected cell monolayers should be scraped and stained as described above. Positive controls must produce syncytia in HEp-2 cells and the resulting cell smears must exhibit a typical staining pattern. Negative controls must stain with the red counterstain and should not exhibit any CPE or specific fluorescent staining.

Preparation of Positive Control Cultures
1. Prepare a 1:1000 dilution of a vigorously glowing RSV isolate or an ATCC culture in GLB or MEM containing 10% FCS. Hold the dilution on ice until ready for inoculation.
2. Remove the growth medium from one 75 cm^2 flask containing an 70-80% confluent HEp-2 cell monolayer.
3. Add 2 ml of the inoculum to the monolayer and allow the virus to adsorb for 1-2 hours at 35-37°C. The flask should be rocked every 15 minutes to assure even distribution of the inoculum and to prevent cell desiccation.
4. Add 10 ml of MEM containing 2% FCS.
5. Incubate the flask at 35-37°C until the CPE is present in 80-100% of the monolayer (usually 3-7 days). Replace the medium after 3 days or if the medium becomes acidic.
6. Scrape the cells into the culture medium.
7. Transfer the cell suspension to a 50 ml polypropylene centrifuge tube and freeze the suspension quickly at -70°C.
8. Thaw the suspension quickly in a 37°C water bath.
9. Centrifuge the cells at 600 x g for 15 minutes to pellet the cell debris.
10. Transfer the supernatant fluids to a sterile 50 ml centrifuge tube and add MEM-20% FCS to the 50 ml mark.

11. Dispense 1 ml of the cell suspension into each of 50 freezing vials and quickly freeze the cells at -70°C or below. Store the vials 2 years at -70°C or 5 years in liquid nitrogen.

Centrifugation Enhanced (Shell Vial) Method

Rapid detection of respiratory syncytial virus is essential for the timely institution of antiviral therapeutic measures and to monitor nosocomial virus transmission. The centrifugation-enhanced (shell vial) method described by Gleaves, et al. (28), has been used to isolate RSV as early as 16 hours after inoculation (18, 20-22). In this procedure, the cell line is not as critical as it is in tube culture isolations because the coverslips are stained before the appearance of CPE. HEp-2, MRC-5, HeLa, HL, A549, and RMK cells have been used in shell vial cultures. Although centrifugation enhanced methods are rapid and specific they are not perfect and tube cultures should also be inoculated to recover slow-growing RSV isolates and for the recovery of other viruses.

Procedure
1. Appropriately label two HEp-2 shell vial cultures for each specimen.
2. Aseptically remove the culture medium from the vials.
3. Inoculate each vial with 0.2-0.5 ml of the specimen.
4. Centrifuge the vials at 700 x g for 40 minutes at room temperature.
5. Allow the virus to adsorb for 1 hour at 35-37°C.
6. Add 1 ml of MEM containing 2% FCS and antibiotics to each vial.
7. Incubate the vials at 35-37°C. Stain one vial after 16-24 hours and the other vial after 48 hours.

Staining of Coverslips

Identification of RSV in shell vial systems is accomplished by staining the coverslips with monoclonal antibodies to RSV. Most specimens will produce 1-10 small foci after 16 hours of culture. Fixation and staining should be done as directed by the manufacturer of the monoclonal antibody or as described below.

1. Carefully remove the culture fluids from the vial and wash the coverslip twice by adding 1-2 ml of PBS and soaking for 5 minutes.
2. Remove the PBS and add 2-3 ml of acetone to each vial. Fix the coverslips for 10 minutes at room temperature.
3. Remove the acetone and allow the coverslip to air dry.
4. Rinse the coverslips briefly with PBS. The moisture trapped between the coverslip and the shell vial will keep the stain from wicking under the coverslip.
5. Add enough fluorescein-labeled monoclonal antibody to the vial to cover the coverslip (100-150 μl) and incubate 30 minutes at 35-37°C. The shell vials should be covered or incubated in a humidified chamber to prevent the antibody from drying.
6. Wash the coverslips twice as previously described (step 1) and remove the PBS.
7. Place a small drop of mounting fluid on a slide.
8. Remove the coverslips from the vial and lay them cell side down on the mounting fluid.
9. Immediately examine the coverslips at 100-300X using a fluorescence microscope.

Interpretation of Results

Coverslips that have individual cells or small syncytia that exhibit a brilliant, apple-green cytoplasmic fluorescence (Fig. 3) are positive for RSV. Coverslips without specific staining are considered negative. Some specimens may trap the monoclonal antibody between the cell monolayer and inoculum cells. Careful examination of cell morphologies can distinguish this type of reaction from RSV-specific reactions. In addition, nonspecific staining can occur if the monoclonal reagents are allowed to dry on the coverslip. In this case, the entire cell sheet may stain apple-green.

Positive Test. The presence of RSV is indicated by the presence of brilliant, apple-green fluorescence in the cytoplasm of infected cells.

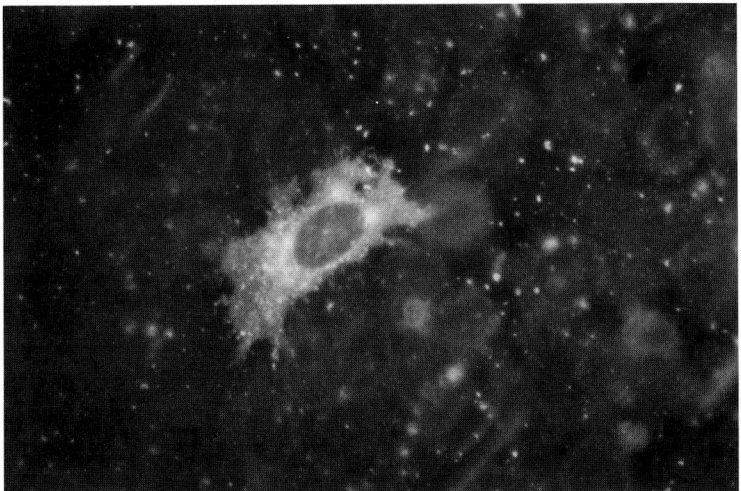

FIG. 3. Respiratory syncytial virus infected A549 shell vial culture stained with monoclonal antibodies 24 hours after infection.

REPORT: Respiratory syncytial virus isolated.

Negative Test. The absence of viable RSV is indicated by the lack of specific fluorescence described above. The entire cell sheet should be stained red by the counterstain.
REPORT: Respiratory syncytial virus not isolated.

QC Procedures.
A 1:10 dilution of the positive control cultures (described above) should be inoculated with each batch of RSV shell vial cultures. Uninfected shell vial cultures serve as negative controls. Infected and uninfected shell vial cultures should be processed and stained as described for the patient cultures. Positive controls must exhibit the typical fluorescence patterns described above. Negative controls must not exhibit specific fluorescent staining.

DIRECT SMEARS

Introduction
Although centrifugation-enhanced (shell vial) methods can reduce the RSV isolation times to as little as 16 hours, this time interval is often too lengthy when antiviral therapy is contemplated. In an effort to reduce the detection time to a clinically useful interval, a number of investigators have shown that direct staining of acetone-fixed nasopharyngeal smears provides a rapid and relatively sensitive method for the detection of RSV (20-22,25). In this procedure, the nasopharynx is vigorously swabbed and a smear is made on an acetone-cleaned microscope slide. The slide is fixed in acetone and stained with monoclonal antibodies to RSV.

DFA methods are more sensitive than culture and they can detect RSV infections even after virus isolation is no longer possible (30). DFA methods are especially valuable in remote laboratories where cold-chain specimen transport is not easily accomplished. However, DFA methods are not without problems. In addition, excess mucus can interfere with antibody binding. Mucus can cause false positive reactions by nonspecifically trapping the fluorescent antibody reagents and false negative reactions by causing the cells to clump thereby, masking the RSV antigens. However, skilled technicians can usually distinguish these reactions from RSV-specific reactions.

Exclusive use of DFA staining can cause false negative reactions when specimens contain viruses other than RSV. In addition, DFA staining will not detect a new infectious agent that might appear in the community. For these reasons, many laboratories employ a combination of DFA and culture methods for the diagnosis of RSV.

Specimen Collection

The success of the direct fluorescent antibody (DFA) procedure depends upon the preparation of a well made cell smear. Smears that are too thick or "lumpy" can cause nonspecific trapping of the DFA reagents. Smears that are grossly contaminated with red blood cells can cause interpretation difficulties due to red cell autofluorescence. The principal reasons why direct smears fail to detect RSV infections are (a) inadequate specimen collection and (b) the presence of too few cells on the slide. Specimen adequacy and its definition are significant sources of laboratory-to-laboratory variation and an area of increasing regulatory concern. Many laboratories require at least 50 cells per smear while others require 1-2 cells/high power field. Whatever the number, too many cells can cause nonspecific trapping of antibody and with too few cells the laboratory may not be able to find an infected cell.

Specimen Processing

Nasal Aspirates, Nasal Washes, and Throat Washes
1. Centrifuge the specimen at 1000 - 1500 x g to pellet the cells.
2. Remove all but 100-200 μl of the fluid.
3. Suspend the cell pellet in the remaining fluid.
4. Place one drop of the cell suspension onto an acetone cleaned slide.
5. Allow the smears to air dry at room temperature. Slides may be stored for up to 48 hours at 2-8°C.

Swabs in Viral Transport Medium
1. Vortex the specimen vigorously to release as many cells from the swab as possible.
2. Remove the swab from the viral transport medium and firmly roll it against the inside of the tube to remove as much fluid as possible.
3. Centrifuge the specimen at 1000 - 1500 x g to pellet the cells.
4. Remove all but 100-200 μl of the viral transport medium.
5. Suspend the cell pellet in the remaining fluid.
6. Place one drop of the cell suspension on an acetone cleaned slide.
7. Allow the smears to air dry at room temperature. Slides may be stored for up to 48 hours at 2-8°C.

Test Procedure

For optimum performance, slides should be fixed and stained within 1-2 hours of specimen collection. Alternatively, unfixed slides may be stored for up to 48 hours at 2-8°C. Fixed slides may be stored at 2-8°C for 1 week or frozen at -20°C for up to 1 year. Storing slides under desiccated conditions will decrease the background staining and minimize antigen degradation.

1. Upon arrival in the laboratory, examine the smear at 100-300X magnification. Ideally, a minimum of 50 cells should be visible before the specimen is considered adequate for further processing.
2. The slide should be processed as directed by the manufacturer of the fluorescein-labeled RSV antibody or as described below.
3. Fix the cells by immersing the slides in cold acetone for 10 minutes. Allow the slides to air dry.
4. Place enough of the fluorescein-labeled monoclonal antibody on each well to cover the cell smear (15-30 μl).
5. Incubate the slide for 30 minutes at 35-37°C in a covered, humidified chamber. Do not allow the antibody to dry. Drying can cause nonspecific antibody binding.
6. Remove the excess antibody with a gentle stream of PBS. Do not direct the stream directly at the cell smear as this could dislodge the cells.
7. Soak the slide in PBS for 5 minutes, shake off the excess fluid, and allow the slide to air dry.
8. Carefully add a small drop of FA mounting fluid to the center of the well.
9. Place a number 1 coverslip on the mounting fluid and carefully remove all the air bubbles.
10. Immediately examine the slide using a

 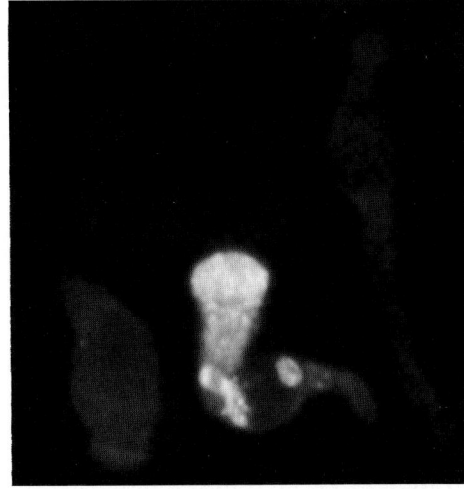

FIG. 4. Respiratory syncytial virus infected respiratory epithelial cell (left) and goblet cell (right) from a nasopharyngeal smear stained with monoclonal antibodies.

fluorescence microscope. For optimum clarity, use 200-300X magnification for screening and 400X for confirmation of cell morphology.

Interpretation of Results

Positive Test. The presence of respiratory syncytial virus is indicated by brilliant apple-green fluorescence in the cytoplasm of exfoliated respiratory epithelial cells (Fig. 4; see Color Plates 3A and 3B following Chapter 7). Only intact cells should be examined because cell debris and clumps of normal cells may exhibit a dull, yellow-green fluorescence which could be misinterpreted as a positive reaction.
REPORT: Respiratory syncytial virus detected.

Negative Test. The absence of respiratory syncytial virus is indicated by the lack of specific staining described above.
REPORT: Respiratory syncytial virus not detected.

Inconclusive Test. Specimens with fewer than 25 cells (or 1-2 cells/HPF) may give inconclusive results.
REPORT: Unacceptable specimen - too few cells.

QC Procedures.

Positive and negative controls should be stained at least once each day to assure that the antibody reagent is performing properly. Positive cells should stain intensely as described above. Negative cells should stain red and should not exhibit specific apple-green fluorescence. Because the same antibody reagent is used for confirmation of tube cultures, shell vial cultures, and direct smears, culture confirmation testing can be used to demonstrate that the reagent is working properly. In laboratories that do not perform RSV isolations or laboratories where no isolations are ongoing, prepared slides should be stained whenever direct smears are stained. Control slides can be purchased commercially from a number of vendors or they can be prepared as described below.

Preparation of Control Slides

1. Add 20 μl of the positive control culture (above) to 2 ml of sterile GLB. Hold the diluted virus on ice until used to inoculate the monolayer.
2. Remove the growth medium from one 75 cm^2 flask of 80-90% confluent HEp-2 cells.
3. Rinse the monolayer twice with 10 ml of serum-free MEM.
4. Add 2 ml of the diluted virus to the flask and

allow the virus to adsorb for 1-2 hours at 35-37°C. Note: The flask should be rocked every 15 minutes to assure even distribution of the virus and to prevent monolayer desiccation.
5. Add 10 ml of MEM containing 2% FCS and antibiotics.
6. Incubate the flask at 35-37°C until the CPE involves 50-60% of the monolayer (2-5 days).
7. Trypsinize the cells and remove them from the flask.
8. Centrifuge the cells at 250 x g for 5 minutes.
9. Resuspend the cell pellet in 5 ml of PBS. Perform a cell count and adjust the cell concentration to 2-5 x 10^6 cells/ml.
10. Dispense 3-10 μl of the cell suspension onto each slide and allow the suspension to air dry.
11. Fix the slides in acetone at room temperature for 10 minutes and allow the slides to air dry.
12. Store the slides at -20°C or below for up to one year. Slides should be stored under desiccated conditions to minimize antigen degradation and background fluorescence.

NOTE: Slides prepared in this manner will contain both positive and negative cells. Therefore, a single slide can be used for QC purposes. When slides are made from flasks with 100% CPE, two slides must be used - one containing a positive cell smear and one containing an uninfected cell smear.

REFERENCES

1. Hall CB. Respiratory syncytial virus. In: Mandell GL, Douglas RG Jr, Bennett JE, eds. *Principles and practice of infectious diseases*, Third Edition. New York: Churchill Livingstone 1990;1265-1279.
2. Toms GL. Respiratory syncytial virus: Virology, diagnosis, and vaccination. *Lung* 1990;Suppl:388-395.
3. Treuhaft MW, Soukup JM, Sullivan BJ. Practical recommendations for the detection of pediatric respiratory syncytial virus infections. *J Clin Microbiol* 1985;22:270-273.
4. Parrott RH, Kim HW, Arrobio JO, et al. Epidemiology of respiratory syncytial virus infection in Washington, D.C. II. Infection and disease with respect to age, immunologic status, race, and sex. *Am J Epidemiol* 1973;98:289-300.
5. Hart RJC. An outbreak of respiratory syncytial virus infection in an old people's home. *J Infect* 1984;8:259-261.
6. Sorvillo FJ, Huie SF, Strassburg MA, Butsumyo A, Shanders WX, Fannin SL. An outbreak of respiratory syncytial virus pneumonia in a nursing home for the elderly. *J Infect* 1984;9:252-256.
7. Mandel SK, Joglekav VM, Kahn AS. An outbreak of respiratory syncytial virus infection in a continuing-care geriatric ward. *Age Aging* 1985;14:184-186.
8. Englund JA, Sullivan CJ, Jordon MC, Dehner LP, Vercellotti GM, Balfour HH. Respiratory syncytial virus infection in immunocompromised adults. *Ann Intern Med* 1988;109:203-208.
9. Watkins RPF, Grover S, Eastwood JB, Crompton MR. Respiratory syncytial virus pneumonia. *J Clin Pathol* 1989;42:1224.
10. Takimoto CH, Cram DL, Root RK. Respiratory syncytial virus infections on an adult medical ward. *Arch Intern Med* 1991;151:706-708.
11. Agius G, Dindinaud G, Biggar RJ, Peyre R, Vaillant V, Ranger S, Poupet JY, Cisse MF, Castets M. An epidemic of respiratory syncytial virus in elderly people: Clinical and serological findings. *J Med Virol* 1990;30:117-127.
12. Englund JA, Anderson LJ, Rhame FS. Nosocomial transmission of respiratory syncytial virus in immunocompromised adults. *J Clin Micrbiol* 1991;29:115-119.
13. Vikerfors T, Grandien M, Olcen P. Respiratory syncytial virus infections in adults. *Am Rev Respir Dis* 1987;136:561-564.
14. Englund JA, Sullivan CJ, Jordan MC, Dehner LP, Vercellotti GM, Balfour HH. Respiratory syncytial virus infection in immunocompromised adults. *Annals Internal Med* 1988;109:203-208.
15. Agah RA, Cherry JD, Garakian AJ. Respiratory syncytial virus (RSV) infection rate in personnel caring for children with RSV infections. *AJDC* 1987;141:695-697.
16. Issacs D, Dickson H, O'Callaghan C, Sheaves R, Winter A, Moxon ER. Handwashing and cohorting in prevention of hospital acquired infections with respiratory syncytial virus. *Arch Dis Child* 1991;66:227-231.
17. Groothuis JR, Woodin KA, Katz R, Robertson

AD, McBride JT, Hall CB, McWilliams BC, Lauer BA. Early ribavirin treatment of respiratory syncytial viral infection in high-risk children. *J Pediatr* 1990;117:792-798.
18. Arens MQ, Swierkosz EM, Schmidt RR, Armstrong T, Rivetna KA. Enhanced isolation of respiratory syncytial virus in cell culture. *J Clin Microbiol* 1986;23:800-802.
19. Bromberg K, Tannis G, Daidone B. Early use of indirect immunofluorescence for the detection of respiratory syncytial virus in HEp-2 cell culture. *Am J Clin Pathol* 1991;96:127-129.
20. Halstead DC, Todd S, Fritch G. Evaluation of five methods for respiratory syncytial virus detection. *J Clin Microbiol* 1990;28:1021-1025.
21. Johnston SLG, Siegel CS. Evaluation of direct immunofluorescence, enzyme immunoassay, centrifugation culture, and conventional culture for the detection of respiratory syncytial virus. *J Clin Microbiol* 1990;28:2394-2397.
22. Matthey S, Nicholson D, Ruhs S, Alden B, Knock M, Schultz K, Schmuecker A. Rapid detection of respiratory viruses by shell vial culture and direct staining by using pooled and individual monoclonal antibodies. *J Clin Microbiol* 1992;30:540-544.
23. Sturgill MA, Hughes JH. Use of high-speed rolling to detect respiratory syncytial virus in cell culture. *J Clin Microbiol* 1989;27:577-579.
24. Swierkosz EM, Flanders R, Melvin L, Miller JD, Kline MW. Evaluation of the Abbott TESTPACK RSV enzyme immunoassay for detection of respiratory syncytial virus in nasopharyngeal swab specimens. *J Clin Microbiol* 1989;27:1151-1154.
25. Waner JL, Whitehurst NJ, Todd SJ, Shalaby H, Wall LV. Comparison of Directigen RSV with viral isolation and direct immunofluorescence for the identification of respiratory syncytial virus. *J Clin Microbiol* 1990;28:480-483.
26. Bruhn FW, Yeager AS. Respiratory syncytial virus in early infancy. *Am J Dis Child* 1977;131:145-148.
27. Toms GL, Scott R. Respiratory syncytial virus and the infant immune response. *Arch Dis Child* 1987;62:544-546.
28. Gleaves CA, Smith TF, Shuster EA, Pearson GR. Rapid detection of cytomegalovirus in MRC-5 cells inoculated with urine specimens by using low-speed centrifugation and monoclonal antibody to an early antigen. *J Clin Microbiol* 1984;19:917-919.
29. Frayha H, Castriciano S, Mahony J, Chernesky M. Nasopharyngeal swabs and nasopharyngeal aspirates equally effective for the diagnosis of viral respiratory disease in hospitalized children. *J Clin Microbiol* 1989;27:1387-1389.
30. Anderson LJ. Paramyxoviridae: Respiratory syncytial virus. In: Balows A, Hausler WJ Jr, Lennette EH, eds. *Laboratory diagnosis of infectious diseases - principles and practice*, Volume II. New York: Springer-Verlag, 1988;540-570.

CHAPTER 25

Human Retroviruses

INTRODUCTION

Retroviruses are unique in that they contain an RNA-dependent DNA polymerase (reverse transcriptase) and replicate their RNA genome by creating double-stranded DNA intermediate. These double-stranded DNA intermediates (provirus) can integrate into the host chromosome, causing transformations and other genetic changes in the host cell. Only four human retroviruses have been isolated to date: human T-cell lymphotropic virus type I (HTLV-I), HTLV-II, human immunodeficiency virus type 1 (HIV-1, originally HTLV-III/LAV) and HIV-2. All of these viruses are members of the *Retroviridae* and as such they are enveloped, negative-stranded RNA viruses. HTLV-I and HTLV-2 are type C oncornaviruses and their 9 kilobase RNA genome codes for approximately 14 polypeptides. HTLV virions are 100 nm in diameter and appear spherical in electron micrographs. HIV-1 and HIV-2 are members of the lentivirus subfamily. In electron micrographs, HIV virions are roughly spherical and approximately 130 nm in diameter. Mature particles have a characteristic electron dense bar-shaped nucleoid structure. Human retroviruses are relatively unstable and are quickly inactivated by alcohols, quaternary ammonium compounds, detergents, and 0.5% sodium hypochlorite solutions.

HTLV-I

HTLV-1 was first isolated from the T-lymphocytes of a 28 year old male with a cutaneous lymphoma in 1980 (1). Since that time, much has been learned about the biology, genetic replication, and geographic distribution of the virus. Epidemiologic data have demonstrated a clear association of HTLV-I and adult T-cell leukemia (ATL). HTLV-I may also play an indirect role in the pathogenesis of some cases of B-cell chronic lymphocytic leukemia (CLL). Like ATL, B-cell CLL has been observed in HTLV-I endemic areas and these patients have high HTLV-I antibody titers. However, the when the leukemic cells from CLL patients were examined, no HTLV-I provirus was detected. Therefore, the link between HTLV-I and B-cell CLL is still considered tenuous.

HTLV-I has also been associated with chronic neurologic disease. In 1985, a group of West Indian patients with tropical spastic paraparesis (TSP) were found to be seropositive for HTLV-I (2). This condition (also called HAM or HTLV-associated myelopathy) has been described in all areas of the world known to be endemic for HTLV-I. The pathogenesis of this condition is especially intriguing because HTLV-I antibodies have been found in the cerebral spinal fluid of HAM patients. In contrast with ATL where patients acquire the disease 20-30 years after HTLV-I infection, HAM developed in some patients only a few years after being transfused with HTLV-I infected blood (3).

HTLV-I is primarily endemic in southwestern Japan and the Caribbean, however, parts of sub-Saharan Africa and Central and South America also have significant seroprevalence. While the overall HTLV-I seroprevalence in the United States is low, increased antibody seropositivity has been observed in the Southwestern United States; in areas possessing large Caribbean emigrant populations; and increasingly, in intravenous drug users and sex partners of male IV drug users. In contrast with HIV-1, HTLV-1 appears to have very little genetic variability. Isolates within Japan show 97-99% homology at the sequence level and isolates obtained from the Caribbean and Africa show 96-99% homology with the Japanese isolates.

HTLV-I is poorly contagious and the virus is not readily transmitted by cell-free body fluids.

> **AT A GLANCE...**
>
> ## HTLV-I
>
> **Virus Detection Methods**
> HTLV-I can be isolated by co-cultivation of patient lymphocytes with phytohemagglutinin-stimulated peripheral blood or cord blood lymphocytes.
>
> **Specimen Source**
> HTLV-I can be isolated from peripheral blood lymphocytes and from cerebral spinal fluid of HTLV-I associated myelopathy patients. The virus appears to be highly cell associated and cannot be readily isolated from cell-free specimens.
>
> **Time to Result**
> Positive culture - 2-4 weeks.
> Negative Culture - 4 weeks.
>
> **HTLV-I Serology**
> Most HTLV-I infections are diagnosed serologically. However, current serologic methods cannot adequately distinguish HTLV-I from HTLV-II infections because considerable antigenic similarities exist, particularly within the p24 gag protein. HTLV-I antibodies have been found in the cerebrospinal fluid of patients with HTLV-I associated myelopathy.
>
> **Epidemiology**
> HTLV-I is primarily endemic in southwestern Japan and the Caribbean, however, parts of sub-Saharan Africa and Central and South America also have significant seroprevalence. Increased antibody seropositivity has been observed in the Southwest United States, in the United Kingdom, and in countries with large Caribbean populations. Approximately 0.1% of males and 0.2% of females infected at childhood develop ATL as adults. HTLV-1 appears to have very little genetic variability.
>
> **Transmission/Incubation Period**
> HTLV-I is poorly contagious and transmission can occur via blood transfusion, contaminated needles, sexual contact, and from mother to child through breast feeding. The incubation period appears to be 10-30 years and infection is assumed to be lifelong. In HTLV-I associated myelopathy, sometimes develops just a few years after infection.
>
> **Inactivators and Disinfectants**
> HTLV-1 is relatively unstable virus and is rapidly inactivated by heat, alcohols and organic solvents. Effective disinfectants include 10% household bleach solutions, quaternary ammonium compounds, phenols, iodophor compounds, and glutaraldehyde compounds.

Transmission can occur via blood cell transfusion, contaminated needles, sexual contact, and from mother to child through breast milk. Most infections are acquired during childhood and can be traced to a seropositive mother. Breast milk appears to be the vehicle of transmission. Indeed, virus-positive lymphocytes are abundant in the breast milk of seropositive mothers. Male to female sexual transmission occurs more often than female to male transmission. Semen is thought to be the main vehicle of transmission because virus-positive mononuclear cells have been observed in the semen of a seropositive man.

Although HTLV-I can be isolated from peripheral blood lymphocytes and from CSF (HAM patients), most HTLV-I infections are diagnosed serologically. All ATL patients produce antibodies to HTLV-I *gag* proteins (p15, p24, and p19). Current serologic methods cannot adequately distinguish HTLV-I from HTLV-II infections because considerable cross-reactivity exists between these viruses, particularly within the p24 protein. Several synthetic peptide assays have been described for distinguishing HTLV-I and HTLV-II infections but these tests are not widely available. HTLV-I and HTLV-II infections can be detected by radioimmunoprecipitation, western blot, the polymerase chain reaction, EIA, and competitive RIAs.

HTLV-II

In 1982 a related, but antigenically distinct, type C retrovirus (HTLV-II) was isolated from a patient with a T-cell variant of a hairy cell leukemia (4). Since that time, only four cases of HTLV-II have been documented by virus isolation. Little is known about the geographic distribution, transmission, and natural history of HTLV-II. Research into these areas has been hampered by the extensive serologic cross-reactivity between HTLV-I and HTLV-II. Although recent reports suggest that intravenous drug users in New York and in England have increased prevalence of HTLV-II antibodies, HTLV-II has not been consistently associated with any diseases.

HIV-1

The human immunodeficiency virus type 1 (HIV-1) is the etiologic agent of the acquired immunodeficiency syndrome (AIDS) and the AIDS-related complex (ARC). The World Health Organization estimates that 8-10 million people are currently infected with the virus, worldwide and 8-10 times that number of infections are predicted for the 21st century (5). Although the first report of a cluster of AIDS patients was made by the CDC in 1981 (6), the virus has been present in the human population for years. Indeed, a retrospective analysis revealed that the first documented case of AIDS was seen in a British seaman who died of progressive immunodeficiency in 1959 (7).

The virus is spread by blood, blood products, and body fluids. The major mode of HIV transmission is through sexual intercourse either through heterosexual contact or male-to-male homosexual contact. Parenteral transmission through sharing of needles by intravenous drug users constitutes the next most prevalent route of transmission. In developing countries, parenteral transmission can also occur via reuse of inadequately sterilized needles and syringes. Transfusion of HIV-1 infected blood and blood factor concentrates was a significant source of transmission prior to the licensure of HIV test kits in 1985. An emerging source of HIV transmission is mother-to-child transmission either at, or before birth. Postpartum transmission of HIV may also occur through breast feeding and HIV has been isolated from cell-free human milk. Salivary transmission has also been reported in one study from the Soviet Union where HIV-infected infants were breast fed and mothers had cracked or bleeding nipples (8).

Once transmitted, the virus binds to CD4-positive cells and establishes a persistent infection with integration of the HIV provirus into the host cell DNA. Virus infection results in a progressive depletion of CD4-positive T-lymphocytes, severe immune depletion, and damage to cells of the central nervous system. PCR analyses indicate that HIV DNA is present in a much larger proportion of CD4-positive cells than was previously reported (9,10). The proportion of infected CD4-positive T cells ranges from 1 in 100 to 1 in 10,000 in asymptomatic individuals compared to earlier studies that detected HIV in 1 in 10,000 to 1 to 100,000 cells. By the time patients present with AIDS, at least 1 in 100 CD4-positive T-cells contain HIV DNA (9,10).

The incubation period for HIV exposure is similar to that of hepatitis B virus. HIV antigens are usually detectable 3-8 weeks after exposure followed by the appearance of antibodies to HIV envelope (gp41) and *gag* (p24) proteins. The initial infection is usually asymptomatic although there

> **AT A GLANCE...**
>
> # HTLV-II
>
> **Virus Detection Methods**
> HTLV-II can be isolated by co-cultivation of patient lymphocytes with phytohemagglutinin-stimulated peripheral blood or cord blood lymphocytes.
>
> **Specimen Source**
> HTLV-II has been isolated from peripheral blood lymphocytes of patients with a T-cell variant of a hairy cell leukemia.
>
> **Time to Result**
> Positive culture - 2-4 weeks.
> Negative Culture - 4 weeks.
>
> **HTLV-II Serology**
> HTLV-I and HTLV-II exhibit extensive serologic cross-reactivity, particularly within the p24 gag protein. Current serologic methods cannot adequately distinguish HTLV-I from HTLV-II infections.
>
> **Epidemiology**
> Although recent reports suggest that intravenous drug users in New York and in England have increased prevalence of HTLV-II antibodies, HTLV-II has not been consistently associated with any diseases.
>
> **Transmission/Incubation Period**
> Little is known about the geographic distribution, transmission, and natural history of HTLV-II.
>
> **Inactivators and Disinfectants**
> Although no definitive information is available, HTLV-II inactivation is presumed to be similar to that of HTLV-I. Effective disinfectants are presumed to include 70% alcohol, 10% household bleach solutions, quaternary ammonium compounds, phenols, iodophor compounds, and glutaraldehyde compounds.

have been reports of individuals experiencing fever, malaise, rashes, and occasionally, encephalopathy (11). During the ensuing "latent" period there is often a progressive subclinical decline in CD4-positive cells and sometimes, lymphadenopathy. The median incubation time from HIV infection to AIDS is 8-10 years. However, homosexual men and some neonates may progress to AIDS more rapidly than other groups.

Diagnosis of HIV infection has relied mainly upon serological testing and, to a lesser extent, on virus isolation. Serological tests are considered either screening or confirmatory. Enzyme immunoassay (EIA) methods using inactivated whole virus lysates or recombinant proteins are used most often for serological screening. EIA manufacturers claim these tests have sensitivities of 98.3% to 100% and specificities of 99.2% to 100%. False positive EIA reactions can occur in patients with immunologic abnormalities, neoplasms, alcoholic hepatitis and in multiparous or multiply transfused patients who develop antibodies to the class II HLA antigens that are present on the cell line used to propagate the virus *in vitro*. Confirmatory testing of repeatedly positive EIA screening tests is usually accomplished by Western Blot or indirect fluorescent antibody staining.

HIV-2

Human immunodeficiency virus type 2 was isolated in West Africa in 1986 (12). Since that time, isolations have been reported in Europe and North America. Although a U.S. blood donor carrying HIV-2 was identified through routine HIV-1 screening, HIV-2 is rare in the United States. Most of the known HIV-2 cases have been clustered in West Africa and in West African citizens in other countries.

Although HIV-2 appears to be somewhat less pathogenic than HIV-1, HIV-2 is capable of infecting CD4-positive T-lymphocytes and causing both AIDS and neurologic disease. Little is known about the transmission, natural history, and inactivation characteristics of HIV-2. In the interim, most researchers assume that these characteristics will be similar to those of HIV-1.

Diagnosis of HIV-2 infections rely upon demonstrating seroconversion to HIV-2 antigens or virus isolation from peripheral blood. Currently licensed HIV-1 antibody tests detect HIV-2 antibodies 20-90% of the time (13). New screening tests have recently been licensed by the Food and Drug Administration for the simultaneous detection of HIV-1 and HIV-2 antibodies.

VIRUS ISOLATION

Introduction

Isolation of human immunodeficiency virus (HIV) from blood or other specimens is often important for establishing a diagnosis in patients whose serological patterns are equivocal and in infants who may still possess maternal antibodies. Culture is also important in early HIV infections when the presence of circulating HIV is the only evidence of infection. Virus isolation is also of value in assessing the effect of antiviral agents, monitoring their use, and studying the pathogenesis of the disease.

HIV cultivation methods have improved steadily since the virus was first isolated in 1983 (14). Original estimates of culture sensitivity were 60-90%, no better than commercial antigen assays. Recently, Jackson, et al. (15) reported that HIV-1 can be isolated from 100% of AIDS patients, 99% of ARC patients, and 98% of asymptomatic, antibody positive individuals. In contrast, the sensitivities of serum antigen testing for these same patient groups were 42%, 37%, and 17%, respectively (15). HIV cultures are sensitive and specific but they are expensive, they take weeks before a result is reported, and HIV cultures create the opportunity for exposing laboratory personnel to high concentrations of virus. Therefore, most clinical laboratories will want to critically evaluate their clinical needs before attempting this test.

Despite the recent introduction of a number of novel HIV microculture systems (16-18), most laboratories continue to use macro systems similar to the one described below. The following protocol has been modified from the CDC procedure published in February, 1987 (19). In this procedure, peripheral blood lymphocytes are co-cultivated with normal donor lymphocytes that have been stimulated with phytohemagglutinin for 2-3 days to increase the number of susceptible, CD4-positive cells. The co-cultivation medium contains T-cell growth factor, polybrene to improve the adherence of the virus to the cells, and antibodies to interferon which results in greater virus production.

Specimen Collection

Peripheral blood leukocytes are the specimen of choice. Although HIV-1 can be isolated from other body fluids including CSF, serum, cervical secretions, saliva, semen, plasma, tears, breast milk, and brain tissue, the probability of recovering HIV-1 from these specimens is significantly less than from peripheral blood leukocytes. The clinical laboratory originally required three days notification before a specimen could be submitted so that PHA-stimulated lymphocytes could be prepared. However, freezing the specimens after lymphocyte separation (20) allows the blood drawing at a

AT A GLANCE...

HUMAN IMMUNODEFICIENCY VIRUS TYPE 1

Virus Detection Methods
HIV-1 can be isolated by co-cultivation of patient lymphocytes with phytohemagglutinin-stimulated peripheral blood or cord blood lymphocytes. Cultures may be useful for detecting HIV infections in patients whose antibody test is negative or indeterminate and in neonates born to high risk, antibody positive mothers.

Specimen Source
Peripheral blood is the specimen of choice. HIV-1 can also be isolated from CSF (in neurological HIV), serum, cervical secretions, saliva, semen, plasma, tears, breast milk, and brain tissue. The probability of recovering HIV-1 from these specimens is significantly less than from blood.

Time to Result
Positive culture - 2-4 weeks.
Negative Culture - 4 weeks.

HIV-1 Serology

IgG   First detectable 3-12 weeks after infection in nearly all cases except neonates. Antibody detection forms the basis for most AIDS screening tests. Once established, HIV antibody levels usually persist throughout the lifetime of the patient. The presence of antibody does not imply immunity to the virus but rather, that the patient is assumed to be infected, and infectious.

p24   The p24 core antigen can be detected in some blood samples 3-4 weeks following infection and detectable antigenemia may be present for up to 3-4 months. The disappearance of p24 antigen is probably related to the development of p24 antibodies. Antigenemia often returns late in the clinical course when patients are symptomatic.

Epidemiology
The human immunodeficiency virus type 1 (HIV-1) causes a worldwide pandemic of acquired immunodeficiency syndrome (AIDS) and AIDS-related complex (ARC). An estimated 8-10 million people are currently infected with the virus and 8-10 times that number of infections are predicted for the 21st century. Considerable genetic variability exists among HIV isolates.

Transmission/Incubation Period
HIV-1 is transmitted by blood, blood products, and body fluids. Major modes of HIV-1 transmission include sexual intercourse; parenteral transmission through shared or inadequately sterilized needles; transfusion of HIV-1 infected blood and blood factor concentrates; and mother-to-child transmission either in utero, at birth, or through breast feeding. The median time from HIV infection to AIDS is 8-10 years. Homosexual men and some neonates may progress to AIDS more rapidly than other groups.

Inactivators and Disinfectants
HIV-1 is relatively unstable and quickly inactivated by alcohols, quaternary ammonium compounds, detergents, and 0.5% sodium hypochlorite solutions. Effective disinfectants include 70% alcohol, 10% household bleach solutions, quaternary ammonium compounds, phenols, and glutaraldehyde compounds.

clinically relevant time and it allows the laboratory to test specimens in batches.

Blood Collection
1. Aseptically collect 20 ml of blood in green top vacutainer tubes containing preservative-free heparin.
2. If the patient's WBC count is known, the guidelines listed below can be used. Otherwise, collect 40 ml of blood.

WBC = 5,000	40 ml
WBC = 6-12,000	20 ml
WBC ≥ 12,000	10 ml
WBC unknown	40 ml

3. Once collected, the tubes should be inverted several times to assure that the heparin is evenly dispersed.
4. The specimens may be shipped to the laboratory at room temperature. However, virus is rarely isolated if the specimen is older than 24 hours old.

Specimen Storage
1. Upon receipt in the laboratory, store the blood in the vacutainer tubes at room temperature until they can be processed. Because virus is rarely isolated from specimens greater than 24 hours old, processing should begin as soon as the specimen is received in the laboratory.
2. If the specimen cannot be tested within 24 hours, the lymphocytes should be separated as described below, washed once, and resuspended in 1 ml of RPMI-1640 containing 20% FCS and 10% DMSO. The cells should be slowly frozen at -70°C (Chapter 6) until they can be tested (20).

Lymphocyte Separation
Lymphocytes are separated from red cells and granulocytes by centrifugation through a ficoll-hypaque density gradient. A number of ready-made gradient materials can be purchased commercially (Ficol-Paque from Pharmacia Fine Chemicals, Histopaque from Sigma Chemical Company or Lymphocyte Separating Medium from Organon-Teknika) for this purpose or the gradient materials can be prepared in the laboratory (Appendix C). Because HIV-positive patients are often lymphopenic, relatively large amounts of blood must be used to provide enough lymphocytes for testing. As a result, a large number of tubes may be required during the initial lymphocyte separation phase. Large-scale lymphocyte separation should be done as directed by the manufacturer of the gradient materials or as described below.

1. Dilute the heparinized blood with an equal volume of sterile saline.
2. Add 40 ml of the diluted blood into sterile 50 ml centrifuge tubes.
3. Draw up 10 ml of ficoll-hypaque into a 10 ml syringe. Attach a sterile 15 cm, 18 ga blunt-tip cannula to the syringe.
4. Carefully underlay the blood with 10 ml of ficoll-hypaque by holding the tip of the cannula just off the bottom of the tube and slowly adding the gradient solution to the tube. A clear zone of ficoll-hypaque should form at the bottom of the tube.
5. Centrifuge the tubes at 400 x g in a swinging bucket rotor for 30 minutes at room temperature. Make sure the brake is off.
6. Lymphocytes will band at the plasma-ficoll interface.
7. Carefully remove and discard the clear upper (plasma) layer.
8. Combine the lymphocyte bands from a single patient in a sterile 50 ml centrifuge tube.
9. Add PBS to the 50 ml mark. Gently rock the tube to mix.
10. Centrifuge at 200 x g for 15 minutes at room temperature to pellet the cells.
11. Remove the supernatant fluids and suspend the cell pellet in 30 ml of sterile PBS.
12. Centrifuge the cells at 200 x g for 15 minutes and remove the supernatant fluids.
13. Resuspend the cell pellet in 5 ml of PBS.
14. Perform a cell count and suspend the cells at

> *AT A GLANCE...*
>
> # HUMAN IMMUNODEFICIENCY VIRUS TYPE 2
>
> **Virus Detection Methods**
> HIV-2 can be isolated by co-cultivation of patient lymphocytes with phytohemagglutinin-stimulated peripheral blood or cord blood lymphocytes.
>
> **Specimen Source**
> Peripheral blood leukocytes are the specimen of choice. However, HIV-2 can be isolated from other body fluids
>
> **Time to Result**
> Positive culture - 2-4 weeks.
> Negative Culture - 4 weeks.
>
> **HIV-2 Serology**
> Little is known about the antibody response to HIV-2 infection. It is presumed that the HIV-2 response is similar to that of HIV-1. Diagnosis of HIV-2 infections is most often based upon demonstration of seroconversion to HIV-2 antigens.
>
> **Epidemiology**
> HIV-2 was first isolated in West Africa in 1986 and most of the known HIV-2 cases have been clustered in West Africa and in West African citizens in other countries. Although HIV-2 has been isolated from patients in Europe and in the United States. HIV-2 infections are rare outside Western Africa. HIV-2 appears to be somewhat less pathogenic than HIV-1. However, HIV-2 is still capable of causing AIDS and neurologic disease.
>
> **Transmission/Incubation Period**
> Little is known about the transmission and natural history of HIV-2. For now, most researchers assume that HIV-2 behaves similarly to HIV-1.
>
> **Inactivators and Disinfectants**
> Although no definitive information is available, HIV-2 inactivation is presumed to be similar to that of HIV-1. Effective disinfectants are presumed to include 70% alcohol, 10% household bleach solutions, quaternary ammonium compounds, phenols, iodophor compounds, and glutaraldehyde compounds.

2×10^7 cells/ml in HIV Propagation medium (Appendix C).

15. Cells can be frozen at 2×10^7 cells/ml in RPMI-1640 containing 20% FCS and 10% DMSO for future testing or for reference purposes.

PHA-Stimulated Peripheral Blood Leukocytes
Peripheral blood lymphocytes are separated from the blood of normal (HIV and HBsAg negative), healthy donors by centrifugation through a ficoll-hypaque density gradient. Lymphocyte separation should be done as directed by the manufacturer or as described below.

1. Aseptically collect 20 ml of blood in green top vacutainer tubes containing heparin.
2. Place 3 ml of the ficoll-hypaque in a clear sterile 15 ml centrifuge tube.
3. Dilute the heparinized blood with an equal volume of sterile saline.

4. Carefully overlay 5 ml of heparinized blood onto the ficoll-hypaque.
5. Centrifuge the tubes at 400 x g in a swinging bucket rotor for 30 minutes at room temperature.
6. Lymphocytes will band at the plasma-ficoll interface.
7. Carefully remove and discard the clear upper (plasma) layer.
8. Remove the lymphocyte bands and pool them in a sterile 50 ml centrifuge tube.
9. Add PBS to the 30 ml mark.
10. Centrifuge the cells at 200 x g for 15 minutes and remove the supernatant fluids.
11. Resuspend the cell pellet in 5 ml of PBS.
12. Perform a cell count.
13. Resuspend the cells at 1×10^6 cells/ml in PHA medium (Appendix C) and distribute into an appropriate number of flasks (25 cm² flask = 10 ml; 75 cm² flask = 30 ml; 150 cm² flask = 100 ml).
14. Incubate the flasks in an upright position for 2-3 days at 35-37°C. Contaminated cultures should be discarded.
15. Cells are ready for co-cultivation or feeding. PHA-stimulated cells can be frozen in RPMI-1640 containing 20% FCS and 10% DMSO for several years. Prior to use, frozen cells should be quickly thawed and washed twice to remove the cryoprotectants. Although cryopreserved lymphocytes were found to be quite effective for isolating HIV from AIDS patients, frozen cells were less efficient than fresh cells for isolating HIV from asymptomatic, antibody positive men (21).

Inoculation
1. Appropriately label a 25 cm² flask containing 5×10^6 PHA-stimulated peripheral blood lymphocytes in 10 ml of PHA medium for each patient isolation.
2. Allow the lymphocytes to settle and, without disturbing the cells, remove approximately 5-7 ml of the medium.
3. Inoculate the flask with 0.5 ml (1×10^7 cells) of the patient lymphocyte suspension.
4. Incubate the flasks in an upright position for 1 hour at 35-37°C.
5. Add HIV Propagation (Appendix C) medium to the 10 ml mark.
6. Incubate the cultures for 24 hours at 35-37°C.
7. Remove half the medium and replace it with an equal volume of propagation medium.
8. Incubate the cultures as before for 1 week.
9. One week after initial culture and every 3-4 days thereafter, remove 5 ml of culture medium from the flasks for antigen detection.
10. Add 5 ml of fresh propagation medium.
11. At weekly intervals, add 3×10^6 fresh PHA-stimulated lymphocytes to the flask.
12. Maintain the cultures for 1 month before discarding them as negative. If the culture becomes contaminated, filter the supernatant fluids through a 0.2 μm filter and add the filtrate to fresh PHA-stimulated lymphocytes.

Culture Confirmation

Several methods can be used to detect HIV in cultures including reverse transcriptase assays, immunofluorescence, and commercial EIA kits for p24 core antigen detection. Reverse transcriptase assays are used in most retrovirology laboratories while most clinical virology laboratories use either the commercial p24 core antigen test or IFA. Several studies have shown that these tests are as sensitive and specific as the reverse transcriptase assay. The assay used in any particular laboratory will depend upon the available equipment and the expertise of the laboratory staff.

Reverse Transcriptase Assay

Reverse transcriptase (RT), a particle associated RNA-dependent DNA polymerase, is routinely used to monitor the replication of retroviruses in culture media. A number of similar RT procedures have been published (1,19,22-25). In these procedures, the cells are removed from the culture media by low speed centrifugation and the supernatant fluids are ultracentrifuged at 140,000 x g to pellet the virus. The virus pellet is lysed and assayed for reverse transcriptase. The assay mixture contains a primer-template poly (rA):poly(dT)$_{12-18}$, tritiated thymidine, and magnesium chloride. Although this procedure is reliable, it is not specific for any single human retrovirus. Thus, the result should be confirmed by a specific test such as immunofluorescence or commercial EIA. RT assays have an advantage because they can detect other HIV strains and other retroviruses that antigen assays may miss.

Immunofluorescence Assay (IFA)

Indirect immunofluorescence assays are used extensively to locate HIV-1 antigens in infected cells. In this procedure, the cells are washed, placed on a slide, and fixed with acetone. Monoclonal antibodies to the p17 or p24 proteins are reacted with the fixed smears. After washing away any unbound antibody, fluorescein-labelled antimouse antibodies are added together with an Evans blue counterstain. The smears are washed, dried and examined under a fluorescence microscope. This procedure is simple, inexpensive, sensitive, and specific. In addition, antibodies are available for HIV-1, HIV-2, and HTLV-I/II. However, interpreting the fluorescence patterns requires a skilled technician who can distinguish specific from nonspecific staining.

Slide Preparation
1. Centrifuge the cultures at 600 to 800 x g to pellet the cells.
2. Remove the culture fluids and store at -70°C for future reference.
3. Resuspend the cells in 1 ml of PBS containing 2% FCS.
4. Place 5-10 µl of the suspension in both wells of an acetone-cleaned two-well slide. Prepare several slides and store the remainder for future reference.
5. Allow the smears to air dry.
6. Fix the slides for 10 minutes in acetone.
7. Slides prepared in this manner can be stored desiccated at -20°C for up to 3 years or they can be stained immediately as described below.

IFA Procedure
1. Remove a positive and negative control slide from the freezer and allow them to come to room temperature.
2. Place one drop (10-30 µl) of the monoclonal antibody preparation on one patient smear and one of each of the control smear(s). Place one drop of PBS on the other smears.
3. Incubate the slides for 30 minutes at 35-37°C in a covered, humidified chamber. Do not allow the antibody to dry on the slide as this could cause nonspecific antibody binding.
4. Remove the excess antibody with a gentle stream of PBS. Do not direct the stream directly at the cell smear as this could dislodge the cells.
5. Soak the slide in PBS for 5 minutes and shake off the excess fluid.
6. Place one drop (15-30 µl) of fluorescein-labelled antimouse antibody on all the smears.
7. Incubate the slide for 30 minutes at 33-35°C in a covered, humidified chamber.
8. Remove the excess antibody with a gentle stream of PBS as before.
9. Soak the slide in PBS for 5 minutes, shake off the excess fluid, and allow the slide to air dry.
10. Carefully add a small drop of FA mounting fluid to the center of each well.
11. Place a number 1 coverslip on the mounting fluid and carefully remove all the air bubbles.
12. Immediately examine the slide using a fluorescence microscope. For optimum clarity, use 200-300X magnification for screening and 400X for confirmation of staining pattern.

Interpretation of Results

Positive Test. The presence of HIV or HTLV antigens is indicated by intense apple-green fluorescence in the cytoplasm of infected cells. Other cells will stain red due to the counterstain. Many cultures will contain multinucleated giant cells that can confuse an inexperienced technician. In addition, cell debris and clumps may exhibit a dull fluorescence which can be misinterpreted as a specific, positive reaction.

REPORT: HIV-1 isolated (HIV-1 antibodies).
HIV-2 isolated (HIV-2 antibodies).
HTLV-I/II isolated (HTLV-I/II antibodies).

Negative Test. If all control reactions function properly, the absence of specific cytoplasmic fluorescence indicates the absence of detectable virus. If the cultures are repeatedly negative after one month in culture they should be reported as negative.

REPORT: HIV-1 not isolated (HIV-1 antibodies).
HIV-2 not isolated (HIV-2 antibodies).
HTLV-I/II not isolated (HTLV-I/II Ab).

QC Procedures.

Appropriate controls are necessary because staining patterns and intensities can vary with the type of antibody (core, gag, etc.) used and the type of virus (HIV, HTLV) being detected. Therefore, appropriate control slides or cultures must be stained concurrently with the patient specimens. Positive control smears must exhibit typical apple-green fluorescence. If no fluorescence is observed in the positive controls, the test is invalid and must be repeated. Negative controls must not stain with the antibody. If specific staining occurs in the negative controls, the test is invalid and must be repeated. Extra control slides prepared from positive and negative cultures (above) can be used as controls.

Commercial p(24) Antigen Assays

Commercial HIV-1 antigen EIAs have several advantages over the RT because antigen assays have excellent sensitivity and they require no radioactive isotopes or ultracentrifugation (26-30). Consequently, many laboratories use the p24 antigen EIA tests to confirm HIV-1 infections.

REFERENCES

1. Poiesz BJ, Fuscetti FW, Gazdar AF, Bunn A, Minna JD, Gallo RC. Detection and isolation of type C retrovirus particles from fresh and cultured lymphocytes of a patient with cutaneous T-cell lymphoma. *Proc Natl Acad Sci USA* 1980;77:7415-7419.
2. Gessian A, Vernant JC, Maurs L, et. al. Antibodies to human T-lymphotropic virus type 1 in patients with tropical spastic paraparesis. *Lancet* 1985;ii:407-409.
3. Osame M, Usuku D, Izumo S, et al. HTLV-I associated myelopathy, a new clinical entity. *Lancet* 1986;1:1301-1302.
4. Kalyanaraman VA, Sarngadharan MG, Robert-Guroff M, et al. A new subtype of human T-cell leukemia virus (HTLV-II) associated with a T-cell variant of hairy cell leukemia. *Science* 1982;218:571-573.
5. Blattner WA. HIV epidemiology: past, present, and future. *FASEB J* 1991;5:2340-2348.
6. Centers for Disease Control. Pneumocystis pneumonia - Los Angeles. *MMWR* 1981;30:250.
7. Crobitt G, Bailey AS, William G. HIV infection Manchester 1959. *Lancet* 1990;1:51.
8. Pokrovsky BB, Kuznetsova I, Eramova I. Transmission of HIV-infection from an infected infant to his mother by breast-feeding. (Abstr. Th.C.48) Sixth International Conference on AIDS, San Francisco, CA. 1990.
9. Schnittman SM, Psallidolopoulos MC, Lane HC, et al. The reservoir for HIV-1 in human peripheral blood is a T cell that maintains expression of CD4. *Science* 1989;245:305-308.
10. McElrath MJ, Steinman RM, Cohn ZA. Latent HIV-1 infection in enriched populations of blood monocytes and T cells from seropositive patients. *J Clin Invest* 1991;87:27-30.
11. Cooper DA, Gold J, Maclean P, et al. Acute AIDS retrovirus infection: Definition of clinical illness associated with seroconversion. *Lancet* 1985;1:537-540.
12. Clavel F, Guetard D, Brun-Vezinet F, et al. Isolation of a new human retrovirus from West African patients with AIDS. *Science* 1987;233:343-346.
13. Hirsch MS, Curran J. Human Immunodeficiency viruses. In: Fields BN, Knipe DM, Chanock RM, Hirsch MS, Melnick JL, Monath TO, Roizman B, eds. *Virology*, Second Edition. New York: Raven Press, 1990;1545-1570.
14. Barre-Sinoussi F, Chermann JC, Rey F, et al. Isolation of a T-lymphotropic retrovirus from a patient at risk for acquired immune deficiency syndrome (AIDS). *Science* 1983;220:868-871.
15. Jackson JB, Kwok SY, Sninsky JJ, et al. Human immunodeficiency virus type 1 detected in all seropositive symptomatic and asymptomatic individuals. *J Clin Microbiol* 1990;28:16-19.

16. Chesebro B, Wherly K. Development of a sensitive quantitative focal assay for human immunodeficiency virus infectivity. *J Clin Microbiol* 1988;62:3779-3788.
17. Hanson CV, Crawform-Miksza L, Sheppard HW. Application of a rapid microplaque assay for determination of human immunodeficiency virus neutralizing antibody titers. *J Clin Microbiol* 1990;28:2030-2034.
18. Dimitrov DH, Melnick JL, Hollinger FB. Microculture assay for isolation of human immunodeficiency virus type 1 and titration of infected peripheral blood mononuclear cells. *J Clin Microbiol* 1990;28:734-737.
19. Velleca WM, Palmer DF, Feorino PM, Warfield DT, Krebs J, Forrester B, Phillips S. *Isolation, culture, and identification of human T-lymphotropic virus type III/lymphadenopathy-associated virus.* U.S. Department of Health and Human Services, Centers for Disease Control 1987.
20. Gallo D, Kimpton J, Johnson PJ. Isolation of human immunodeficiency virus from peripheral blood lymphocytes stored in various transport media and frozen at -60°C. *J Clin Microbiol* 1989;27:88-90.
21. Balachandran R., Thampatty P, Rinaldo C, Gutpa P. Use of cryopreserved normal peripheral blood lymphocytes for isolation of human immunodeficiency virus from seropositive men. *J Clin Microbiol* 1988;26:595-597.
22. Levy JA, Shimabukuro J. Recovery of AIDS-associated retroviruses from patients with AIDS or AIDS-related conditions and from clinically healthy individuals. *J Infect Dis* 1985;152:734-738
23. Asjo B, Morfeldt-Manson L, Albert J, et al. Replicative capacity of human immunodeficiency virus from patients with varying severity of HIV infection. *Lancet* 1986;2:660-662.
24. Rey MA, Spire B, Dormont D, Barre-Innoussi F, Montaigner L, Chermann JC. Characterization of the RNA dependent DNA polymerase of a new human T lymphotropic retrovirus (lymphadenopathy associated virus). *Biochem Biophys Res Commun* 1984;121:126-133.
25. Gallo D, Kimpton JS, Dailey PJ. Comparative studies on use of fresh and frozen peripheral blood lymphocyte specimens for isolation of human immunodeficiency virus and effects of cell lysis on isolation efficiency. *J Clin Microbiol* 1987;25:1291-1294.
26. Gupta P, Balachandran R, Grovit K, Webster D, Rinaldo C Jr. Detection of human immunodeficiency virus by reverse transcriptase assay, antigen capture assay, and radioimmunoassay. *J Clin Microbiol* 1987;25:1122-1125.
27. Healey DS, Maskill WJ, Neate EV, Beaton F, Gust ID. Comparison of enzyme immunoassay and reverse transcriptase assay for detection of HIV in culture supernates. *J Virol Methods* 1987;17:237-245.
28. Land D, Beaton F, McPhee DA, Gust ID. Comparison of core antigen (p24) assay and reverse transcriptase activity of detection of human immunodeficiency virus type 1 replication. *J Clin Microbiol* 1989;27:486-489
29. Jackson JB, Coombs RW, Sannerud K, Rhame FS, Balfour HH Jr. Rapid and sensitive viral culture for human immunodeficiency virus type 1. *J Clin Microbiol* 1988;26:1416-1418.
30. Feorino P, Forrester B, Schable C, Warfield D, Schochetman G. Comparison of antigen assay and reverse transcriptase assay for detecting human immunodeficiency virus in culture. *J Clin Microbiol* 1987;25:2344-2346.

CHAPTER 26

Rhinovirus

INTRODUCTION

The genus rhinovirus comprises one of the four genera belonging to the family *Picornaviridae*. Rhinoviruses are small (28-34 nm) non-enveloped icosahederal viruses. Their 7.2 kilobase, positive sense, single-stranded RNA genome codes for a single polycystronic message that is post-translationally modified into four structural proteins (VP1-4). Rhinoviruses can be distinguished from their enterovirus cousins by their inability to replicate at 35-37°C and their sensitivity to pH 3 conditions. Rhinoviruses are extremely hardy and can persist for hours to days on environmental surfaces (1). Rhinoviruses are resistant to chloroform, ether, 70% ethanol, 5% phenol, and most detergents.

Rhinoviruses are found throughout the world and more than 100 rhinovirus serotypes have been identified. Epidemiologic studies have shown that a number of different serotypes usually circulate within any given geographic area. However, the prevalent serotypes vary from area to area. Rhinovirus infections occur throughout the year and peak incidences are observed in the fall and spring. The average individual experiences 1-6 rhinovirus infections per year with the highest incidence occurring in children under 5 years of age.

Collectively, rhinoviruses are the most common infectious agents of man and they are the principal cause of man's most common illness - the common cold (2). There is also evidence that rhinoviruses play a role in acute sinus infections (3), exacerbations of chronic bronchitis (4,5), and wheezing in asthmatic children (6,7). Rhinoviruses are transmitted directly through inhalation of virus-laden particles generated when infected individuals cough or sneeze. Indirect transmission can occur through contamination of the hands with respiratory secretions and subsequent autoinoculation of the nose and eyes. Indirect transmission is probably the most important means of transmitting the virus in the general population as rhinoviruses can persist on the skin for several hours (1). Transmission occurs most often in the home and school children are the most frequent means for introducing infection into the home (8-10).

The incubation period is 1-4 days (11) and once transmitted, rhinoviruses quickly establish infections within the nasal epithelial cells. Virus spreads rapidly from cell to cell until the acute phase of illness when nearly all of the nasal epithelium is infected. The mechanism of rhinorrhea is poorly understood but it is thought to caused by the direct cytocidal effect of virus replication in the nasal epithelial cells. This destruction of the nasal epithelium also causes the edema, nasal obstruction, sneezing, and sore throat commonly associated with colds. In some patients, these symptoms may be accompanied by headache, cough, malaise, and occasionally, lower respiratory symptoms (12,13). Upper respiratory tract symptoms generally last for 2-5 days and patients are infectious from 1 day before to 5 days after onset of symptoms. In rare cases, rhinovirus may be isolated from nasal secretions for up to 2 weeks after the onset of symptoms. However, the virus titer is extremely low in these individuals.

Laboratory diagnosis of rhinovirus infections depends primarily upon viral isolation and identification. Serological diagnosis is usually not appropriate because of the large number of rhinovirus serotypes and the relative lack of cross-reactivity among these viruses. However, several serological methods can be used if the rhinovirus serotype is known or if the clinical isolate is available for testing.

AT A GLANCE...

RHINOVIRUS

Virus Detection Methods
Tube cultures of human diploid fibroblasts, WI-38, WI-26, HeLa M and MRC-5 cells.

Specimen Source
Specimen of choice is a nasal swab, aspirate or wash within 72 hours after the onset of symptoms. Rhinovirus is rarely isolated from throat swabs or washes.

Time to Result
Standard culture: Positive culture - 1 day to 2 weeks.
                    Negative culture - 2 weeks.

Rhinovirus Serology
Serological diagnosis is usually not appropriate because of the large number of rhinovirus serotypes and the relative lack of cross-reactivity among these viruses.

Epidemiology
Rhinoviruses are found throughout the world and more than 100 rhinovirus serotypes have been identified. A number of different serotypes usually circulate within any given geographic area. However, the prevalent types vary from area to area. Rhinovirus infections occur throughout the year and peak incidences are observed in the fall and spring. The average individual experiences 1-6 rhinovirus infections per year with the highest incidence occurring in children under 5 years of age. The annual number of rhinovirus infections tends to decrease with age.

Transmission/Incubation Period
Rhinoviruses are transmitted directly through inhalation of virus-laden particles generated when infected individuals cough or sneeze or cough. Indirect transmission can occur through contamination of the hands with respiratory secretions and subsequent autoinoculation of the nose and eyes. Indirect transmission is probably the most important means of transmitting the virus in the general population as rhinoviruses can persist on the skin for several hours. Transmission occurs most often in the home and school children are the most frequent means for introducing infection into the home. Patients are infectious from 1 day before to 5 days after onset of symptoms.

Inactivators and Disinfectants
Rhinoviruses are extremely hardy and can persist for hours to days on environmental surfaces. Rhinoviruses are resistant to chloroform, ether, 70% ethanol, 5% phenol, and most detergents. Effective disinfectants include 0.3% formaldehyde solutions, 10% household bleach solutions (0.3-0.5 ppm free residual chlorine), 0.1 N HCl, 2% sodium hydroxide, and 2% glutaraldehyde solutions. The presence of organic material can protect rhinoviruses from inactivation.

VIRUS ISOLATION

Introduction

Rhinoviruses can be isolated in HFF, MRC-5, WI-38, HeLa M, and WI-38 cells. However, the culture conditions must be closely controlled for efficient rhinovirus isolation. Variations in culture conditions can cause significant differences in rhinovirus susceptibility. To counter this problem, some laboratories use more than one cell type (i.e., HFF and MRC-5 cells) for rhinovirus isolations.

Inoculation of multiple tubes also improves the rhinovirus isolation frequency. The condition of the culture medium must also be closely controlled. Most rhinovirus isolation media are well buffered to minimize pH shifts that could inactivate the virus.

The optimal temperature for rhinovirus replication is 33-35°C, the temperature of human nasal passages and large airways. At 37°C, rhinovirus yields are only one-tenth of the normal 33-35°C levels (14). Roller drum incubation also provides improved isolation rates, presumably through increased oxygenation of the medium.

Rhinovirus CPE in HFF, WI-38, and MRC-5 cells is usually detectable 2-6 days after inoculation. Rhinovirus CPE consists of focal areas of large and small rounded cells that possess prominent, refractile plasma membranes and pyknotic nuclei. CPE spreads slowly through the monolayer and the CPE may subside unless the virus is passed to fresh cell cultures. Because the CPE produced by enteroviruses and rhinoviruses are often indistinguishable, pH 3 sensitivity testing is usually done to differentiate these viruses. Rhinoviruses are rapidly inactivated at pH 3 while enteroviruses are resistant to these conditions.

Many laboratories also inoculate at least one HEp-2 and one RMK tube whenever a rhinovirus infection is suspected. These extra tubes allow the laboratory to detect other viral agents (e.g., myxoviruses and adenoviruses) that can cause cold-like symptoms. Rhinoviruses grow slowly or not at all in RMK and HEp-2 cells.

Specimen Collection and Storage

Rhinovirus can be isolated from nasopharyngeal swabs, washes and aspirates. Virus is rarely isolated from throat washes, throat swabs, or sputa and these specimens are not recommended. After collection, specimens should be placed in viral transport medium and held on wet ice (2-8°C) until they are inoculated onto cell cultures. Because rhinoviruses are pH labile, special care must be taken to maintain the pH at or near neutrality during specimen transport. After collection, specimens can be stored for up to 24 hours at 2-8°C. If inoculation must be delayed beyond 24 hours, specimens should be rapidly frozen on dry ice and stored at or below -70°C. **DO NOT FREEZE SPECIMENS AT -20°C.** Storage at -20°C and repeated freeze/thaw cycles can destroy virus infectivity.

Specimen Preparation
Swabs
1. Vortex the specimen with several glass beads for 20-30 seconds to release any bound cells or virus.
2. Remove the swab from the viral transport medium and firmly roll it against the inside of the tube to remove as much fluid as possible.

Nasal Wash and Aspirate Specimens
1. Nasal aspirate specimens should be diluted to 3 ml with viral transport medium.
2. Vortex specimens with several glass beads for 20-30 seconds.
3. Centrifuge the suspension at 200 x g for 5 minutes to remove any mucus.
4. Carefully remove the mucus and use the remainder of the specimen for inoculation into cell cultures.

Standard Tube Culture
Procedure
1. Appropriately label two freshly confluent HFF, WI-38, or MRC-5 tubes for each specimen.
2. Remove the medium from the cells.
3. Add 2 ml of maintenance medium containing 2% FCS and antibiotics to each tube.
4. Add 0.2-0.5 ml of specimen to each tube.
5. Incubate the tubes on a roller drum (10-15 rph) at 33-35°C for 14 days. During this time, the culture medium should be replaced every third day or whenever it becomes acidic (yellow).
6. Observe the cells at least every other day for CPE.
7. Tubes displaying rhinovirus CPE should be scraped and transferred to fresh culture tubes. Passaging rhinovirus is important because

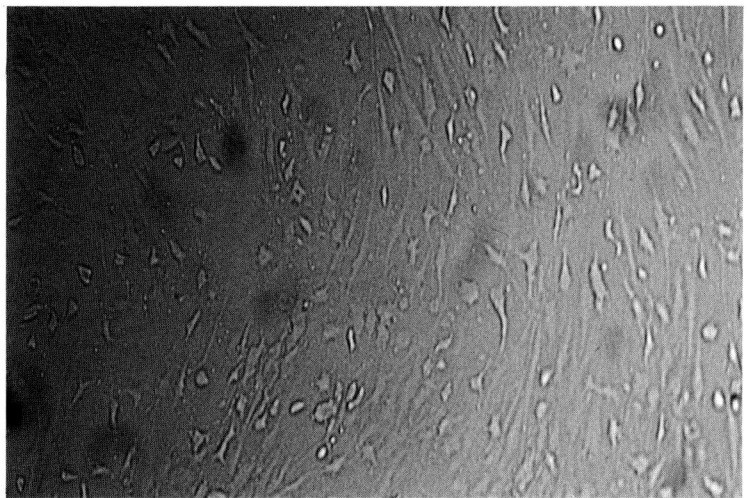

FIG. 1. Rhinovirus CPE in HFF cells.

rhinovirus CPE can regress if left in culture too long. In addition, passage will increase the virus concentration and facilitate acid stability pH 3 inactivation testing.
8. Incubate the virus passage tube until the CPE involves more than 50% of the monolayer.

Interpretation of Results
Positive test. Rhinovirus CPE is usually evident 2-6 days after inoculation. In human diploid fibroblast cells, rhinovirus CPE appears as focal areas of large and small round cells with prominent, refractile plasma membranes and pyknotic nuclei (Fig. 1). Positive cultures must be tested for acid stability before reporting results.
REPORT: No report should be made based upon CPE alone.

Negative Test. The absence of characteristic CPE after 14 days indicates the absence of viable and culturable rhinovirus.
REPORT: Rhinovirus not detected.

pH 3 Inactivation Test
1. Appropriately label two sterile glass tubes for each specimen.
2. Place 0.9 ml of pH 3 medium into one sterile glass tube and 0.9 ml of maintenance medium (pH 7.4) in the second tube.
3. Add 0.1 ml of the culture supernatant fluids to each tube.
4. Incubate at room temperature for 1-3 hours.
5. Add sterile 1N NaOH to the pH 3 tube until the medium is about pH 7.
6. Appropriately label two freshly confluent HFF, WI-38, or MRC-5 tubes for each specimen.
7. Remove the medium from the fresh culture tubes and replace it with 2 ml of maintenance medium.
8. Add 0.5 ml of the pH 3-treated virus to one culture tube and 0.5 ml of the pH 7-treated virus to the other culture tube.
9. Incubate the cultures on a roller drum at 33-35°C for 7 days.
10. Examine the cultures at least every other day for evidence of CPE.

Interpretation of Results
Positive Test. The pH 3 inactivation test is considered positive if the pH 3 medium neutralizes the infectivity of the supernatant fluids and if the pH 7 tube produces CPE. Virus isolates are considered positive for rhinovirus if they produce characteristic CPE and if the pH 3 inactivation test is positive.
REPORT: Rhinovirus isolated.

Negative Test. The CPE caused by enteroviruses and rhinoviruses are often indistinguishable. Therefore, cultures that produce characteristic CPE at 32-33°C and are not inactivated at pH 3.0 are considered enteroviruses. Further characterization should be done as described in Chapter 12.

REPORT: No report should be made until the virus is identified.

QC Procedures.

Because cell cultures may have unexplained day-to-day variations in rhinovirus susceptibility, rhinovirus clinical isolates must be inoculated with each batch of rhinovirus isolations. This procedure is essential for assuring that the cells remain susceptible to the rhinovirus infection. Virus isolates should produce characteristic CPE within 1-7 days after inoculation and should be neutralized by pH 3 treatment. Failure to produce these results invalidates that batch of cultures and virus specimens should be reinoculated onto fresh monolayers as soon as possible.

Rhinovirus isolates and enterovirus isolates must be tested for pH 3 inactivation at the same time as the patient specimen. Enteroviruses must produce CPE after pH 3 treatment and rhinovirus isolates must not produce CPE after treatment. If the patient control (pH 7) tube does not produce CPE, the virus titer in the specimen may be too low for pH 3 testing. If this occurs, the virus should be passaged to improve the virus concentration and tested again for pH 3 stability.

Preparation of Positive Control Cultures
1. Add 20 µl of a high-titer rhinovirus isolate to 2 ml of GLB.
2. Remove the growth medium from one 75 cm^2 flask of freshly confluent HFF or MRC-5 cells.
3. Add 2 ml of the inoculum to the flask and allow the virus to adsorb for 1-2 hours at 33-35°C. The flask should be rocked every 15 minutes to assure even virus distribution and to prevent monolayer desiccation.
4. Add 20 ml of maintenance medium to the flask.
5. Incubate the flask at 33-35°C until CPE involves 50-70% of the cell monolayer.
6. Scrape the cells from the flask into the cell culture medium.
7. Transfer the cell suspension to a 50 ml polypropylene centrifuge tube and freeze at -70°C.
8. Thaw the medium quickly in a 37°C water bath.
9. Centrifuge the tube at 600-800 x g to pellet the cell debris.
10. Transfer the supernatant fluids to a sterile vessel containing 80 ml growth medium.
11. Dispense 1.0 ml of the diluted supernatant fluids into each of 50 freezing vials and freeze the vials at -70°C or below. Vials can be stored in liquid nitrogen for up to 10 years or at -70°C for up to 5 years.

REFERENCES

1. Hendley JO, Wenzel RP, Gwaltney JM Jr. Transmission of rhinovirus colds by self-inoculation. *N Eng J Med* 1973;288:1361-1363.
2. Couch RB. Rhinoviruses. In: Schmidt NJ, Emmons RW, eds. *Diagnostic procedures for viral, rickettsial, and chlamydial infections*, Sixth edition. Washington DC: American Public Health Association 1989;709-729.
3. Evans RO Jr, Sydnor JB, Moore WFC. Sinusitis of the maxillary antrum. *N Eng J Med* 1975;293:735-739.
4. McNamara MJ, Phillips IA, Williams OB. Viral and *Mycoplasma pneumoniae* infections in exacerbations of chronic lung disease. *Am Rev Respir Dis* 1969;100:10-24.
5. Stenhouse AC. Rhinovirus infection in acute exacerbations of chronic bronchitis: A controlled prospective study. *Br Med J* 1967;3:461-463.
6. Minor TE, Dick EC, Baker JW, Quellette JJ, Cohen M, Reed CE. Rhinovirus and influenza type A infections as precipitant of asthma. *Am Rev Respir Dis* 1976;113:149-153
7. Minor TE, Baker JW, Dick E, et al. Greater frequency of viral respiratory infection in asthmatic children as compared with their nonasthmatic siblings. *Pediatrics* 1974;85;472-477.
8. Dick EC, Inhorn SL. Rhinoviruses. In: Feigin RD, Cherry JD, eds. *Textbook of Pediatric*

Infectious Diseases. Philadelphia: W. B. Saunders Co. 1987;1539-1558.
9. Dick EC, Jennings LC, Mink KA, Wartgow CD, Inhorn SL. Aerosol transmission of rhinovirus colds. *J Infect Dis* 1987;156:442-448.
10. Casey JM, Dick, EC. Acute respiratory infections. In: Casey JM, Foster C, and Hixson EG, eds. *Winter Sports Medicine.* Philadelphia: F. A. Davis Co. 1990;112-128.
11. Douglas RG Jr, Cate TR, Gerone JP, Couch RB. Quantitative rhinovirus shedding patterns in volunteers. *Am Rev Respir Dis* 1966;94:159-167.
12. Cate TR, Douglas RG Jr, Couch RB. Interferon and resistance to upper respiratory virus illnesses. *Proc Soc Exp Biol Med* 1969;131:631-636.
13. Fridy WW Jr, Ingram RH Jr, Hierholzer JC, Coleman MT. Airways function during mild respiratory illness. *Ann Intern Med* 1974;80:150-155.
14. Gwaltney JM Jr. Rhinovirus. In: Mandell GL, Douglas RG, Jr., Bennett JE, eds. *Principles and practice of infectious disease*, Third Edition. New York: Churchill Livingstone, Inc., 1990;1399-1404.

CHAPTER 27

Rotavirus

INTRODUCTION

Rotaviruses are segmented, double-stranded RNA viruses and members of the *Reoviridae*. Rotavirus virions are icosahederal, nonenveloped, and 55-75 nm in diameter. The 11 double-stranded RNA segments that comprise the rotavirus genome are surrounded by two capsid layers. In electron micrographs, rotavirus particles resemble wheels (hence the name) with a wide hub, short spokes, and a narrow rim. Six distinct human rotavirus serotypes have been identified (1). Four serotypes are endemic within the United States.

Rotaviruses are the principal cause of severe diarrheal illness of infants and young children worldwide. In the United States, rotaviruses are responsible for one-third of all hospitalizations for diarrhea in children under 5 years of age (2). Although diarrheal diseases are not a predominant cause of mortality in the United States (approximately 500 deaths/year), the toll from diarrheal diseases in developing countries is staggering (5-18 million deaths/year). Rotavirus diarrhea occurs in all age groups with peak occurrences in children 6-24 months of age. This age group typically experiences an abrupt onset of explosive, watery diarrhea that lasts for 5-8 days. Diarrhea may be preceded by vomiting for 1-3 days. Other common clinical features include isotonic dehydration, compensated metabolic acidosis, and low-grade fever. Newborn infants are often asymptomatic and when diarrhea occurs, it is usually mild. However, severe illnesses have been reported among premature infants (3). Most rotavirus infections in adults tend to be mild or asymptomatic. However adults who care for young children, adult travelers, military personnel, and elderly, institutionalized adults often have symptomatic rotavirus infections.

Transmission occurs via the fecal-oral route, most commonly by person-to-person transmission and by contaminated fomites. Nosocomial infections in pediatric wards also appear to be common. High infection rates have also been reported in preschools and orphanages. Rotaviruses are quite stable to drying. In addition, the presence of fecal material increases the stability of these viruses in the environment and increases their resistance to disinfectants. On human hands, the virus is infective for at least 4 hours (4) and is resistant to several hygienic handwashing agents. Effective disinfection of contaminated material and careful hand washing are the most important measures for controlling the spread of rotavirus infection, especially in a hospital or institutional setting. Food and water have also been suggested as potential sources of transmission. Rotaviruses have been reported to survive in potable and recreational waters for weeks. Although rotaviruses have been detected in respiratory tract secretions, respiratory transmission has not been proven. The mean incubation period is 2 days with a range of 1-3 days. Following infection, patients are infectious from 1 day before, to 8-10 days after, the onset of symptoms. Immunocompromised patients may shed virus for 2-3 weeks.

Although rotaviruses can be detected intermittently throughout the year, seasonal rotavirus outbreaks usually occur from November through April in the United States. There is no seasonal variation in rotavirus diarrheas in tropical areas. In the United States, the peak month of rotavirus activity varies from region to region. Seasonal peaks in rotavirus detection appear in a regular sequence beginning in the Southwest (Nov-Dec) and followed by the Northwest (Dec-Jan), the Midwest (Jan-Feb), the East (Feb-Mar), and New England (Mar-Apr) (5,6).

AT A GLANCE...

ROTAVIRUS

Virus Detection Methods
Cell culture isolations are not clinically efficacious. However, rotavirus can be isolated in standard tube cultures of primary African green monkey kidney, MA104, LLC-MK2, CV-1, Vero, and primary cynomolgus monkey kidney cells. Centrifugation enhanced (shell vial) cultures using AGMK, LLC-MK2, CV-1, Vero, or MA104 cells have also been used. Most laboratories use enzyme immunoassays and latex agglutination tests for rotavirus detection.

Specimen Source
Stool, collected during the first 3-5 days of illness, is the specimen of choice. Stool specimens collected 8 days after the onset of symptoms rarely contain virus. Rectal swabs do not provide enough virus for most procedures and are not recommended.

Time to Result
Standard Culture: Positive Culture - 3-10 days. Negative Culture - 10 days.
Shell Vial Method: Positive Culture - 24-48 hours. Negative Culture - 48 hours.

Rotavirus Serology

IgM Detectable in the serum and feces 5 days after onset of symptoms. Antibody levels usually peak 1-2 weeks and persists for approximately 4 months.

IgG Because of maternal IgG, it is not possible to demonstrate seroconversion in children less than 3 months of age. IgG persists for at least a year.

IgA Serum IgG and IgA coproantibody are detectable from day 4 postinfection. Serum IgA persists for 4-12 months.

Epidemiology
Rotavirus infections occur throughout the year. However, in temperate climates, sharp seasonal outbreaks occur each winter (November - April). Infants less than 3 months and adults are often have only mild symptoms. Nosocomial infections occur frequently.

Transmission/Incubation Period
Rotavirus is highly contagious and transmission occurs predominantly through the fecal-oral route. Waterborne outbreaks are infrequently reported. Although rotaviruses have been detected in respiratory tract secretions, respiratory transmission has not been proven. The mean incubation period is 2 days with a range of 1-3 days. Following infection, patients are infectious from 1 day before, to 8-10 days after, the onset of symptoms. Immunocompromised patients may shed virus for 2-3 weeks.

Inactivators and Disinfectants
Rotaviruses are resistant to many disinfectants including chlorheximidine gluconate and providine-iodine solutions. The presence of fecal material increases the stability of these viruses in the environment and increases their resistance to disinfectants. Phenols, formalin, iodophor compounds, and 10% household bleach appear to be effective disinfectants. However, human rotaviruses appear to be less sensitive to chlorine than animal rotaviruses.

The laboratory diagnosis of rotavirus infections can be accomplished by isolating the virus in cell culture, electron microscopy, enzyme immunoassay (EIA), latex agglutination, and membrane-based EIA methods. Serological tests are not clinically efficacious. Cell culture isolations are not clinically useful because isolations do not provide information in a clinically relevant timeframe and because rotavirus isolation efficiencies are relatively low. Electron microscopy is currently the gold standard for rotavirus detection (7). However, few hospitals have EM facilities. Most laboratories, therefore, use either EIA or latex agglutination methods for rotavirus detection. More than a dozen rotavirus detection kits are commercially available and rotavirus detection can be accomplished in 3 minutes (latex tests) to 3 hours (EIA tests). Latex agglutination assays are usually faster and easier to use than EIA methods. However, latex assays are usually less sensitive than EIA methods.

VIRUS ISOLATION

Introduction

Because of their fastidious nature, human rotaviruses were nearly impossible to cultivate from clinical specimens until 1980, when Wyatt et al. successfully adapted a single virus strain to grow in primary African green monkey kidney cells (8). Since that time, rotaviruses have been isolated in roller tube cultures and in centrifugation-enhanced (shell vial) cultures. A number of cell lines including primary cynomolgus monkey kidney (CMK), primary African green monkey kidney (AGMK), Vero, LLC-MK2, CV-1, and MA104 cells have been used to propagate human and animal rotaviruses. However, primary AGMK appear to be more sensitive than other cells for the isolation of human rotaviruses from clinical specimens (8,10). A distinctive feature of rotavirus growth in cell culture is that virus infectivity can be increased by, and in some cases is dependent upon, trypsin treatment (11,12). Indeed, successful isolation of human rotaviruses from clinical specimens depends upon treatment of the specimen with trypsin and virus propagation in roller tube cultures containing serum-free, trypsin-supplemented medium (9). Primary isolations generally require 5-10 days for detection. After one passage through primary AGMK cells human rotaviruses generally do not need additional trypsin treatments although trypsin-supplemented medium is still required. After two passages in AGMK, many human rotaviruses can be propagated in continuous cultures such as CV-1 and MA104 cells (11)

Specimen Collection and Storage

Stool, collected during the first 3-5 days of symptoms, is the specimen of choice. At least 1 gram of stool should be placed in a leakproof, screwtop plastic or glass container. With liquid stools, the cap should also be sealed with tape and the container should be placed in a sealed plastic bag containing absorbent material. Rectal swabs are generally not adequate because they usually do not contain enough virus. Specimens may be stored at 2-8°C for up to 24 hours and should be transported to the laboratory on wet ice. If specimens cannot be inoculated within 24 hours of collection, they should be frozen at -70°C. Repeated freeze/thaw cycles can inactivate the virus and should be avoided.

Sample Preparation (tube and shell vial cultures)
1. Prepare a 10-15% suspension of feces in serum-free MEM containing antibiotics.
2. Vortex the suspension vigorously to extract the virus.
3. Centrifuge the suspension at 700-1500 x g for 15 minutes to pellet the solids. Transfer the supernatant fluids to a clean tube.
4. Filter the supernatant fluids through a low protein binding 0.45 µm filter.
5. Add TPCK trypsin to produce a 10 µg/ml final concentration.
6. Incubate the specimen for 30 minutes at 35-37°C.

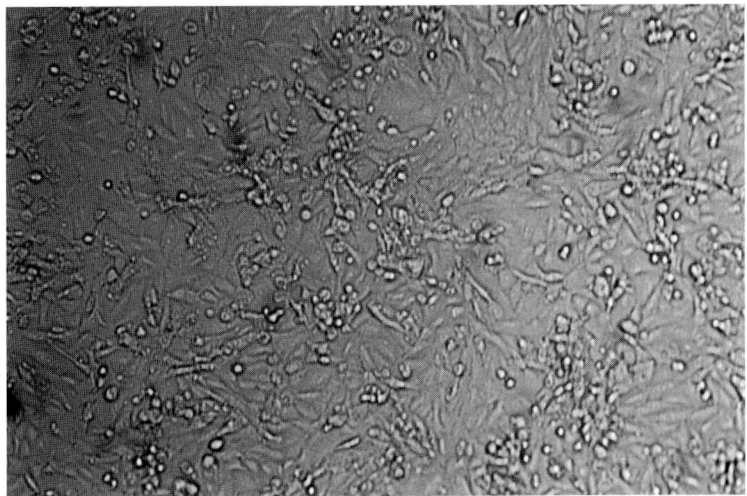

FIG. 1. Rotavirus-infected Vero cells.

Standard Tube Culture
Procedure
1. Appropriately label one tube containing freshly confluent primary AGMK cells.
2. Remove the medium from the cells and wash the monolayer twice with HBSS or serum-free MEM containing antibiotics. These wash steps are important because serum contains trypsin inhibitors.
3. Add 0.2 ml of treated specimen to the tube and allow the specimen to adsorb to the cells for 2 hours at 35-37°C in a roller rack (10-15 rph).
4. Add 2 ml of serum-free MEM containing antibiotics and 0.5-2.0 µg/ml TPCK trypsin.
5. Return the tube to the roller drum and incubate as before for up to 5 days.
6. Examine the cells at least every other day for CPE (Fig. 1).
7. Perform fluorescent antibody testing (see below) on any monolayer that exhibits CPE or on day 5, whichever is earlier.
8. If the fluorescent antibody test is negative, scrape the cells from the tube, transfer a portion of the cells to a new AGMK cell monolayer, and incubate as before for 5 days.

Interpretation of Results
Some isolates may not produce CPE within 5 days and blind passage is sometimes required before CPE is evident. Rotavirus CPE is characterized by cell rounding and granulation (Fig. 1). Eventually, the entire monolayer will slough off the tube wall and disintegrate. Because CPE-negative cultures may contain virus, no report should be made based upon CPE alone.

Culture Confirmation Procedure
1. Remove all but a few drops of medium from the cell culture tube.
2. Scrape the cells from the tube surface and resuspend the cells in the remaining medium.
3. Vortex the tube to break up some of the cell clumps.
4. Use a pipet to spot cells onto an acetone-cleaned glass slide, saving some of the cells for passage onto fresh AGMK cells if the staining is negative.
5. The slide should be processed as directed by the manufacturer of the rotavirus antibody or as described below.
6. Allow the slide to air dry then fix it in acetone for 10-15 min. Fixed slides may be stored for up to one week at 2-8°C in a moisture free container before staining.
7. Place enough of the rotavirus antibody on each well to cover the cell smear (approximately 15-30 µl).

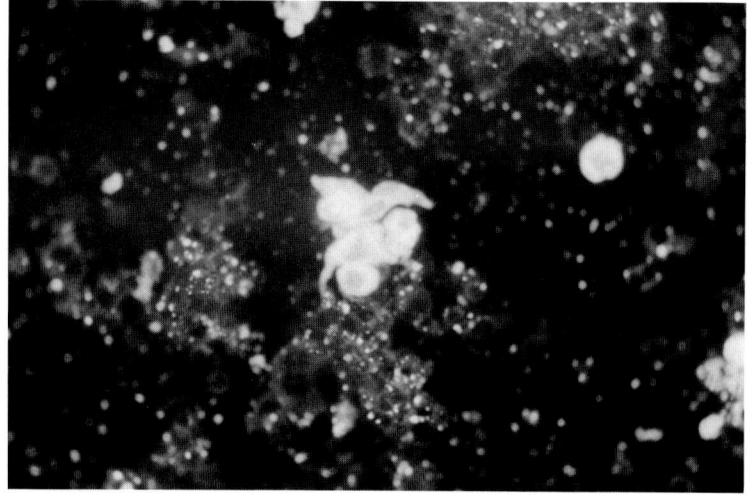

FIG. 2 Fluorescent antibody staining of rotavirus-infected Vero cells. Cells were scraped from the tube and stained.

8. Incubate the slide for 30 minutes at 35-37°C in a covered, humidified chamber. Do not allow the antibody to dry on the slide as this could cause nonspecific antibody binding.
9. Remove the excess antibody with a gentle stream of PBS. Do not direct the stream directly at the cell smear as this could dislodge the cells.
10. Soak the slide in PBS for 5 minutes and shake off the excess fluid.
11. Add 15-30 μl of the appropriate fluorescein-labelled antispecies antibody (conjugate) to the well and incubate for 30 minutes at 35-37°C as before.
12. Remove the excess antibody with a gentle stream of PBS. Do not direct the stream directly at the cell smear as this could dislodge the cells.
13. Soak the slide in PBS for 5 minutes, shake off the excess fluid, and allow the slide to air dry.
14. Carefully add a small drop of FA mounting fluid to the center of each well.
15. Place a number 1 coverslip on the mounting fluid and carefully remove all the air bubbles.
16. Immediately examine the slide using a fluorescence microscope. For optimum clarity, use 200-300X magnification for screening and 400X for confirmation of cell morphology.

Interpretation of Results

Positive Test. The presence of rotavirus is indicated by the presence of characteristic apple-green fluorescence in the cytoplasm of the infected cells on the slide (Fig. 2; see Color Plate 2C following Chapter 7). Only individuals experienced in reading FA reactions should examine the cell smears because cell debris and cell clumps may exhibit a dull fluorescence which can be misinterpreted as a positive reaction by inexperienced technologists.
REPORT: Rotavirus isolated.

Negative Test. The absence of rotavirus virus is indicated by the lack of intense apple-green fluorescence in the infected cells. If the cultures were CPE-, and FA-negative after one blind passage, they may be reported as negative. However, failure to isolate rotavirus does not preclude the presence of rotavirus in the specimen.
REPORT: Rotavirus not isolated.

QC Procedures.
Subpassages of human rotavirus isolates or ATCC cultures should be inoculated into cell

culture tubes with each batch of isolations. Uninfected cell cultures serve as negative controls. Infected and uninfected cell monolayers are scraped and stained as described above. Positive controls must exhibit CPE and intense apple-green cytoplasmic fluorescence. Negative controls must not exhibit specific fluorescent staining.

Preparation of Positive Control Cultures
1. Add 20 µl of a recent human rotavirus isolate (or an ATCC culture) to 2 ml of serum-free MEM.
2. Add TPCK trypsin to produce a 10 µg/ml final concentration.
3. Incubate the specimen for 30 minutes in a 35-37°C water bath.
4. Remove the medium from a 75 cm² flask of freshly confluent AGMK cells and wash the monolayer twice with HBSS or serum-free MEM. This wash step is important because serum contains trypsin inhibitors.
5. Add 2 ml of the treated virus to the flask of AGMK cells and allow the specimen to adsorb to the cells for 2 hours at 35-37°C. Note: The flask should be rocked every 15 minutes to assure even distribution of the virus and to prevent monolayer desiccation.
6. Add 20 ml of serum-free MEM containing antibiotics and 0.5-2.0 µg/ml TPCK trypsin.
7. Incubate the flask at 33-35°C until CPE involves 80-90% of the cells and monolayer sloughs off the flask.
8. Transfer the medium to a 50 ml polypropylene centrifuge tube and freeze/thaw the cells twice to release any intracellular virus.
9. Centrifuge the tube at 600-800 x g to pellet the cell debris.
10. Transfer the supernatant fluids to a sterile vessel containing 80 ml of sterile serum-free MEM.
11. Dispense 0.5 ml of the diluted supernatant fluids into each of 200 freezing vials and freeze the vials at or below -70°C. Store the vials in liquid nitrogen for up to 10 years or at -70°C for up to 3 years.

Centrifugation Culture (Shell Vial) Method
Centrifugation-enhanced (shell vial) cultures can decrease the time-to-result to 24 hours. Although shell vial methods are more rapid than conventional tube cultures, these methods are inherently less sensitive. Because shell vials are stained before CPE develops, a variety of cell lines, including MA104, LLC-MK2, CV-1, Vero, and primary African green monkey kidney (AGMK) cells may be used.

Procedure
1. Appropriately label two shell vial cultures containing freshly confluent primary AGMK cells, LLC-MK2, CV-1, Vero, or MA104 cells.
2. Remove the medium from the cells and wash the monolayer twice with HBSS or serum-free MEM containing antibiotics. This wash step is important because serum contains trypsin inhibitors.
3. Add 0.2 ml of treated specimen to the vial and centrifuge at 3000 x g for 1 hour at 25-35°C.
4. Add 1 ml of serum-free MEM containing antibiotics and 0.5-2.0 µg/ml TPCK trypsin.
5. Incubate at 35-37°C.
6. Stain one coverslip at 24 hours and the other after 48 hours.

Staining of Coverslips
Identification of rotavirus in shell vial systems is accomplished by staining the coverslips with rotavirus antibodies. Staining should be accomplished as directed by the manufacturer of the antibodies or as described below.
1. Carefully remove the culture fluids from the vials and wash the coverslip twice by adding 1-2 ml of PBS and soaking for 5 minutes.
2. Remove the PBS and add 2-3 ml of acetone to each vial. Fix the coverslip for 10 minutes at room temperature.
3. Remove the acetone and allow the coverslip to air dry in the vial.
4. Rinse the coverslip briefly with PBS. The moisture trapped between the coverslip and the vial will keep the stain from wicking

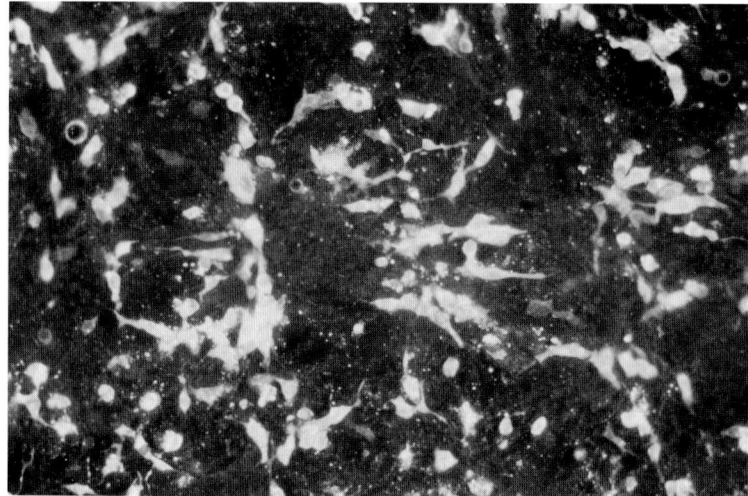

FIG. 3. Fluorescent antibody-stained shell vial (coverslip) cultures of rotavirus-infected CV-1 cells stained with a polyclonal antibody.

under the coverslip.
5. Add enough antibody to the vial to cover the coverslip (100-150 µl) and incubate for 30 minutes at 33-35°C. The shell vials should be covered and incubated in a humidified chamber to prevent the antibody from drying.
6. Wash the coverslip twice as previously described (step 1).
7. Remove the PBS and add enough fluorescein-labelled antispecies antibody (conjugate) to cover the coverslip (100-150 µl).
8. Incubate for 30 minutes as before.
9. Wash the coverslip twice as previously described (step 1).
10. Place a small amount of mounting fluid on a slide. Remove the coverslips from the vials and lay them on the mounting fluid, cell side down.
11. Immediately examine the coverslips at 100-300X using a fluorescence microscope.

Interpretation of Results

Coverslips that have individual cells or small groups of cells that exhibit brilliant apple-green cytoplasmic fluorescence are considered positive for rotavirus (Fig. 3; see Color Plate 4A following Chapter 7). Coverslips without the specific fluorescence described above are considered negative. Some specimens may nonspecifically trap the monoclonal antibodies between the cell monolayer and inoculum cells. Careful examination of the cell morphology can distinguish this type of reaction from specific reactions. Nonspecific staining can also occur if the antibody reagents are allowed to dry on the coverslip. In this case, the entire cell sheet may stain apple-green.

Positive Test. The presence of rotavirus is indicated by the presence of brilliant apple-green cytoplasmic fluorescence (Fig. 3).
REPORT: Rotavirus isolated.

Negative Test. A negative test is indicated by the lack of specific fluorescence described above. The entire cell sheet should be stained red by the counterstain. Failure to isolate rotavirus does not preclude the presence of rotavirus in the specimen.
REPORT: Rotavirus not isolated.

QC Procedures.
A 1:10 dilution of recent human rotavirus isolate or an ATCC culture (see Preparation of Positive Control Cultures, above) should be inoculated with each batch of shell vial cultures. Uninfected shell vial cultures serve as negative controls. Infected and uninfected shell vial cultures should be

processed and stained as described for the patient cultures. Positive controls must exhibit the typical fluorescence patterns. Negative controls must not exhibit specific fluorescent staining.

REFERENCES

1. Kapikian AZ, Chanock RM. Rotaviruses. In: Fields BN, Knipe DM, Chanock RM, Hirsch MS, Melnick JL, Monath TP, Roizman B, eds. *Virology,* Second Edition. New York: Raven Press, 1990;1353-1404.
2. Ho MS, Glass RI, Pinsky PF, Anderson LJ. Rotavirus as a cause of diarrheal morbidity and mortality in the United States. *J infect Dis* 1988;158:1112-1116.
3. Dearlove J, Lathan P, Dearlove B, et al. Clinical range of neonatal rotavirus gastroenteritis. *Brit Med J* 1983;286:1473-1475.
4. Ansari, SA, Springthorpe VS, Sattar SA. 1991. Survival and vehicular spread of human rotaviruses: Possible relation to seasonality of outbreaks. *Rev Infect Dis* 1991;13:448-461.
5. LeBaron CW, Lew J, Glass RI, et al. Annual rotavirus epidemic patterns in North America: results of a 5-year retrospective survey of 88 centers in Canada, Mexico, and the United States. *JAMA* 1990;264:983-988.
6. Centers for Disease Control. Rotavirus surveillance - United States, 1989-1990. MMWR 1991;40:80-87.
7. Dennehy, PH, Gauntlett DR, Spangenberger SE. Choice of reference assay for the detection of rotavirus in fecal specimens: Electron microscopy versus enzyme immunoassay. *J Clin Microbiol* 1990;28:1280-1283.
8. Wyatt RG, Janes WD, Bohl EH, et al. Human rotavirus type 2: cultivation *in vitro. Science* 1989;207:189-191.
9. Wyatt RG, James HD Jr.,Pittman AL, Hoshino Y, Greenberg HB, Kalica AR, Flores, J, Kapikian AZ. Direct isolation in cell culture of human rotaviruses and their characterization into four serotypes. *J Clin Microbiol* 1983;18:310-317.
10. Ward RL, Knowlton DR, Pierce MJ. Efficiency of human rotavirus propagation in cell culture. *J Clin Microbiol* 1984;19:748-753.
11. Barnett, BB, Spendlove RS, Clark ML. Effect of enzymes on rotavirus infectivity. *J Clin Microbiol* 1979;10:111-113.
12. Clark SM, Roth JR, Clark ML, Barnett BB, Spendlove RS. Trypsin enhancement of rotavirus infectivity: mechanism of enhancement. *J Virol* 1981;39:816-822.

CHAPTER 28

Rubella Virus

INTRODUCTION

Rubella is a positive-stranded RNA virus and a member of the *Togaviridae*. Rubella virus is relatively small (60-70 nm), and virions are roughly spherical. Individual virions are characterized by a 30 nm electron-dense core surrounded by a lipid envelope. The envelope contains 5-6 nm surface projections that have hemagglutinin activity. The RNA genome consists of a single-stranded (11 kilobase) RNA molecule that codes for a 110 kD precursor polyprotein and three structural proteins. Rubella virus is relatively unstable and is rapidly inactivated by lipid solvents, trypsin, formalin, and extremes in pH and heat.

Rubella occurs throughout the world. In temperate climates, seasonal outbreaks of rubella usually occur in the spring. Prior to widespread vaccine usage, minor rubella epidemics occurred every 6-9 years while major epidemics occurred at 30 year intervals. Rubella infections were most commonly seen in children 5-9 years of age. Today, rubella is increasingly seen in older age groups (1). Disease is characterized by rash, fever, and a characteristic cervical and occipital lymphadenitis that usually appears a few days before the rash. Typically, rubella rash is macular or maculopapular and closely resembles rashes caused by enteroviruses, human parvovirus, and some arboviruses. The rash initially appears on the face and rapidly spreads to the trunk, arms, and legs. The rash may be inapparent, particularly in dark-skinned individuals. After its initial appearance, the rash usually disappears in 1-3 days. Although adults frequently develop transient polyarthralgia and arthritis after natural infection, other complications such as thrombocytopenia and encephalopathy occur infrequently.

Rubella infections during pregnancy can frequently have disastrous effects on the fetus, causing fetal death, premature delivery, and an array of congenital defects (2). The effects on the fetus depend upon the time of infection. Generally, the younger the fetus, the more severe the resulting disease. During the first 8 weeks of gestation, the fetus has a 40-60 percent chance of spontaneous abortion or of acquiring multiple congenital defects. At 12 weeks, the fetus has a 30-35 percent chance of developing a single defect such as deafness or congenital heart disease. Fetal infection during the 16th week carries a 10 percent risk of a single congenital defect. Occasionally, fetal damage (deafness) has been observed when rubella infections occur as late as the 20th week of gestation (3). Infants with congenital rubella may excrete high titers of virus in nasopharyngeal and other secretions until they are nearly 2 years of age. Therefore, contact with these children posses a risk for susceptible pregnant women.

Rubella virus is transmitted through inhalation of virus-laden aerosol droplets. Once transmitted, rubella virus replicates in the mucosa and the lymphoid tissue of the nasopharynx. The characteristic prodromal enlargement of the posterior and occipital lymph nodes is caused by virus replication in these tissues. Rubella virus is spread via the lymphatics and/or through a transient viremia. Viremia is no longer detectable 16-21 days after exposure and coincides with the production of rubella antibodies and the appearance of the maculopapular rash. Rash usually resolves within 12 hours or it may persist for up to 5 days.

Because rubella-infected patients usually present with mild illness and nonspecific symptomatology, rubella is often difficult to diagnose clinically. Rubella has been confused with many other infections including scarlet fever, mild measles, infectious mononucleosis, toxoplasmosis, roseola,

AT A GLANCE...

RUBELLA VIRUS

Virus Detection Methods
Virus isolations are rarely done except to confirm a diagnosis of congenital rubella and to determine the duration of virus excretion in infants. Isolation is usually done in tube cultures of AgMK, RK-13, SIRC, some Vero strains, and in primary vervet monkey kidney cells.

Specimen Source
Throat and NP swabs collected up to 5 days after the onset of rash are the specimens of choice for the detection of active infection. Cerebrospinal fluid is an appropriate specimen for suspected rubella meningoencephalitis. In cases of suspected congenital rubella virus can be isolated from NP and throat swabs and urine for several months after birth. Rubella may be isolated from tissues including brain, lung, liver and kidney, fetal tissue, amniotic fluids, and placenta. Viremia is rarely detectable after the rash appears.

Time to Result
Positive culture - 3-7 days. However, 1-2 secondary passages are often required.
Negative Culture - 3 weeks.

Rubella Serology

IgM First detectable 1-5 days after the rash appears. Peak titers are observed 1-4 weeks after infection and IgM antibody titers often drop to undetectable levels after 6-12 weeks.

IgG Rubella IgG is first detectable 1-3 days after onset of symptoms. Antibody levels peak after 1-2 weeks and remain detectable throughout the lifetime of the patient. Demonstration of rubella IgG in a single serum specimen indicates that the patient is immune to subsequent infections. Seroconversion or demonstration of a four-fold increase in antibody titer is indicative of a recent infection. IgG subtype prevalence: IgG1>>IgG3>IgG4.

IgA First detectable 1 week after the onset of symptoms with peak levels occurring 2-3 weeks postinfection. Rubella serum IgA levels usually decline thereafter. In some patients, serum IgA titers can persist for years. The IgA response appears to be limited to IgA1.

Epidemiology
Most rubella cases occur in the spring and early summer (February - May) with peak incidence in March and April. Prior to 1965, rubella epidemics occurred at irregular 6-9 year intervals and major epidemics occurred at 30 year intervals. However, the incidence of epidemic rubella has been dramatically reduced by immunization programs. Only one rubella serotype has been identified.

Transmission/Incubation Period
Rubella is mildly contagious and the disease is transmitted through droplet spread and subsequent inhalation of virus-laden aerosols. The incubation period is 16-21 days with a mean of 18 days. The incubation period may be prolonged in patients receiving gamma globulin. Patients are infectious for 1 week before and 7-10 days after the rash appears. Although rubella virus can be isolated from the pharynx of persons who have received the rubella vaccine, these individuals do not appear to transmit rubella to others.

Inactivators and Disinfectants
Rubella virus is very thermolabile and is quickly inactivated by heat, alcohols, and organic solvents. Effective disinfectants include 70% alcohol, 10% household bleach solutions, quaternary ammonium compounds, phenols, iodophor compounds, and glutaraldehyde compounds.

erythema infectiosum, and some enteroviral infections (4,5). Routine laboratory studies are not always helpful for diagnosis because they may only reveal a leukopenia and atypical lymphocytes. Therefore, a definitive diagnosis of rubella can only be made by specific laboratory tests such as isolation of rubella virus or demonstration of seroconversion. The diagnosis of confirmed congenital rubella cannot be established solely on the basis of clinical findings and requires (a) the direct isolation of rubella virus, (b) the demonstration of rubella virus-specific IgM in cord blood or neonatal serum, or (c) the persistence of virus-specific IgG in the infant's serum for more than 6 months.

Serology remains the principal for establishing a clinical diagnosis of acquired rubella and congenital infections. In many laboratories, traditional tests such as complement fixation, hemagglutination inhibition (HAI), neutralization, and infected cell surface immunofluorescence assays are rapidly being replaced by highly sensitive and specific enzyme immunoassays and latex agglutination tests. Despite the availability of more sensitive tests, the rubella HAI test, which was introduced in 1965 (9), remains the reference standard for antibody detection (10). The HAI test is technically more demanding than many of the newer tests and significant interpretation errors and laboratory-to-laboratory variations have been observed. Furthermore, there are rare individuals who are HAI negative but EIA positive and who fail to develop HAI antibody on repeated vaccination but who show rising anti-rubella virus antibody titers by EIA and whose sera are capable of immunoprecipitation by all three structural polypeptides of rubella virus (2). Serological diagnosis of acquired rubella can be made by demonstrating seroconversion, a four-fold increase in rubella IgG in acute and convalescent sera, or by the presence of rubella-specific IgM. Diagnosis of congenital rubella is aided by the demonstration of rubella virus-specific IgM in cord blood or neonatal serum, or the persistence of virus-specific IgG in the infant's serum for more than 6 months.

VIRUS ISOLATION

Introduction

Rubella virus was first isolated in cell culture in 1962 (6,7). In the clinical laboratory, rubella isolations are rarely done except to confirm the diagnosis of congenital rubella or to determine the duration of virus excretion in infants. Primary or secondary African green monkey kidney (AgMK) cells appear to be the most sensitive cells for primary isolations. However, a number of other cells including RK-13, SIRC, some Vero strains, BHK-21, and primary vervet monkey kidney cells have been used for this purpose. Rubella virus rarely produces CPE in primary cell cultures and the cytopathic effects produced in continuous cell lines (RK-13, SIRC, and Vero) are capricious, requiring carefully controlled pH, cell growth, and passage history conditions. Preliminary identification of rubella virus-infected monolayers has traditionally been accomplished by the virus interference test. Verification of rubella virus infection requires either virus neutralization with specific antisera or fluorescent antibody staining.

Specimen Collection

The specimens of choice for the detection of active infection are a nasopharyngeal and a throat swab. Rubella isolations are more likely when the specimen is collected within 5 days after the onset of rash. Cerebrospinal fluid is an appropriate specimen for suspected rubella meningoencephalitis. In cases of suspected congenital disease, virus can be recovered for up to 20 months from NP and throat swabs and from urine. After that time, throat swabs and CSF become the most appropriate specimens. Autopsy material including brain, lung, liver and kidney often yield virus. Fetal tissue, amniotic fluids, and placenta also yield virus after congenital infection. Although virus has been isolated from serum, circulating lymphocytes, and from skin, these sites do not yield virus often enough for routine diagnostic purposes (2).

Rubella virus is very thermolabile and specimens should be inoculated onto cell cultures immediately upon receipt in the laboratory. If specimens cannot be immediately inoculated, they should be stored at 2-8°C for up to 3 days. If longer delays are anticipated, specimens should be stored at or below -70°C. Rubella is rarely isolated from specimens that have been transported at ambient temperature for more than 3 days (4).

Specimen Preparation
Swabs
1. Vortex the viral transport medium, the swabs, and several glass beads for 20-30 seconds to release any bound cells or virus.
2. Remove the swabs from the transport medium and firmly roll them against the inside of the tube to remove as much fluid as possible.

Tissue and Autopsy Specimens

Tissue specimens should be weighed, minced with sterile scissors, and homogenized using a disposable dounce homogenizer. Add enough viral transport medium to the homogenizer to produce a 5% (w/v) suspension. Remove the suspension from the homogenizer and vortex vigorously to extract as much antigen as possible. Centrifuge the specimen at 300-600 x g but do not separate the tissue from the supernatant fluids. Carefully remove 300 μl aliquots for inoculation onto cell cultures.

Test Procedure
1. Appropriately label 2 tubes containing freshly confluent AgMK cells.
2. Add 0.2 to 0.3 ml of specimen to each tube.
3. Place the tubes in a roller drum and incubate at 33-35°C and 10-15 rph.
4. Examine the cells at least every other day for cytotoxicity and CPE. If toxicity is evident, remove the medium and refeed the cells with cell culture maintenance medium.
5. Tubes should be tested (interference test or FA) on day 7 or 8 postinoculation. If the interference test is negative, the remaining tube should be frozen quickly, thawed slowly, and centrifuged at 600 x g to pellet the cell debris. The supernatant fluids are used to inoculate 2 fresh AgMK tubes. A minimum of 2 blind passages are required before cultures are considered negative. Alternatively, cells can be scraped and stained with antibodies to rubella virus.

Interference Test

The interference test is based upon the observation that rubella virus-infected monolayers inhibit the replication of enteroviruses such as ECHO 11 and coxsackie A9 virus. In addition to their sensitivity to rubella virus, primary AgMK cells also have a second advantage in that they are also susceptible to enteroviruses. Therefore, AgMK cells can be used for both the initial isolation of rubella virus and the subsequent interference test without requiring subpassages onto other cell lines. The interference test is not specific for rubella virus and positive specimens should be subpassaged and examined by immunofluorescence for the presence of rubella antigen.

Test Procedure
1. Remove media from the inoculated tubes and freeze at -70°C or below.
2. Rinse cells once with 2-5 ml of sterile HBSS.
3. Add 1 ml of maintenance medium containing 100 $TCID_{50}$ of ECHO 11 to each tube.
4. Incubate the tubes at 35-37°C.
5. Examine the challenge and control tubes daily for the presence of CPE.

Interpretation of Results

After 24 hours, little or no CPE should be evident in the (No Rubella) control tube. If heavy (3-4+) CPE is present, the challenge inoculum contained > 100 $TCID_{50}$ units of ECHO 11 virus and the results may not be valid. ECHO-11 CPE should be evident in the (No Rubella) control within 2-4 days.

Positive Result. If (a) no CPE is present in the patient tube; (b) rubella control tube shows no ECHO 11 CPE; and (c) the No Rubella control shows ECHO 11 CPE, the test is positive for an interfering agent.
REPORT: Interfering agent isolated.

Negative Result. If ECHO-11 CPE occurs in the patient tube at about the same level as in the dilution control, the result is negative for an interfering agent and suggests that no rubella virus is present in the specimen. Negative results should not be reported until after the second passage. If desired, a third challenge can be done.
REPORT: Rubella not isolated after 21 days.

QC Procedures.

Positive (rubella virus) and negative (no-rubella) AgMK control tubes must be challenged each time a patient tube is challenged. Rubella control tubes must interfere with ECHO 11 virus replication and the uninoculated tube must exhibit 1-3+ CPE within 2-4 days. If heavy (3-4+) CPE is present in the no-rubella control tube after 24 hours, the test is invalid and must be repeated. Likewise, if no CPE is evident in the no-rubella control tube after 4 days, the test is invalid and must be repeated.

Preparation of ECHO 11 Virus Challenge
1. Inoculate one freshly confluent RMK tube culture with 0.2 ml of a 1:100 dilution of virus stock (Chapter 12).
2. Incubate the tube at 35-37°C for 24-48 hours until the tube shows 1+ CPE (25% of the cell sheet affected). At this point, the supernatant fluids will contain 10^4 to 10^5 virions per ml. A 1:100 or a 1:1000 dilution will produce a suspension containing 100 $TCID_{50}$ units/ml.
3. Virus dilution should be prepared in MEM containing 10% FCS. Challenge virus should be aliquotted and frozen at -70°C or below for up to 3 years. Alternatively, the challenge virus may be stored in liquid nitrogen for up to 7 years.

Culture Confirmation
1. Remove all but a few drops of medium from the cell culture tube. Store the supernatant fluids at or below -70°C for future passages.
2. Scrape the cells from the surface of the tube.
3. Resuspend the cells in the medium remaining in the tube.
4. Use a pipet to spot cells onto an acetone cleaned glass slide.
5. The slide should be processed as directed by the manufacturer of the antibody to rubella virus or as described below.
6. Allow the slide to air dry then fix it in acetone for 10-15 min. Fixed slides may be stored for up to a week at 2-8°C in a moisture free container before staining.
7. Add enough rubella antibody (approximately 15-30 µl) to cover the cell smear.
8. Incubate the slide for 15-30 minutes at 35-37°C in a covered, humidified chamber. Do not allow the antibody to dry on the slide as this could cause nonspecific antibody binding.
9. Remove the excess antibody with a gentle stream of PBS. Do not direct the stream directly at the cell smear as this could dislodge the cells.
10. Soak the slide in PBS for 5 minutes, shake off the excess fluid.
11. Add enough fluorescein-labelled antispecies antibody (approximately 15-30 µl) to cover the cell smear.
12. Incubate the slide for 15-30 minutes at 35-37°C in a covered, humidified chamber. Do not allow the antibody to dry on the slide.
13. Remove the excess antibody with a gentle stream of PBS as before.
14. Soak the slide in PBS for 5 minutes, shake off the excess fluid, and allow the slide to air dry.
15. Carefully add a small drop of FA mounting fluid to the center of each well.
16. Place a number 1 coverslip on the mounting fluid and carefully remove all the air bubbles.
17. Immediately examine the slide using a fluorescence microscope. For optimum clarity, use 200-300X magnification for screening and 400X for confirmation of cell morphology.

Interpretation of Results
Positive Test. The presence of rubella virus is indicated by the presence of speckled apple-green fluorescence in the cytoplasm of the infected cells

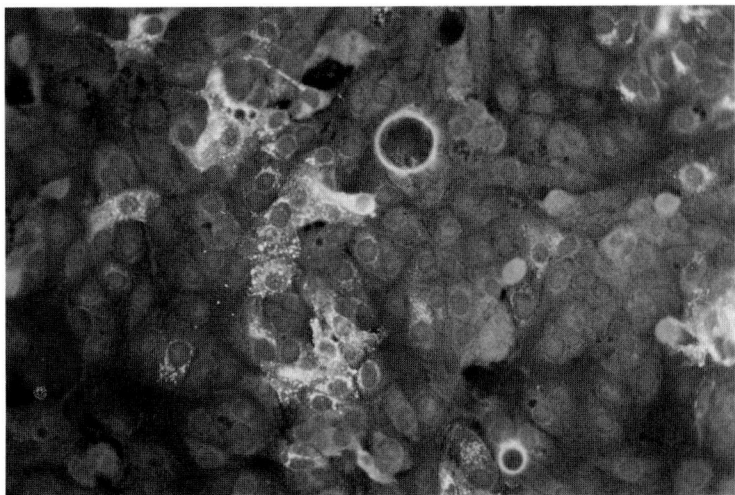

FIG. 1. Fluorescent antibody staining of rubella-infected Vero cells.

(Fig. 1). Only individuals experienced in reading FA reactions should examine the cell smears because cell debris and cell clumps may exhibit a dull fluorescence which can be misinterpreted as a specific, positive reaction by inexperienced technologists.
REPORT: Rubella virus isolated.

Negative Test. The absence of rubella virus is indicated by the lack of intense apple-green fluorescence in the infected cells (Fig. 1).
REPORT: Rubella virus not detected.

QC Procedures.

Subpassages of clinical rubella isolates or ATCC cultures should be inoculated into cell culture tubes with each batch of rubella isolations. Uninfected cell cultures serve as negative controls. Infected and uninfected cell monolayers are scraped and stained as described above. Positive controls must exhibit typical cytoplasmic apple-green fluorescence. Negative controls must not exhibit specific fluorescent staining.

Preparation of Positive Control Cultures

1. Add 20 µl of a rubella virus isolate or an ATCC culture (ATCC VR 315) to 2 ml of sterile MEM containing 2% FCS. Hold the diluted virus on ice until used to inoculate the monolayer.
2. Remove the growth medium from one 75 cm^2 flask of an 80-90% confluent BHK-21 cells.
3. Add 2 ml of the diluted virus to the monolayer and allow the virus to adsorb for 1-2 hours at 33-35°C. Note: The flask should be rocked every 15 minutes to assure even distribution of the virus and to prevent monolayer desiccation.
4. Add 20 ml of MEM containing 2% FCS
5. Incubate the flask at 33-35°C for 7-10 days.
6. Remove the cell culture medium and place it in a 50 ml polypropylene centrifuge tube.
7. Centrifuge the tube at 600-800 x g to pellet any cell debris.
8. Transfer the supernatant fluids to a sterile vessel containing 80 ml of MEM containing 10% FCS.
9. Dispense 1.0 ml of the diluted supernatant fluids into each of 100 freezing vials and freeze the vials for up to three years at -70°C. Vials can be stored in liquid nitrogen for up to 10 years.

REFERENCES

1. Centers for Disease Control. Increase in rubella and congenital rubella syndrome- United States, 1988-1990. *MMWR* 1991;40:93-99.

2. Wolinski JS. Rubella. In: Fields BN, Knipe DM, Chanock RM, Hirsch MS, Melnick JL, Monath TP, Roizman B, eds. *Virology*, Second Edition. New York: Raven Press 1990;815-838.
3. Gershon A. Rubella virus (German measles). In: Mandell GL, Douglas RG Jr., Bennett JE, eds. *Principles and practice of infectious disease*, Third Edition. New York: Churchill Livingstone, Inc., 1990;1242-1247.
4. Cherry JD. Newer viral exanthems. *Prog Med Virol* 1973;16;269-283.
5. Bell EF, Ross CA, Grist NR. ECHO 9 infection in pregnant women with suspected rubella. *J Clin Pathol* 1975;28:267-269.
6. Parkman PD, Buescher RL, Arnstein MS. Recovery of rubella virus from army recruits. *Proc Soc Exp Biol Med* 1962;111:225-230.
7. Weller TH, Neva FA. Propagation in tissue culture of cytopathic agents from patients with rubella-like illness. *Proc Soc Exp Biol Med* 1962;111:215-225.
8. Best JM, O'Shea S. *Togaviridae*: Rubella virus. In: Balows A, Hausler WJ Jr, Lennette EH. *Laboratory diagnosis of infectious diseases - Principles and practice*. New York: Springer-Verlag 1988;435-450.
9. Server JL, Hardy JB, Nelson KB, Gilkeson MR. Rubella in the collaborative perinatal research study. II. Clinical and laboratory findings in children. *Am J Dis Child* 1969;118:123-132.
10. Herrmann KL. Available rubella serologic tests. *Rev Infect Dis* 1985;7:S108-S112.

CHAPTER 29

Varicella-Zoster Virus

INTRODUCTION

Varicella-zoster virus (VZV) is a human herpesvirus and a member of the Alphaherpesvirinae subfamily (1). VZV is an enveloped virus and virions are generally round or polygonal. The VZV nucleocapsid is approximately 100 nm in diameter while the intact virus particles are 180-200 nm. VZV has an extremely large double-stranded linear DNA genome that codes for more than 30 polypeptides (2).

As the name implies, varicella-zoster virus causes two distinct clinical diseases - varicella (chickenpox) and herpes zoster (shingles). Chickenpox is a seasonal childhood disease characterized by low grade fever, malaise, and a generalized vesicular rash. In contrast, shingles exhibits no seasonality and it usually occurs in immunocompromised patients and in older adults. Herpes zoster is characterized by painful vesicular eruptions that are generally limited to a single dermatome. Chickenpox may occur after exposure to either chickenpox or shingles and is a manifestation of primary VZV infection. During primary infection, VZV establishes a latent infection and infection in the dorsal root ganglia. Shingles results from the reactivation of this latent infection. The incidence and frequency of reactivation depends upon a balance of virus and host factors and is not associated with exposure to exogenous VZV.

Chickenpox has a worldwide distribution and, in temperate climates, chickenpox occurs more frequently in the winter and spring months. Although chickenpox is usually a benign, self-limiting disease, chickenpox causes approximately 100 deaths per year in the United States. While the mortality rate for immunocompetent children is less than 2 per 100,000 cases, the mortality rate for adults is 15 times the childhood rate (3). Perinatal varicella can occur when a susceptible mother contracts chickenpox within 5 days of delivery or 48 hours postpartum. Under these circumstances, infant mortality can be as high as 30% because of immune system immaturity and lack of protective maternal antibodies.

Several infections can be mistaken for chickenpox including impetigo, herpes simplex virus, and enterovirus infections. Individuals with chickenpox usually present with a generalized vesicular rash, low-grade fever (100-103°F), and malaise lasting for 3-5 days. The hallmark of this infection is a rash that typically consists of maculopapules, dew drop-like vesicles, and scabs in varying stages of evolution. The rash usually appears on the trunk and face and then spreads rapidly to other areas of the body. Successive crops of vesicles will generally appear over a 2-4 day period. Vesicles initially contain clear fluid but quickly pustulate. Most vesicles are small, 5-13 mm in diameter, and possess an erythematous base (3).

Chickenpox in the immunocompromised patient can be life-threatening. Immunocompromised children, particularly those with leukemia, have more lesions, and the lesions often have a hemorrhagic base. Healing takes nearly three times longer than in normal children (4). Immunocompromised patients are more likely to have progressive complications of visceral organs and mortality rates can be as high as 15-18% (4).

The dermatomal rash that is characteristic of shingles has been recognized as a distinct clinical entity even in the early medical literature. The first suggestion that chickenpox and herpes zoster were caused by the same agent was made by von Bokay (5) who observed that children exposed to family members with shingles often contracted chickenpox. The incidence of herpes zoster is sporadic and the

Chapter 29

> *AT A GLANCE...*
>
> # VARICELLA-ZOSTER VIRUS
>
> **Virus Detection Methods**
> Tube cultures of human diploid fibroblasts, HFF, WI-38, CV-1, and MRC-5 cells.
> Centrifugation-enhanced (shell vial) cultures using MRC-5 or CV-1 cells.
> Direct fluorescent antibody staining of vesicular cell smears.
>
> **Specimen Source**
> Specimen of choice is vesicle fluid (up to 3 days after vesicle appears).
> Blood - Viremia is rarely detectable after onset of rash except in immunocompromised patients.
> Cerebrospinal Fluid - For suspected VZV encephalitis.
> Oropharyngeal Swabs - Virus isolation from this site is rare and therefore not recommended.
>
> **Time to Result**
> Standard Culture: Positive Culture - 3 days to 3 weeks. Negative Culture - 3-4 weeks.
> Shell Vial Method: Positive Culture - 2-4 days. Negative Culture - 4 days.
> Direct Fluorescent Antibody Method: 30-45 minutes.
>
> **VZV Serology**
>
> IgM First detectable 2-5 (varicella) or 8-10 (herpes zoster) days after rash. Peak titers are observed 8-11 days (varicella) or 18-19 (herpes zoster) days after rash develops. IgM antibodies often drop to undetectable levels within 5-6 weeks.
>
> IgG First detectable 4-6 days after onset of rash, peaking after 4-8 weeks. Antibody titers remain elevated for 6-8 months then decline 4-8 fold. IgG antibodies will normally persist at low levels indefinitely.
>
> IgA First detectable 4-6 days after onset of rash with peak levels 1-3 weeks later. VZV IgA antibody responses remain elevated for 1-2 months then decline to undetectable levels within 4 months.
>
> **Epidemiology**
> Most cases occur in winter and early spring (January through April) with peak incidence occurring in March and April. Epidemics occur at irregular 5 year intervals. There is only one known antigenic type.
>
> **Transmission/Incubation Period**
> VZV is highly contagious and transmission is via direct contact and through droplets. Patients are infectious 1-2 days before the rash appears and for 3-5 days thereafter. The incubation period is 11-21 days with a mean of 13-17 days.
>
> **Inactivators and Disinfectants**
> VZV is very labile and is quickly inactivated by heat, alcohols, organic solvents, and repeated freeze/thaw cycles. Effective disinfectants include 70% ethanol, quaternary ammonium compounds, phenols, iodophor compounds, and glutaraldehyde compounds.

disease occurs in all age groups. Overall, shingles afflicts about 10 percent of the total population (6) with the highest incidence of disease (5-10 cases per 1000) occurring in people over 60 years of age. Approximately 4% of patients with shingles will have a recurrence of the disease.

Shingles is characterized by unilateral vesicular eruptions that appear along a single dermatome. Although thoracic and lumbar dermatomes are most frequently affected, shingles can involve the ophthalmic branch of the trigeminal nerve, leading to a sight-threatening condition called zoster ophthalmicus. Disease onset is often heralded by pain along the dermatome that precedes the skin lesions by 48-72 hours. In the immunocompetent host, lesions continue to form for 3-5 days and the total duration of the disease is usually 10-15 days. After the disease abates, it may take as long as a month before the skin returns to normal.

Herpes zoster in the immunocompromised host is more severe than in normal individuals. In these patients, lesion formation continues for up to 2 weeks and scabbing may not occur for 3-4 weeks (7). Patients with lymphoproliferative malignancies are at increased risk for cutaneous dissemination and visceral involvement that can result in varicella pneumonitis, hepatitis, and meningoencephalitis. However, disseminated herpes zoster is rarely fatal, even in the immunocompromised patient. Chronic herpes zoster can also occur in the immunocompromised host, particularly in individuals with AIDS. Chronic herpes zoster is hallmarked by sustained new lesion formation with an absence of lesion healing. This syndrome has been associated with an acyclovir-resistant VZV isolate (3).

Serological testing for VZV antibodies is usually done to confirm a recent infection or to determine the immune status of the patient. For diagnostic purposes, demonstration of a four-fold rise in titer between acute and convalescent sera is generally diagnostic of VZV infection. However, the problems associated with heterotypic IgG responses have caused many laboratories to shift from this traditional IgG testing protocol to the use of a single IgM antibody test for active disease. The presence of VZV IgM is diagnostic of a recent VZV infection. In addition, several studies have found correlations between the presence of VZV antibody in spinal fluid and VZV-associated encephalitis (8,9).

Immune status testing of high risk individuals and hospital staff is of increasing importance in the laboratory. Because VZV can be spread before the rash appears, infected health care workers can unknowingly spread varicella to susceptible individuals within the hospital. Immune status testing is also important for identifying VZV vaccine candidates before they are treated with chemotherapeutic agents. A number of techniques are available to measure VZV-specific antibody levels and demonstration of VZV antibody in a single serum specimen by any of the methods listed below indicates immunity to VZV. However, only the FAMA, ELISA, IFA and virus neutralization tests are sensitive enough to detect VZV antibodies in subjects who had VZV infections more than 10 years previously. FAMA remains the gold standard for VZV antibody testing and early studies questioned whether IFA and ELISA tests could accurately predict VZV immunity (10,11). More recent literature indicates that the IFA and the ELISA tests are equivalent to the FAMA test (12). Thus, a number of test formats are available for immune status testing and the test of choice will depend upon testing volume and technical preference of the individual laboratory.

Immunocompetent patients will usually have detectable IgM titers 2-5 days after the onset of symptoms. IgM levels will generally peak 8-11 days later at which time some sera will have IFA or EIA titers of 1:2560. An IgM response can be expected in 100% of chickenpox cases (13) and up to 80% of shingles cases (14, 15). IgM antibody responses often drop to undetectable levels within a few weeks after the exanthem subsides. Demonstration of VZV-specific IgM is usually diagnostic of disease.

IgG antibodies usually appear 4-6 days after the onset of symptoms and titers reach peak levels 4-8 weeks later. VZV IgG concentrations generally remain elevated for 6-8 months and then decline 4- to 8-fold over several years. Thereafter, IgG levels generally persist at low levels indefinitely. IgG3 subclass antibodies are predominant in chickenpox whereas VZV-specific IgG1 is the dominant subclass in shingles (16).

Antibody levels and profiles can be significantly different in immunocompromised and immunosuppressed patients. Arvin, et al. reported that the median IgG titer for immunocompromised patients with acute varicella was 1:9 while the median IgG titers of immunocompetent individuals was 1:1024 (17).

Heterotypic booster immune responses can further confuse the IgG antibody picture. In this phenomenon, the antigenic similarities between VZV and HSV cause some patients with previous exposure to VZV to experience a four-fold rise in VZV IgG titers following primary HSV infection (18-20). These heterotypic booster immune responses appear to affect all VZV antibody assays but fortunately, the effect appears to be limited to the IgG response (10).

IgA antibodies usually appear 3-5 days after the onset of rash and peak levels are observed 1-3 weeks after the exanthem subsides. VZV IgA antibody responses generally remain elevated for 1-2 months then decline rapidly. VZV IgA antibodies are usually undetectable four months after the infection subsides. However, approximately 10% of adults who had childhood varicella will have low levels (<1:10) of VZV IgA (21).

VIRUS ISOLATION

Introduction

Virus isolation is the gold standard by which all other VZV detection methods are measured. However, traditional cell culture methods have limited clinical utility and because physicians may not receive isolation results for 3-4 weeks. To decrease isolation times, many laboratories have begun to use the centrifugation-enhanced (shell vial) culture methods originally described by Gleaves, et. al. for CMV (22). Shell vial methods can accommodate the same clinical specimens as traditional tube cultures and can reduce VZV isolation time from 21 days to 2-5 days (23). Although shell vial methods are faster, they are slightly less sensitive than traditional tube cultures. The extreme lability of the virus places further demands upon the sensitivity of the cell culture system. As a result, shell vials have not completely replaced traditional tube culture systems and most laboratories inoculate both tube and shell vial cultures for VZV isolations.

VZV grows readily in human diploid fibroblast cell lines and virus isolations are most often accomplished in MRC-5, human foreskin fibroblasts (HFF), CV-1, or WI-38 cells. Isolation times and procedures for standard tube cultures vary from lab-to-lab. Some laboratories perform blind passages on negative cultures after 14 days while others incubate the original culture for 14-28 days without subpassage. In tube cultures, CPE may be visible the second day after inoculation, however, most isolates require 5-7 days of cultivation to produce visible foci. Maximum virus recovery is obtained with newly confluent cell monolayers that are in an active growth phase. It is therefore important to grow these cells and maintain these cell monolayers in a medium containing 10% FCS.

VZV specimens should be inoculated onto two fibroblast culture tubes and 2 MRC-5 shell vials. The addition of two MRC-5 tubes and 1 RMK tube must be used to rule out enterovirus or HSV infections. An A549 tube can be added to rule out adenovirus infection.

Specimen Collection and Storage.

Fresh vesicular fluid obtained from a nonpurulent lesion is the specimen of choice. Selecting a lesion containing clear vesicle fluid is important because viral recovery rates decline as lesions pustulate. Attempts to recover virus from crusted lesions are a waste of time (14). VZV can be isolated from chickenpox lesions if the specimen is collected within the first 3 days of the exanthem. In shingles, VZV can be isolated 7-10 days after the rash appears. Successful isolations beyond these times are rare unless the patient is immunosuppressed.

Although VZV has been isolated from the blood of immunocompromised individuals suffering from VZV pneumonia (24-26), viremia is difficult to detect under normal circumstances (5). VZV has also been isolated from cerebrospinal fluids (5) and from autopsy specimens (27). Attempts to recover VZV from throat swabs or washings, oropharyngeal secretions, and urine are rarely successful and should not be attempted.

VZV is extremely labile and specimens should be inoculated onto cell cultures immediately or stored at -70°C. Specimens should not be frozen at -20°C. Levin, et al. demonstrated that freezing VZV at -20°C for 24 hours resulted in a 99% reduction in viral titer versus a 10-30% reduction when the samples were frozen at -70°C (28).

Specimen Preparation
Swabs
1. Vortex the viral transport medium, the swabs, and several glass beads for 20-30 seconds to release any bound cells or virus.
2. Remove the swab from the transport medium and firmly roll it against the inside of the tube to remove as much fluid as possible.

Tissues

Tissue specimens should be weighed, minced with sterile scissors, and homogenized using a disposable tissue homogenizer. Add enough viral transport medium to the homogenizer to produce a 10-20% (w/v) suspension. Remove the suspension from the homogenizer and vortex vigorously to extract as much antigen as possible. Centrifuge the specimen at 300-600 x g for 10 minutes at 4°C but do not separate the tissue from the supernatant fluids. Carefully remove the required amount of specimen for inoculation.

Standard Tube Culture
Procedure
1. Appropriately label two human fibroblast cell culture tubes.
2. Remove the maintenance medium from the cells and add 2 ml of growth medium.
3. Add 0.2-0.5 ml of the specimen to each tube.
4. Place the tubes in a roller rack (10-15 rph) and incubate at 35-37°C.
5. Cell cultures should be examined for at least every other day for 21 days. During this time the culture medium should be replaced if the pH changes drastically (i.e., red to yellow). Cell cultures that become toxic should be reinoculated onto new cell cultures.

Interpretation of Results

Presumptive identification of VZV in human diploid fibroblast cultures is based upon the presence of characteristic focal CPE and narrow host range. Culture confirmation is usually done using specific antibodies to VZV (29,30). VZV is highly cell associated and spreads slowly from cell-to-cell (29). This pattern of growth produces the focal lesions that are characteristic of VZV--infected monolayers (Fig.1). In human diploid fibroblast cultures, CPE develops slowly and is characterized by groups of refractile, irregularly shaped cells in an otherwise undisturbed cell sheet (Fig. 1). Under optimal conditions, the foci will enlarge to involve the entire cell sheet. Most specimens will produce from 1-15 foci within the 21 day incubation period.

Culture Confirmation Procedure
1. Once CPE develops, remove all but a few drops of culture medium from the tube.
2. Scrape the cells from the surface of the tube and suspend them in the remaining fluid.
3. Using a pipette, place a drop of the cell suspension onto an acetone-cleaned slide. Retain the remainder of the cell suspension for passage into new culture tubes if necessary.
4. Allow the slide to dry at room temperature.
5. The slide should be processed as directed by the manufacturer of the monoclonal antibody or as described below.
6. Fix the slide in cold acetone for 10-15 minutes and allow the slide to air dry. For optimum performance, slides should be stained immediately after fixation. Alternatively, fixed slides may be stored for up to 1 week at 2-8°C or one year at -20°C or below. Slides should be stored under desiccated conditions in order to minimize antigen degradation and background staining.
7. Add enough of the fluorescein-labelled antibody to cover the cell smear (15-50 μl).

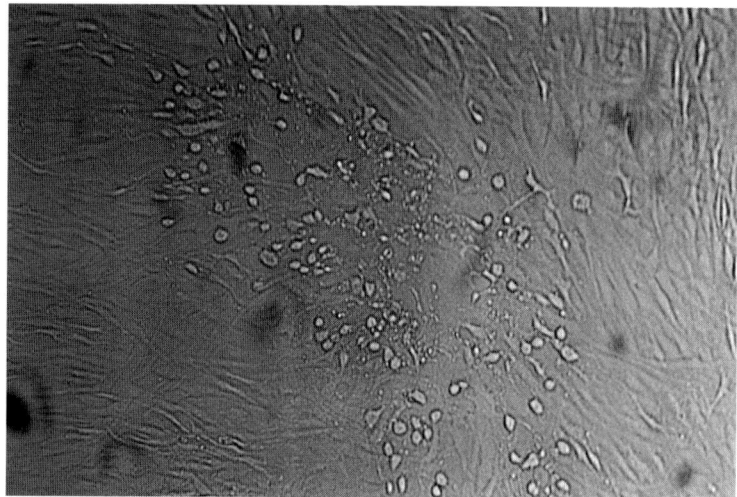

FIG. 1. Cytopathic effects produced in VZV-infected HFF cells 5 days after infection.

8. Incubate the slide for 30 minutes at 35-37°C in a covered, humidified chamber. Do not allow the antibody to dry as this could cause nonspecific antibody binding.
9. Remove the excess antibody with a gentle stream of PBS. Do not direct the stream directly at the cell smear as this could dislodge the cells.
10. Soak the slide in PBS for 5 minutes, shake off the excess fluid, and allow the slide to air dry.
11. Carefully add a small drop of FA mounting fluid to the center of each smear.
12. Place a number 1 coverslip on the mounting fluid and carefully remove all the air bubbles.
13. Immediately examine the slide using a fluorescence microscope. For optimum clarity, use 200-300X magnification for screening and 400X for confirmation of cell morphology.

Interpretation of Results
Positive Test. The presence of VZV is indicated by the appearance of characteristic CPE in the tube culture **and** intense apple-green fluorescence in the cytoplasm of the infected cells (Fig. 2; see Color Plate 2B following Chapter 7). Cell debris and cell clumps may exhibit a dull fluorescence which can be misinterpreted as a positive reaction.

REPORT: Varicella-zoster virus isolated.

Negative Test. The absence of VZV virus is indicated by the uniform red coloration of the cells and a lack of specific staining.
REPORT: Varicella-zoster virus not isolated.

QC Procedures.
Subpassages of VZV clinical isolates or ATCC cultures should be inoculated with each batch of VZV isolations. Uninfected cell cultures can serve as negative controls. Infected and uninfected cell monolayers should be scraped and stained as described above. Positive controls must have typical CPE and the cell smears must contain cells exhibiting intense apple-green fluorescence. Negative controls must stain with the red counterstain and should not exhibit any CPE or specific fluorescent staining.

Preparation of Positive Control Inocula
1. Infect one 75 cm^2 flask containing newly confluent HFF cells with 1.0 ml of a VZV isolate or an ATCC culture.
2. Allow the virus to adsorb for 1-2 hours at 35-37°C. The flask should be rocked every 15 minutes to assure even virus distribution

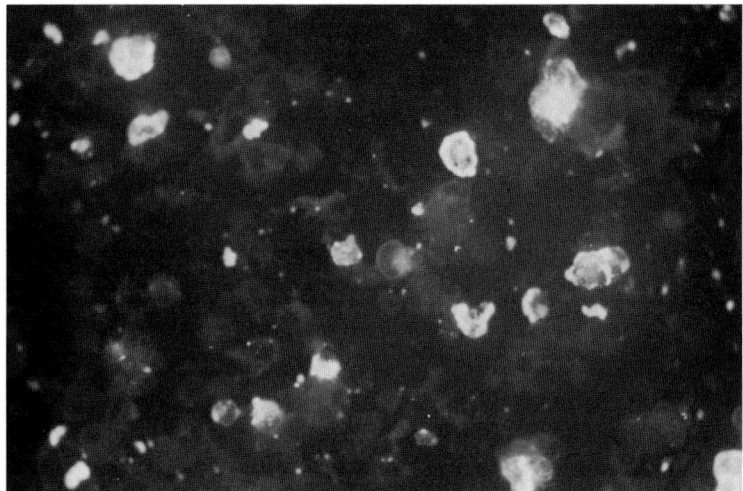

FIG. 2 Fluorescent antibody staining of VZV-infected human foreskin fibroblast after the cells were scraped from the culture tubes.

and to prevent monolayer desiccation.
3. Add 10 ml of DMEM containing high glucose, HEPES, sodium pyruvate, glutamine, and 10% fetal calf serum.
4. Incubate the flask at 35-37°C until the CPE involves 80-100% of the monolayer. Replace the medium after 5-7 days or if it becomes acidic.
5. Trypsinize the cells and remove them from the flask. Centrifuge the cells at 250 x g for 5 minutes.
6. Resuspend the cell pellet in 20 ml of PBS.
7. Slowly add 20 ml of 2X Freezing Medium (Appendix C) to the cell suspension.
8. Dispense 1 ml of the cell suspension into each of 40 freezing vials and freeze the cells as described in Chapter 3.
9. Store the vials in liquid nitrogen for up to 5 years or 18 months at -70°C.

Centrifugation Enhanced (Shell Vial) Method
Procedure
1. Appropriately label two MRC-5 or CV-1 shell vial cultures for each specimen.
2. Aseptically remove the culture medium from the vials.
3. Inoculate each vial with 0.2-0.5 ml of the specimen.
4. Centrifuge the vials at 700 x g for 40 minutes at room temperature.
5. Add 1 ml MEM-10% FCS and antibiotics to each vial.
6. Incubate the vials at 35-37°C. Stain one vial after 48 hours and the other vial after 96 hours of incubation.

Interpretation of Results
Identification of VZV in shell vial systems is accomplished by staining the coverslips with monoclonal antibodies to VZV. Most specimens will produce 1-20 small foci after 48 hours.

Staining of Coverslips
1. Carefully remove the culture fluids from the vial and wash the coverslip twice by adding 1-2 ml of PBS and soaking for 5 minutes.
2. Remove the PBS and add 2-3 ml of acetone to each vial. Fix the coverslips for 10 minutes at room temperature.
3. Remove the acetone and allow the coverslip to air dry in the vial.
4. Rinse the coverslip briefly with PBS. (The moisture trapped between the coverslip and the shell vial will keep the stain from wicking under the coverslip.)

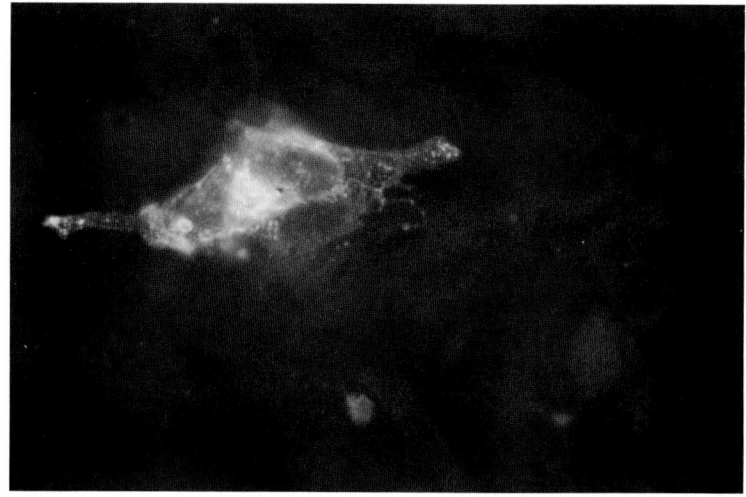

FIG. 3. VZV-infected CV-1 coverslip culture 48 hours after infection. Coverslip was stained with monoclonal antibodies to VZV.

5. Add enough monoclonal antibody to the vial to cover the coverslip (100-150 µl) and incubate 30 minutes at 35-37°C. The shell vials should be covered or incubated in a humidified container to prevent the antibody from drying.
6. Wash the coverslips twice as previously described (step 1).
7. Place a small amount of mounting fluid on a slide.
8. Remove the coverslips and lay them cell side down on the mounting fluid.
9. Immediately examine the coverslips at 100-300X using a fluorescence microscope.

Interpretation of Results
Positive Test. The presence of varicella-zoster virus is indicated by brilliant apple-green fluorescence in cells that are arranged in discrete foci (Fig. 3).
REPORT: Varicella-Zoster Virus Isolated.

Negative Test. The absence of VZV isolation is indicated by the lack of specific fluorescence described above. The entire cell sheet should be stained red by the counterstain. Because of the lability of the virus, a negative result does not preclude VZV infection.

REPORT: Varicella-Zoster Virus Not Detected.

QC Procedures.
Subpassages of VZV clinical isolates or ATCC cultures should be inoculated with each batch of VZV isolations. Uninfected cell cultures can serve as negative controls. Infected and uninfected cell monolayers are scraped and stained as described above. Positive controls must exhibit the typical focal fluorescence patterns. Negative controls must not exhibit specific fluorescent staining. Positive control specimens can be prepared as described above.

DIRECT SMEARS

Introduction
Although virus isolation is the gold standard for the laboratory diagnosis of varicella-zoster virus infections, virus lability and long isolation times limit the diagnostic utility of most culture systems. Several studies have shown that the detection of VZV by direct fluorescent antibody (DFA) methods are faster and more sensitive than traditional culture (13, 30-33). In contrast with 2-14 day isolation times, DFA staining of cell smears and tissue specimens can usually be accomplished in less than

an hour. DFA methods that employ monoclonal antibodies are very specific and VZV antigens can be detected in vesicles and tissues for a longer period of time than culture. Realizing this, some laboratories routinely centrifuge exfoliated cells from the viral transport medium, make smears, and stain the cells with DFA reagents. Although rapid, sensitive, and specific, the DFA methodology is not a panacea. DFA reagents can only detect specific antigens. Therefore, DFA could miss specimens containing more than one virus unless multiple smears are prepared and tested with a multiple antibodies. Exclusive use of DFA reagents could also miss new infectious agents that might appear in the community. For these reasons, most laboratories employ a combination of DFA and culture methods for the diagnosis of VZV.

Specimen Collection

The success of the DFA procedure depends upon the submission of a well made cell smear. Smears that are too thick or "lumpy" can cause nonspecific trapping of the DFA reagent. Smears that are grossly contaminated with red blood cells can also cause interpretation difficulties due to red cell autofluorescence. Finally, cell smears containing too few cells could cause the laboratory to report a false-negative result because no positive cells were observed. Specimen adequacy and its definition are significant sources of laboratory-to-laboratory variation and an area of increasing regulatory concern. Many laboratories require at least 50 cells per smear while other laboratories require 1-2 cells/high power (400X) field. Whatever the number, too many cells can cause nonspecific trapping of the conjugate and with too few cells, the laboratory may not be able to find an infected cell.

Specimen Preparation

Swabs in Viral Transport Medium

1. Vortex the specimen vigorously to release as many cells from the swab as possible.
2. Remove the swab from the viral transport medium and firmly roll it against the inside of the tube to remove as much fluid as possible.
3. Centrifuge the specimen at 1,000-1,500 x g to pellet the cells.
4. Remove all but 100-200 μl of the viral transport medium.
5. Suspend the cell pellet in the remaining fluid.
6. Spread one drop of the cell suspension over one well of an acetone-cleaned slide.
7. Allow the smears to air dry at room temperature.

Test Procedure

For optimum performance, slides should be fixed and stained within 1-2 hours of specimen collection. Alternatively, unfixed slides may be stored for up to 48 hours at 2-8°C. Fixed slides may be stored at 2-8°C for 1 week or frozen at -20°C for up to 1 year. Storing slides under desiccated conditions will decrease the background staining and minimize antigen degradation.

1. Upon arrival in the laboratory, examine the slide wells at 100-300X magnification. Ideally, a minimum of 50 cells should be visible on each well before the specimen is considered adequate for further processing.
2. The slide should be processed as directed by the manufacturer of the labelled antibody or as described below.
3. Fix the cells by immersing the slides in cold acetone for 10 minutes. Allow the slides to air dry.
4. Place enough of the fluorescein-labelled monoclonal antibody on one well to cover the cell smear (15-30 μl). The second well can be stained with labelled antibodies to VZV, HSV, or another virus as appropriate.
5. Incubate the slide for 30 minutes at 35-37°C in a covered, moist chamber. Do not allow the antibody to dry. Drying could cause nonspecific antibody binding.
6. Remove the excess antibody with a gentle stream of PBS. Do not direct the stream directly at the cell smear as this could dislodge the cells.
7. Soak the slide in PBS for 5 minutes, shake off the excess fluid, and allow the slide to air dry.

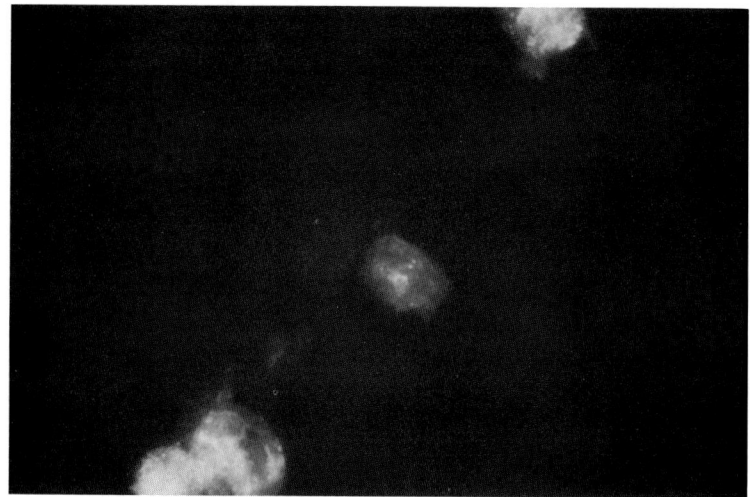

FIG. 4. Vesicular cell smear that was stained with monoclonal antibodies to VZV.

8. Carefully add a small drop of FA mounting fluid to the center of each well.
9. Place a number 1 coverslip on the mounting fluid and carefully remove all the air bubbles.
10. Immediately examine the slide using a fluorescence microscope. For optimum clarity, use 200-300X magnification for screening and 400X for confirmation of cell morphology.

Interpretation of Results

Positive Test. The presence of VZV is indicated by intense apple-green fluorescence in the cytoplasm of the exfoliated cells (Fig. 4; see Color Plate 3C following Chapter 7). Only intact cells should be examined because cell debris and clumps of normal cells may exhibit a dull fluorescence which could be misinterpreted as a positive reaction.
REPORT: Varicella-zoster virus detected.

Negative Test. The absence of VZV is indicated by the lack of specific staining.
REPORT: Varicella-zoster virus not detected.

Inconclusive Test. Specimens with fewer than 25 cells on each well (and no positive cells) may give erroneous results.
REPORT: Unacceptable specimen - too few cells.

QC Procedures.

Positive and negative controls should be stained at least once each day to assure that the antibody reagent is performing properly. Positive cells should stain intensely as described above. Negative cells should only be stained by the counterstain and should not exhibit specific fluorescence. Because the same antibody reagent is used for confirmation of tube cultures, shell vial cultures, and direct smears, culture confirmation testing can be used to demonstrate that the reagent is working properly. In laboratories that do not perform VZV isolations or laboratories where no isolations are ongoing, prepared slides should be stained whenever direct smears are stained. Control slides can be purchased commercially from a number of vendors or they can be prepared in the laboratory as described below.

Preparation of Control Slides

1. Infect one 75 cm^2 flask of newly confluent HFF cells with 1.0 ml of a VZV isolate or an ATCC culture.
2. Allow the virus to adsorb for 1-2 hours at 35-37°C.
3. Add 12 ml of complete medium containing DMEM (high glucose) and 10% fetal calf serum.

4. Incubate the flask at 35-37°C until the CPE involves 40-60% of the monolayer. Replace the medium after 3-4 days if the medium becomes acidic.
5. Trypsinize the cells and remove them from the flask. Centrifuge the cells at 250 x g for 5 minutes.
6. Resuspend the cell pellet in 5 ml of PBS. Perform a cell count and adjust the cell concentration to 2-5 x 10^6 cells/ml.
7. Dispense 3-10 μl of the cell suspension onto each slide and allow the suspension to air dry.
8. Fix the slides in acetone at room temperature for 10 minutes and allow the slides to air dry.
9. Store the slides at -20°C or below for up to one year. Slides should be stored under desiccated conditions to minimize antigen degradation and background fluorescence.

NOTE: Slides prepared in this manner will contain both positive and negative cells. Therefore, a single slide can be used for QC purposes. When slides are made from flasks with 100% CPE, two slides must be used - one containing a positive cell smear and a second slide containing an uninfected cell smear.

REFERENCES

1. Roizman B, Carmichael LE, Deinhardt F, et. al. Herpesviridae: Definition, provisional nomenclature, and taxonomy. *Intervirology* 1981;16;201-217.
2. Gelb LD. Varicella-zoster virus. In: Fields BN, Knipe DM, Chanock RM, Hirsch MS, Melnick JL, Monath TP, Roizman B, eds. *Virology*, Second Edition. New York: Raven Press, 1990;2011-2054.
3. Whitley RJ. Varicella-zoster virus. In: Mandell GL, Douglas RG, Jr., Bennett JE, eds. *Principles and practice of infectious disease*, Third Edition. New York: Churchill Livingstone, Inc., 1990;1153-1159.
4. Feldman S, Hughes WT, Daniel CB. Varicella in children with cancer; seventy-seven cases. *Pediatrics* 1975;56:388-397.
5. Gold E. Serologic and virus isolation studies of patients with varicella or herpes-zoster infection. *N Eng J Med* 1966;274:181-185.
6. Ragizzino MW, Melton LJ III, Kurland LT, et. al. Population-based study of herpes zoster and its sequelae. *Medicine* (Baltimore) 1982;51:310-316.
7. Hope-Simpson RE. The nature of herpes zoster: A long-term study and a new hypothesis. *Proc R Soc Med* 1965;58:9-
8. Andiman WA, White-Greenwal M, Tinghitella T. Zoster encephalitis: Isolation of virus and measurement of varicella-zoster-specific antibodies in cerebrospinal fluid. *Am J Med* 1982;73:769--772.
9. Gershon AA, Steinberg S, Greenberg S, Taber L. Varicella-zoster-associated encephalitis: detection of specific antibody in cerebrospinal fluid. *J Clin Microbiol* 1980;12:764-767.
10. Schmidt NJ, Gallo D. Class-specific antibody responses to early and late antigens of varicella and herpes simplex viruses. *J Med Virol* 1984;13;1-12.
11. Williams V, Gershon AA, Brunell PA. Serologic response to varicella-zoster membrane antigens measured by immunofluorescence. *J Infect Dis* 1974;130:699-672.
12. Landry ML, Cohen SD, Mayo DR, Fong CYK, Andiman WA. Comparison of fluorescent antibody to membrane antigen test, indirect immuno-fluorescence assay, and a commercial enzyme-linked immunosorbent assay for determination of antibody to varicella-zoster virus. *J Clin Microbiol* 1987;25:832-835.
13. Preblud SR. Varicella: Complications and costs. *Pediatrics* 1986;78:728-735.
14. Sundqvist, VA. Frequency and specificity of varicella zoster virus IgM response. *J Virol Methods* 1982;5:219-227.
15. Schmidt, NJ, and Arvin, AM. Sensitivity of different assay systems for immunoglobulin M responses to Varicella-Zoster virus in reactivated infections (Zoster). *J Clin Microbiol* 1986;23:9-78-979.
16. Sundqvist VA, Linde A, Wahren B. Virus-specific immunoglobulin G subclasses in herpes simplex and varicella-zoster virus infections. *J Clin Microbiol* 1984;20:94-98.
17. Arvin, AM, Koropchak CM, Williams BRG, Grumet FC, Foung SKH. Early immune response in healthy and immunocompromised subjects with

primary varicella-zoster virus infection. *J Infect Dis* 1986;154:422-429.
18. Gallo D, Schmidt NJ. Comparison of anti-complement immunofluorescence and fluorescent antibody-to-membrane antigen tests for determination of immunity status to varicella-zoster and for serodifferentiation of varicella-zoster and herpes simplex virus infections. *J Clin Microbiol* 1981;14:539-543.
19. Schmidt NJ. Further evidence for common antigens in herpes simplex and varicella-zoster viruses. *J Med Virol* 1982;9:27-36.
20. Schmidt NJ, Lennette EH, Magoffin RL. Immunological relationship between herpes simplex and varicella-zoster viruses demonstrated by complement fixation, neutralization, and fluorescent antibody tests. *J Gen Virol* 1969;4:321-328.
21. Tovi F, Hadar T, Sidi J, Sarov B, Sarov I. The significance of specific IgA antibodies in the serum in the early diagnosis of zoster. *J Infect Dis* 1985;152:230-
22. Gleaves CA, Smith TF, Shuster EA, Pearson GR. Comparison of standard tube and shell vial cell culture techniques for the detection of cytomegalovirus in clinical specimens. *J Clin Microbiol* 1985;21:217-221.
23. Schirm J, Meulenberg JJ, Pastoor GW, van Voorst Vader PC, Schroder FP. Rapid detection of varicella-zoster virus in clinical specimens using monoclonal antibodies on shell vials and smears. *J Med Virol* 1989;28:1-6.
24. Feldman S, and Epp E. Isolation of varicella-zoster virus from blood. *J Pediatr* 1976;88:265-267.
25. Feldman S, and Epp E. Detection of viremia during incubation of varicella. *J Pediatr* 1979;94:746-748.
26. Myers MG. Viremia caused by varicella-zoster virus: Association with malignant progressive varicella. *J Infect Dis* 1979;140:229-233.
27. Cheatham WJ, Weller TH, Dolan TF, Dower JC. Varicella: report of two fatal cases with necropsy, virus isolation and serologic studies *Am J Pathol* 1956;32:1015-1035.
28. Levin MJ, Leventhal S, Masters HA. Factors influencing quantitative isolation of varicella-zoster virus. *J Clin Microbiol* 1984;19:880-883.
29. Weller TH. Varicella-zoster virus. In: Lennette EH, Schmidt NJ, eds. *Diagnostic procedures for viral, rickettsial, and chlamydial infections*, 5th edition. Washington, DC: American Public Health Association, 1979:375-398.
30. Schmidt NJ, Lennette EH, Woodie JD, Ho HH. Immunofluorescent staining in the laboratory diagnosis of varicella-zoster virus infections. *J Lab Clin Med* 1965;66:403-412.
31. Olding-Stenkvist E, Grandien M. Early diagnosis of virus-caused vesicular rashes by immunofluorescence on skin biopsies. I. Varicella, zoster, and herpes simplex. *Scand J Infect Dis* 1976;8:27-35.
32. Gleaves CA, Lee CF, Bustamante CI, Meyers JD. Use of murine monoclonal antibodies for laboratory diagnosis of varicella-zoster virus infection. *J Clin Microbiol* 1988;26:1623-1625.
33. Rawlinson WD, Dwyer DE, Gibbons VL, Cunningham AL. Rapid diagnosis of varicella-zoster virus infection with a monoclonal antibody based direct immunofluorescence technique. *J Virol Methods* 1989;1:13-18.

APPENDIX A

Glossary of Commonly Used Terms

A549	A continuous cell line derived from a lung adenocarcinoma.
ACIF	Anticomplement immunofluorescence test. An amplified fluorescent antibody method commonly used to detect Epstein-Barr virus nuclear antigens.
Ad	Abbreviation for adenovirus.
AGMK	Primary African green monkey kidney cells.
ATCC	American Type Culture Collection.
Blind Passages	Transferring all or part of a cell culture to a new culture irrespective of whether any CPE or signs of infection are present.
BGM	Buffalo green monkey kidney cells. These cells support the growth of enteroviruses and Chlamydia.
BHK	Baby hamster kidney cells.
BJAB	A transformed lymphoid cell line that does not produce Epstein-Barr virus nuclear antigens.
BME	Basal medium (Eagle).
BSC-1	A continuous cell line derived from African green monkey kidney cells.
BSS	Balanced or buffered salt solution: Isotonic solution of inorganic salts commonly used as a diluent or media base.
Cell Culture	The process of propagating dissociated cells *in vitro*. Sometimes erroneously referred to as tissue culture.
Cell Line	A cell culture that can be propagated indefinitely.
CHO	Chinese hamster ovary cells.
CMK	Primary cynomolgus monkey kidney cells.

Complete Medium	Medium containing all the additives necessary to promote cell growth. Commonly used to denote medium containing serum, L-glutamine, growth factors, antibiotics, and buffers.
Confluent	A condition where cells throughout the culture are in contact with each other creating what appears to be a continuous sheet (or monolayer) of cells.
Contact inhibition	The property of some cells to cease growing when they abut other cells. Cells <u>without</u> contact inhibition tend to pile up in culture.
Continuous cell line	A cell culture that can be propagated indefinitely.
CPE	Cytopathic effect. Changes in cellular structure resulting from external agents. Common cytopathic effects include cell destruction, syncytia formation, cell rounding, vacuole formation, and formation of inclusion bodies.
CV-1	A continuous cell line derived from African green monkey kidney cells.
DFA	Direct fluorescent antibody
DMEM	Dulbecco's modified EMEM. This medium contains a higher concentration of vitamins, amino acids, and glucose than EMEM and is used to propagate primary human cells *in vitro*.
DMSO	Dimethylsulfoxide, a solvent used as a cryoprotectant when freezing mammalian cell cultures.
EDTA	Ethylenediaminetetraacetic acid (Versene), a chelating agent used to remove divalent cations.
EIA	Enzyme immunoassay, a generic name for enzyme-linked immunosorbent assay.
ELISA	Enzyme linked immunoassay, an assay for antigen or antibody that utilizes a solid support coated with capture reagents and an enzyme-labeled reagents that react with the immobilized antigen or antibody. The presence of the enzyme label causes a chemical reaction, the products of which can be detected visually, spectrophotometrically, or with a fluorimeter.
EMEM	Eagle's minimum essential medium, a commonly used medium suitable for a wide variety of anchorage-dependent cells.
FA	Fluorescent antibody. Usually used to denote an assay system utilizing antibodies labeled with a fluorescent compound.
FAMA	Fluorescent antibody to membrane antigen. A common method for detecting antibodies to varicella-zoster virus.

FBS	Fetal bovine serum, used synonymously with fetal calf serum (FCS).
FCS	Fetal calf serum, a nonspecific media additive used to promote cell growth.
FITC	Fluorescein isothiocyanate, a fluorescent label commonly used in fluorescent antibody reagents.
Fomites	Inanimate objects that can be used to transmit disease.
FRhK-4	A continuous cell line derived from fetal rhesus monkey kidney cells and commonly used for isolation of Hepatitis A virus.
GLB	Gelatin-lactalbumin hydrolysate, an inexpensive and relatively inert solution for preparing virus dilutions and for stabilizing fragile viruses before inoculation.
HA	Hemagglutination, a process whereby an external agent (i.e., virus or antibody) can prevent the formation of a red blood cell button in a round bottom tube or well. This process is due to the ability of the agent to link two (or more) red cells thereby forming a lattice that will not pack neatly into the bottom of the tube or well.
HAd	Hemadsorption, the process whereby the presence of viral proteins in the outer membrane of the host cell can cause red blood cells to adsorb or stick to the mammalian cell. This method is commonly used to identify some virus-infected cells before CPE is present.
HBSS	Hank's balanced salt solution, a physiologic solution commonly used for dilutions and for washing cells.
HEK	A primary culture of human embryonic kidney cells.
HEL	A primary culture of human embryonic lung cells.
HeLa-229	A rapidly growing cervical adenocarcinoma cell line commonly used to grow Chlamydia. A subline of HeLa cells.
HEp-2	Human laryngeal carcinoma cell line commonly used to isolate respiratory syncytial virus.
HEPES	N-2-hydroxyethylpiperazine-N'-2-ethanesulfonic acid, an organic buffer used to maintain the pH of the culture medium at about pH 7.5.
HFF	Human foreskin fibroblast cells. These are human diploid fibroblasts with a finite life span.
HL	A continuous human cell line derived from HeLa cells and commonly used to isolate *Chlamydia pneumoniae* (TWAR).

HPF	High powered field, usually 400-800X.
HSV	Herpes simplex virus
Humidity chamber	Any type of chamber (e.g., a petri dish) containing a moistened sponge or paper towel. This chamber is commonly used to prevent antibody reagents from drying on FA slides or coverslip cultures.
IFA	Immunofluorescence assay or indirect fluorescence assay. Immunoassay methods for detecting antigen or antibodies using antibodies that are conjugated to a fluorescent reporter molecule.
IgA	Immunoglobulin A.
IgG	Immunoglobulin G.
IgM	Immunoglobulin M.
KB cells	A continuous cell line derived from an epidermoid carcinoma of the mouth.
Graham 293 cells	A continuous cell line derived from embryonic human kidney and used to isolate fastidious enteric adenoviruses.
Laminar flow hood	A special type of machine that produces an ultraclean environment for cell culture by providing a unidirectional flow of air through high efficiency particulate air (HEPA) filters and onto the work surface. Some laminar flow cabinets can serve as biological safety cabinets while others (type 1 hoods) cannot.
LLC-MK$_2$	A continuous cell line derived from rhesus monkey kidney.
McCoy	A mouse cell line similar to the L-cells that are commonly used to isolate *Chlamydia trachomatis*.
Medium	
Growth medium	Culture medium containing serum and other additives designed to promote cell growth.
Maintenance medium	Culture medium with reduced concentrations of serum (usually 1-2%) designed to support cellular metabolism but not active cell growth.
Freezing medium	Cell culture medium containing cryoprotectants designed to stabilize the cells during their transition into and out of, the frozen state.
Serum free medium	Cell culture medium without serum. Usually used when serum may be toxic to cells or when serum components may inhibit virus replication.

Transport medium	A proteinaceous fluid designed to maintain the viability of viruses from the time they are collected until they can be inoculated onto cell cultures.
MEM	Minimum essential medium, a culture medium commonly used to propagate anchorage-dependent cultures.
MDCK Cells	Madin-Darby canine kidney cells, a canine cell line used to isolate and grow influenza A and influenza B virus.
ML	Mink lung cells. Commonly used to isolate herpes simplex virus and for cytomegalovirus shell vial cultures.
Mounting fluid	A glycerol-based fluid used in FA procedures. Mounting fluid has approximately the same refractive index as glass and thereby prevents loss of signal strength due to refraction.
MRC-5 Cells	A well characterized primary human diploid fibroblast cell strain. These cells support the growth of a number of viruses including herpesviruses and enteroviruses.
MK Cells	Monkey kidney cells. Generally denotes primary cells from cynomolgus or rhesus monkey kidneys.
Microtiter plate	A polystyrene plate consisting of an 8 x 12 well matrix of 0.3-0.4 ml wells. These plates may have round, "V", or flat bottoms and are commonly used for hemagglutination, EIA, and cell culture assays.
NDV	Newcastle disease virus.
Nt	Neutralization test. A test that measures the ability of a specific agent to interrupt the virus to replication cycle.
Neut	Neutralization test.
NP	Nasopharynx or nasopharyngeal.
Passage	The process of dividing a cell monolayer and transplanting it to a new flask for continued growth. This term is often used to denote the relative age of a cell line or strain (passage level).
PBS	A phosphate-buffered physiological saline solution.
PMK	Primary monkey kidney cells.

Primary cells/ Primary cultures	Cells that are usually diploid and have a finite life span in culture. Most primary cultures will not grow beyond 50 passages. Primary cultures are used because they do not cause tumors and because they are closer to the physiologic and genetic norm than continuous cell lines.
PRK	Primary rabbit kidney cells.
Raji cells	A human lymphoblastoid cell line derived from a Burkitts lymphoma. These cells continuously produce Epstein-Barr virus nuclear antigens but not viral capsid antigens.
RBC	Red blood cells.
RD cells	A continuous human cell line derived from a malignant embryonic rhabdomyosarcoma of the pelvis.
RK cells	A continuous cell line derived from rabbit kidney cells.
RK-13	A continuous cell line derived from rabbit kidney cells.
RhMK	Primary rhesus monkey kidney cells.
RPMI-1640	A culture medium formulated at the Roswell Park Memorial Institute for the propagation of lymphoblastoid cells.
Rubber policeman	A scraping device used to remove cells from flasks or tubes.
RMK	Primary rhesus monkey kidney cells.
Roller rack	A slanted, slowly rotating (1-3 rpm) apparatus used for incubating tube cultures.
RSV	Respiratory syncytial virus.
Saline	A physiological salt solution containing 0.85% sodium chloride.
Shell vials	Short flat-bottomed glass tubes that will accommodate a 12 mm coverslip. Commonly used for centrifugation-enhanced culture methods.
SIRC	A continuous cell line derived from rabbit cornea and used for rubella virus isolations.
Split ratio	The ratio at which a cell monolayer is divided to inoculate new cultures. For instance, a culture that received one quarter of the original monolayer did so through a 1:4 split.
SVC	Shell vial culture, usually used to denote the centrifugation-enhanced culture method.

Tissue culture	The process used to propagate entire tissues *in vitro*. This term is often erroneously used to denote methods for *in vitro* propagation of dispersed cells (cell culture).
WI-38	A well characterized human diploid cell culture.
Vero	A continuous cell line derived from African green monkey kidney cells.
Vortex	To mix using a vortex mixer.
VTM	Viral transport medium, a protein-containing fluid designed to preserve the infectivity of viruses from the time they are collected until they can be inoculated onto cell cultures.
VZV	Varicella zoster virus.

APPENDIX B

Reagent Resources

INTRODUCTION

This listing is provided as general guide for locating reagents commonly used in the clinical virology laboratory. This listing is not complete and obviously some vendors will be inadvertently omitted from the list. The authors make no warranties that the reagents listed below are suitable for clinical virology laboratories, merely that such reagents are available for sale. For a more extensive listing, we recommend the Lindscotts Directory for Biological Reagents (707) 544-9555.

Adenovirus
   Control Slides: Bion, Gull
   Monoclonal Antibodies: API, Bartels, Chemicon, Gull
   Virus Controls: ATCC

Biological Safety Hoods
   Baker, Labconco, NuAire

Cell Lines
   Frozen: ATCC, BioWhittaker
   Ready-to-use: Bartels, BioWhittaker, ViroMed

Chlamydia trachomatis
   Control Slides: Bion, Gull, Kallestad, Ortho, Syva
   Monoclonal Antibodies: API, DPC, Difco, Chemicon, Gull, Kallestad, Ortho, Syva, ViroStat
   Controls: ATCC

Chlamydia pneumoniae
   Monoclonal Antibodies: Washington Research Foundation
   Controls: Washington Research Foundation, ATCC

Coverslips
   CMS, Baxter, Fisher

Culture Tubes
   CMS, Baxter, Fisher

Cell Culture Media
   BioWhittaker, Fisher, Gibco, KC Biologicals, Specialty Media

Clostridium difficile
   Antisera: Bartels, TechLab

Cell Cultures: Bartels, Viromed
Controls: Bartels, TechLab

Cytomegalovirus
    Control Slides: Bion, Gull, Kallestad, Syva
    Monoclonal Antibodies: Chemicon, Dupont, Gull, Syva, ViroStat
    Virus Controls: ATCC

Enterovirus
    Typing Reagents: World Health Organization
    Viruses: ATCC

Epstein-Barr Virus
    Control Slides (EBNA): Bion, Gull, Hillcrest, Zeuss/Wampole
    Guinea Pig Complement: Rockland, Cedar Lane, Cappel
    Monoclonal Antibodies: Chemicon, ViroStat
    Virus Controls: ATCC

Guinea Pig Red Blood Cells
    BioWhittaker, Brown Laboratory, Kroy Medical

Hepatitis A Virus
    Monoclonal Antibodies: Chemicon

Herpes Simplex Virus
    Control Slides: Bion, Gull, Ortho, Syva
    Monoclonal Antibodies: API, Chemicon, Dako, DPC, Kallestad, Ortho, Syva, ViroStat
    Virus Controls: ATCC

Human Herpesvirus 6 (HHV-6)
    Control Slides: Stellar Biosystems

Incubators, CO_2
    Forma, Lab-Line

Influenza Virus
    Control Slides: Bion, Bartels, Gull
    Monoclonal Antibodies: API, Bartels, Gull, ViroStat
    Virus Controls: ATCC

Measles
    Control Slides: Bion, Gull
    Monoclonal Antibodies: Chemicon
    Virus Controls: ATCC

Mumps

Control Slides: Bion, Gull
Monoclonal Antibodies: Chemicon, ViroStat
Virus Controls: ATCC

Parainfluenza Virus
Control Slides: API, Bartels, Bion, Gull
Monoclonal Antibodies: API, Bartels, Chemicon, Gull, ViroStat
Virus Controls: ATCC

Human Papilloma Virus
Test kits: Enzo, Gibco, Orgenics, Orion, Oncor

Respiratory Syncytial Virus
Control Slides: API, Bartels, Bion, Genetic Systems, Gull, Ortho
Monoclonal Antibodies: API, Bartels, Chemicon, Genetic Systems, Gull, Ortho, ViroStat, Vitek
Virus Controls: ATCC

Human Retroviruses
Monoclonal Antibodies: Chemicon

Rotavirus
Antibodies: Chemicon, Dako, Silenus, Virostat
Virus Controls: ATCC

Rubella Virus
Control Slides: Bion, Gull
Antisera: Chemicon, Flow, ViroStat
Virus: ATCC

Serum, Fetal Calf
BioWhittaker, Gibco BRL, Hyclone, Fisher

Shell Vials
CMS, Baxter, Fisher

Slides
Carlson Scientific, Cell-Line Associates, CMS, Baxter, Erie Scientific, Fisher

Trypsin
Gibco, BioWhittaker, Hyclone, Sigma

Varicella-Zoster Virus
Control Slides: Bion, Gull, Ortho
Monoclonal Antibodies: Chemicon, Ortho

Tissue Culture Rotators
Belco, Lab-Line

API-Analytab Products, Inc.
200 Express Street
Plainview, NJ 11803
(800) 645-7034
(516) 349-4000

Atlantic Antibodies
Incstar Corporation
1990 Industrial Blvd.
P.O. Box 285
Stillwater, MN 55082
(800) 328-1482
(612) 439-9710

American Type Culture Collection
12301 Parklawn Drive
Rockville, MD 20852
(800) 638-6597
(301) 881-2600

The Baker Company
P.O. Drawer E
Sanford, ME 04073
(800) 992-2537
(207) 324-3869

Bartels Diagnostics
P.O. Box 3093
Bellevue, WA 98009
(800) 277-8357
(206) 392-2992

Baxter Healthcare Corporation
Scientific Products Division
1430 Waukegan Rd.
McGaw Park, IL 60085
(708) 689-8410

Bellco Glass, Inc.
P.O. Box B
340 Edrudo Rd.
Vineland, NJ 08360
(800) 257-7043
(609) 691-1075

Brown Laboratory
P.O. Box 424
Topeka, KS 66601
(913) 233-3174

BioWhittaker, Inc.
8830 Biggs Ford Road
Walkersville, MD 21793
(800) 638-3976
(800) 538-3961

Cappel
Organon Teknika Corporation
100 Akzo Avenue
Durham, NC 27704
(800) 523-7620

Carlson Scientific
514 South Third St.
Peotone, IL 60468

Cederlane Laboratories, Ltd.
300 Shames Drive
Westbury, NY 11590
(800) 645-6264
(800) 255-9378
(516) 333-2221

Chemicon International, Inc.
27515 Enterprise Circle West
Temecula, CA 92390
(800) 437-7500
(714) 676-8080

CMS - Curtin Matheson Scientific
9999 Veterans Memorial Drive
Houston, TX 77038
(713) 820-9898

Dako Corporation
22 North Milpas Street
Santa Barbara, CA 93103
(800) 235-5763
(800) 235-5743
(805) 963-9881

DPC - Diagnostic Products Corp.
5700 W. 96th Street
Los Angeles, CA 90045
(800) 421-5116
(213) 776-0180

Difco Laboratories
P.O. Box 331058
Detroit, MI 48232-7058
(800) 521-0851
(313) 462-8500

Erie Scientific
Portsmouth Industrial Park
Portsmouth, NH 03801-5691
(800) 258-0834
(603) 431-8410

Enzo Diagnostics, Inc.
325 Hudson Street
New York, NY 10013
(800) 221-7705
(212) 741-3838

Fisher Scientific
711 Forbes Avenue
Pittsburgh, PA 15219
(412) 562-1394

Flow Laboratories, Inc.
7655 Old Springhouse Road
McClean, VA 22102
(800) 368-3569
(703) 893-5925

Forma Scientific, Inc.
P.O. Box 649
Millcreek Road
Marietta, OH 45750
(800) 848-3080
(614) 373-4763

Gibco BRL
P.O. Box 68
Grand Island, NY 14072-0068
(800) 828-6686

Gull Laboratories
1011 East 4800 South
Salt Lake City, UT 84117
(800) 448-4855
(801) 263-3524

Hillcrest Biologicals
10703 Progress Way
Cypress, CA 90630-4714
(800) 445-0185
(714) 220-1900
(213) 420-2657

Hyclone Laboratories, Inc.
1725 South Hyclone Road
Logan, UT 84321-6212
(800) 492-5663
(801) 753-4584

Kallestad Diagnositc, Inc.
Sanofi Diagnostics Pasteur
1000 Lake Hazeltine Dr.
Chaska, MN 55318
(800) 666-5511
(612) 448-4848

Kirkegaard & Perry Laboratories, Inc.
2 Cessna Court
Gaithersburg, MD 20879
(800) 638-3167
(301) 948-7755

Kroy Medical
Stillwater, MN 55082

Labconco
8811 Prospect Avenue
Kansas City, MO 64132
(800) 821-5525
(816) 333-8811

Lab-Line Instruments, Inc.
15th and Bloomingdale Avenues
Melrose Park, IL 60160
(800) 323-0257
(312) 450-2600

Nuaire, Inc.
2100 Fernbrook Lane
Plymouth, MN 55447
(800) 328-3352
(612) 553-1270

Orgenics
P.O. Box 576, Kings Park
New York, NY 11756
(800) 628-8751
(301) 776-3151

Orion Diagnositca
P.O. Box 83
02101 Espoo
Finland

Oncor
209 Perry Parkway
Gaithersburg, MD 20877
(800) 776-6267
(301) 963-3500

Ortho Diagnostic Systems
Route 202
Raritan, NJ 08869
(800) 526-3875
(908) 218-1300

Sigma Chemical Company
P.O. Box 14508
St. Louis, MO 63178
(800) 325-3010
(314) 771-5750

Silenus Laboratories
P.O. Box 398
Hawthorn, Australia 3122
61-3-819-5000

Specialty Media
245 Bryn Mawr Avenue
Lavallette, NJ, 08735
(201) 830-4008

Syva Company
P.O. Box 10058
Palo Alto, CA 94303
(800) 227-8994
(415) 960-0720

Stellar Biosystems
9075 Guilford Road
Columbia, MD 21046
(800) 962-6790
(301) 381-8850

TechLab, Inc.
Virginia Tech Corporate Research Park
1861 Pratt Drive
Balcksburg, VA 24061
(703) 231-3943

ViroMed Laboratories, Inc.
5100 Gamble Drive
Suite 55
Minneapolis, MN 55416
(612) 545-5277

ViroStat, Inc.
P.O. Box 8522
Portland, ME 04104
(207) 883-1491

Vitek Systems
595 Anglum Drive
Hazelwood, MO 63042
(800) 638-4835

Washington Research Foundation
4225 Roosevelt Way N.E., Suite 303
Seattle, WA 98105
(206) 633-3569

Wellcome Diagnostics
3030 Cornwallis Road
Research Triangle Park, NC 27709
(800) 334-8570
(919) 248-4617

WHO Enterovirus Reference Laboratory
Statens Seruminstitut
80 DK 2300
Copenhagen, DENMARK

Appendix C

REAGENT FORMULATIONS

Table 1. Reagent Formulations

Solution	Ingredients		Preparation and Storage
Alsever Solution (Modified)	20.5 g 4.2 g 8.0 g 0.55 g 1.0 L	Dextrose NaCl Sodium Citrate ($Na_3C_6H_5O_7 \cdot 2H_2O$) Citric Acid ($C_6H_8O_7 \cdot H_2O$) Endotoxin-Free Distilled Water	Dissolve reagents in 700 ml of water. Adjust pH to 6.0-6.2 with 1N HCl or 1 N NaOH as needed. Filter sterilize using a 0.2 µm filter. Store at 2-8°C for up to 2 years unopened and 3 months after opening the bottle.
0.25 mg/ml Amphotericin B Stock Solution	50.0 mg 10.0 ml 190.0 ml	Amphotericin B (Fungizone, Squibb) Sterile Endotoxin-Free Water Hank's Balanced Salts Solution (HBSS)	Reconstitute 1 vial (50 mg) of Amphotericin B with 10 ml of sterile water. Dilute to 200 ml with sterile HBSS. Dispense in convenient aliquots. Store frozen at -20°C for up to 1 year or until the expiration date on the original vial. Thawed vials should be stored at 2-8°C and discarded after 1 week.
Antibiotic Solution for Treating Specimens in Chlamydia Transport Medium	6.0 ml 0.8 ml 23.2 ml	0.25 mg/ml Amphotericin B Stock Solution 10 mg/ml Gentamicin Solution Hank's Balanced Salts Solution (HBSS)	Mix solutions in a sterile tube. Dispense in 2 ml volumes and store at -20°C for up to 1 year but not to exceed the closest expiration date on the original vial. 0.2 ml of this solution is added to specimens in chlamydia transport medium to inhibit bacterial and fungal growth.

Table 1. Reagent Formulations (Continued)

Solution	Ingredients		Preparation and Storage
Antibiotic Solution for Treating Specimens in Viral Transport Medium	6.0 ml 6.0 ml 0.8 ml 7.2 ml	Penicillin/Streptomycin Solution 0.25 mg/ml Amphotericin B Stock Solution 10 mg/ml Gentamicin Solution Hank's Balanced Salts Solution (HBSS)	Mix solutions in a sterile tube. Dispense in 2 ml volumes and store at -20°C for up to 1 year but not to exceed the closest expiration date on the original vial. 0.2 ml of this solution is added to specimens in viral transport medium to inhibit bacterial and fungal growth.
ATV Trypsin	8.0 g 0.4 g 1.0 g 0.58 g 0.5 g 0.2 g 0.5 ml 1.0 L	NaCl KCl Dextrose NaHCO$_3$ Trypsin (Porcine parvovirus tested) Versine (disodium ethylenediamine tetraacetate) Phenol red stock solution Distilled water	Dissolve all the reagents except the trypsin and phenol red in 800 ml of water. Some heating may be required to dissolve the versine. Cool to room temperature. Add trypsin and phenol red. Add remaining water to the 1 liter mark. Filter sterilize through a 0.22 μm filter. Dispense in convenient aliquots and store at -20°C for up to 1 year.
1X Cell Freezing Medium	77.5 ml 15.0 ml 7.5 ml	Eagle's Minimum Essential Medium Fetal Calf Serum (Heat Inactivated) Dimethylsulfoxide	Use sterile reagents. Store at 2-8°C for up to 3 months.
2X Cell Freezing Medium	27.5 ml 15.0 ml 7.5 ml	Eagle's Minimum Essential Medium Fetal Calf Serum (Heat Inactivated) Dimethylsulfoxide	Use sterile reagents. Store at 2-8°C for up to 3 months.

Table 1. Reagent Formulations (Continued)

Solution	Ingredients		Preparation and Storage
Cell Freezing Medium With Glycerol	80.0 ml 15.0 ml 5.0 ml	Eagle's Minimum Essential Medium Fetal Calf Serum (Heat Inactivated) Sterile Glycerol	Use sterile reagents. Store at 2-8°C for up 3 months.
Cell Growth Medium	450.0 ml 50.0 ml 10.0 ml 5.0 ml 5.0 ml 0.5 ml	Eagle's Minimal Essential Medium (MEM) Fetal Calf Serum (Heat Inactivated) 200 mM L-glutamine 1 M HEPES Penicillin/Streptomycin Solution 10 mg/ml Gentamicin Solution	All solutions should be sterile. A sterility check should be done before the medium is used. Store at 2-8°C for up to 2 months. Additional L-glutamine should be added after 1 month.
Cell Maintenance/Refeed Medium (Chlamydia) and Chlamydia Overlay Medium	450.0 ml 10.0 ml 10.0 ml 5.0 ml 0.5 ml 8.0 ml 25.0 ml 2.0 ml	Eagle's Minimal Essential Medium (MEM) Fetal Calf Serum (Heat Inactivated) 200 mM L-glutamine 1 M HEPES 10 mg/ml Gentamicin Solution 0.25 mg/ml Amphotericin B Stock Solution 8.8% Glucose 200 µg/ml Cycloheximide Stock Solution	All solutions should be sterile. A sterility check should be done before the medium is used. Store at 2-8°C for up to 2 months. Additional L-glutamine should be added after 1 month.
Cell Maintenance/Refeed Medium (RMK Cells)	500.0 ml 10.0 ml 5.0 ml 5.0 ml 0.5 ml 2.0 ml	Eagle's Minimal Essential Medium (MEM) 200 mM L-glutamine 1 M HEPES Penicillin/Streptomycin Solution 10 mg/ml Gentamicin Solution 0.25 mg/m. Amphotericin B Stock Solution	All solutions should be sterile. A sterility check should be done before the medium is used. Store at 2-8°C for up to 2 months. Additional L-glutamine should be added after 1 month.

Table 1. Reagent Formulations (Continued)

Solution	Ingredients		Preparation and Storage
Cell Maintenance/Refeed Medium (Virus Isolations)	450.0 ml 10.0 ml 10.0 ml 5.0 ml 5.0 ml 0.5 ml 2.0 ml	Eagle's Minimal Essential Medium (MEM) Fetal Calf Serum (Heat Inactivated) 200 mM L-glutamine 1 M HEPES Penicillin/Streptomycin Solution 10 mg/ml Gentamicin Solution 0.25 mg/ml Amphotericin B Stock Solution	All solutions should be sterile. A sterility check should be done before the medium is used. Store at 2–8°C for up to 2 months. Additional L-glutamine should be added after 1 month.
Chlamydia Transport Medium (2SP)	68.46 g 2.01 g 1.01 g 100.0 ml 1.0 L	Sucrose K_2HPO_4 KH_2PO_4 Fetal Calf Serum Distilled Water	Combine dry chemicals and add distilled water to the 1 liter mark. The solution may be warmed to dissolve the chemicals. Autoclave at 121°C for 15 minutes. Cool to room temperature and add FCS. Aseptically dispense 3 ml into sterile screw-capped vials. Store at 2–8°C for up to 1 year.
200 µg/ml Cycloheximide Stock Solution	10.0 mg 50.0 ml	Cycloheximide Sterile Distilled Water	Add the water to a vial containing the cycloheximide. Dispense into 1 ml aliquots and store at -20°C for up to 1 year or until the expiration date on the original vial. **CAUTION: CYCLOHEXIMIDE IS EXTREMELY TOXIC. WEAR GLOVES AND A MASK WHEN PREPARING THIS REAGENT.**

Table 1. Reagent Formulations (Continued)

Solution	Ingredients		Preparation and Storage
Ficoll-Hypaque	20.0 ml 6.04 g 80.0 ml	Hypaque, sodium, 50% (Winthrop Laboratories) Ficoll (400,000 MW) (Sigma Chemical Co.) Distilled Water	Filter sterilize through a 0.2 µm filter. Dispense into sterile conical centrifuge tubes. Store at 2-8°C for up to 6 months.
10 mg/ml Gentamicin Solution	10.0 ml 30.0 ml	Gentamicin, 40 mg/ml Hank's Balanced Salts Solution (HBSS)	Mix solutions in a sterile tube. Dispense into convenient amounts. Store at 2-8°C for up to 1 year but not to exceed the expiration date on the original vial.
GLB - Gelatin-Lactalbumin Hydrolysate	5.0 g 2.5 g 1.0 L	Gelatin Lactalbumin hydrolysate (Gibco) Hanks'Balanced Salts Solution	Dissolve the gelatin and lactalbumin hydrolysate in HBSS. Gentle heating may be required to dissolve the gelatin. Dispense 99 ml into each of 10 glass bottles and autoclave at 121°C for 15 minutes. Store at room temperature for up to 2 years. **Before use**: adjust the pH to 7.2-7.4 with sodium bicarbonate and add 1 ml of Penicillin/Streptomycin solution. Store at 2-8°C for up to 1 month after opening.
8.8% Glucose	8.8 g 100.0 ml	Glucose Endotoxin-Free Distilled Water	Stir to dissolve. Filter sterilize using a 0.2 µm filter. Store at room temperature for up to 2 years.

Table 1. Reagent Formulations (Continued)

Solution	Ingredients		Preparation and Storage
Guinea Pig RBCs, 0.5%	0.5 ml 100.0 ml	Packed, washed guinea pig red blood cells PBS or physiologic saline	Place 8 ml of sterile PBS or physiologic saline in a graduated 15 ml centrifuge tube. Add 2 ml of guinea pig RBCs in Alsever's solution. Centrifuge at 700 x g for 10 minutes. Carefully remove the supernatant fluids. Repeat the washes until the supernatant fluids are clear. Prepare a 0.5% suspension as described. Store at 2-8°C for up to 7 days.
1 M HEPES Buffer	23.83 g 100.0 ml	N-2-hydroxyethylpiperazine-N'-2-ethanesulfonic acid (HEPES) Endotoxin-free Distilled Water	Stir to dissolve. Autoclave at 121°C for 15 minutes. Store at 2-8°C for up to 1 year.
1 N Hydrochloric Acid	91.7 ml 8.3 ml	Endotoxin-Free Distilled Water Concentrated (11.6N) HCl	Add the acid to the water. Filter sterilize. Store at room temperature for up to 1 year.
Mounting Fluid for Fluorescent Antibody Tests	18.0 ml 2.0 ml	Glycerol, Spectranalyzed (Fisher Scientific) Phosphate Buffered Saline, pH 8.5	Some types of glycerol will autofluorescence. Therefore, only glycerol grades that have been tested for autofluoresce should be used. Dispense into dropper bottles and store at room temperature for up to 3 months.

Table 1. Reagent Formulations (Continued)

Solution	Ingredients		Preparation and Storage
Penicillin/Streptomycin Solution	1. Vial 2.0 g 200.0 ml	Penicillin, 5 million Units per vial Streptomycin, 1 g per vial Sterile Endotoxin-Free Distilled Water	Dissolve 1 vial of penicillin in 100 ml of sterile distilled water. Dissolve 2 vials of streptomycin in sterile distilled water. Combine the two solutions to produce a stock solution containing 25,000 U/ml of penicillin and 10,000 µg/ml streptomycin. Dispense 5 ml into sterile tubes and store at -20°C for up to 1 year but not to exceed the nearest expiration date on the original vials. The final antibiotic concentration in cell culture medium is 250 U/ml penicillin and 100 µg/ml streptomycin.
Phosphate Buffered Saline (PBS), pH 7.4	40.0 g 1.0 g 5.8 g 1.0 g 5.0 L	NaCl KCl Na_2HPO_4, anhydrous KH_2PO_4 Deionized, endotoxin free, water	Stir to dissolve. Dispense into 500 ml or 1 liter bottles. Autoclave at 121°C for 15 minutes. Store at room temperature indefinitely. Once opened, discard after 2 months.
Sorenson's Phosphate Buffered Saline (PBS), pH 7.4	8.5 g 47.18 g 6.13 g	NaCl $Na_2HPO_4 \cdot 7H_2O$ $NaH_2PO_4 \cdot H_2O$	Stir to dissolve. Dispense into 500 ml or 1 liter bottles. Autoclave at 121°C for 15 minutes. Store at room temperature indefinitely. Once opened, discard after 2 months.

Table 1. Reagent Formulations (Continued)

Solution	Ingredients		Preparation and Storage
pH 3 Medium	10.0 ml 1.0 ml 1.0 ml 2.0 ml 86.0 ml	EMEM Penicillin/Streptomycin Solution Essential Amino Acids (100X) (Gibco) 1N HCl Sterile Physiologic Saline	All reagents should be sterile. Add reagents and check the pH. The pH should be 2.0-2.2 to overcome the buffering capacity of the culture media. Store at 2-8°C for up to 6 months.
Phytohemagglutinin P Stock Solution	50.0 mg 100.0 ml	Phytohemagglutinin P (Difco Laboratories) RPMI-1640	Stir to dissolve. Filter through a 0.22 µm filter. Dispense in convenient aliquots. Store at -20°C for up to two weeks.
PHA Medium (Retrovirus Assay)	380.0 ml 100.0 ml 5.0 ml 8.0 ml 5.0 ml 4.0 ml 5.0 ml	RPMI-1640 (Gibco) Fetal Calf Serum (Hyclone), Heat inactivated Pencillin/Streptomycin Solution 0.25 mg/ml Amphotericin B Solution 200 mM L-Glutamine Solution (Gibco) 7.5% Sodium Bicarbonate Solution Phytohemagglutinin P stock solution	Combine reagents in s sterile 500 ml bottle. Add bicarbonate until pH is 7.2-7.4. PHA medium can be stored for at 2-8°C for two weeks after preparation. For best results, PHA should be thawed and added to the medium just before use.
Phenol Red Stock Solution, 1%	20.0 ml 10.0 g 1.0 L	1 N Sodium Hydroxide Phenol Red (USP) Endotoxin-Free Distilled Water	Add the phenol red to a 1 liter volumetric flask. Add the sodium hydroxide and stir to dissolve. Up to 40 additional ml of 1 N sodium hydroxide may be needed to completely dissolve the phenol red. Add water to the 1 liter mark. Filter sterilize using a 0.45 µm filter. Store at room temperature for up to 1 year.

Table 1. Reagent Formulations (Continued)

Solution	Ingredients		Preparation and Storage
0.1% Polybrene Stock Solution	0.1 g 100.0 ml	Polybrene (hexadimethrene bromide, Sigma) RPMI-1640	Stir to dissolve. Solution will be slightly cloudy. Filter through a 0.2 μm filter. Store at 2-8°C for up to 6 months.
Propagation Medium (Retrovirus Assay)	360.0 ml 100.0 ml 25.0 ml 0.5 ml 1.0 ml 5.0 ml 8.0 ml 5.0 ml 4.0 ml	RPMI-1640 (Gibco) Fetal Calf Serum, (Hyclone) Heat inactivated. Interleukin-2 (Cellular Products, Inc.) Anti-human interferon alpha, 200,000 U/ml (Miles Scientific) Polybrene, 0.1% stock Solution Pencillin/Streptomycin Solution 0.25 mg/ml Amphotericin B stock solution 200 mM L-Glutamine (Gibco) 7.5% Sodium Bicarbonate Solution	Combine the reagents in a sterile 500 ml bottle. Add bicarbonate until pH is 7.2-7.4. Medium without PHA may be stored at 2-8°C for up to 6 weeks. After PHA is added, medium should be used within 2 weeks. Store at 2-8°C.
7.5% Sodium Bicarbonate	7.5 g 100.0 ml	$NaHCO_3$ Endotoxin-Free Distilled Water	Stir to dissolve. Autoclave at 121°C for 15 minutes. Store at room temperature for up to 1 year. Once opened store at 2-8°C for up to 30 days.
1 N Sodium Hydroxide	40.0 g 1.0 L	NaOH pellets Endotoxin-Free Distilled Water	Combine NaOH and water in a glass flask. Stir to dissolve. Autoclave at 121°C for 15 minutes. Transfer the solution to a sterile polypropylene container. Store at room temperature for 6 months.

Table 1. Reagent Formulations (Continued)

Solution	Ingredients		Preparation and Storage
70% Sorbitol	70.0 g	Sorbitol	Dissolve sorbitol in 70 ml of water. Gentle heating can be applied if necessary to dissolve the sorbitol. Add water to the 100 ml mark. Filter sterilize through a 0.2 μm filter. Add phenol red and adjust the pH to 7.0-7.2 with 1 N NaOH or 1 N HCL. Store at 2-8°C for up to 2 years.
	100.0 ml	Endotoxin-Free Distilled Water	
	0.1 ml	1% Phenol Red Stock Solution	
Trypan Blue, 0.4%	0.2 g	Trypan Blue	Dissolve the trypan blue in the normal saline. Filter through a 0.45 μm filter. Store at room temperature for 1 year unopened and 2 months after opening.
	50.0 ml	Physiological Saline	
Viral Transport Medium	2.95 g	Tryptose Phosphate Broth	Stir to dissolve. Some heating may be required to dissolve reagents. Autoclave at 121°C for 15 minutes. Cool and dispense 3 ml into sterile screw-capped vials. Store at 2-8°C for up to 1 year.
	2.06 g	$Na_2HPO_4 \cdot 7H_2O$	
	0.08 g	$NaH_2PO_4 \cdot H_2O$	
	5.0 g	Gelatin	
	1.5 ml	1% Phenol Red Stock Solution	
	1.0 L	Endotoxin-Free Distilled Water	
Viral Transport Medium (HEPES Tryptose Gelatin Medium)	26.0 g	Tryptose Broth	Dissolve gelatin in 100 ml of hot water. Add remaining ingredients and about 500 ml of water. Add 1 N NaOH until the pH is 7.2-7.4. Add water to the 1 liter mark. Autoclave at 121°C for 15 minutes. Dispense 3 ml into sterile screw-top vials. Store at 2-8°C for up to 1 year.
	5.0 g	Gelatin	
	2.0 ml	1% Phenol Red Stock Solution	
	10.0 ml	1 M HEPES	
	1.0 L	Endotoxin-Free Distilled Water	

Table 1. Reagent Formulations (Continued)

Solution	Ingredients		Preparation and Storage
Viral Transport Medium (MEM-Gelatin plus Antibiotics)	100.0 ml 5.0 g 2.0 ml 1.0 ml 1.0 L 0.4 ml	10X Eagle's Minimal Essential Medium Gelatin 10 mg/ml Gentamicin Solution Penicillin/Streptomycin Solution Endotoxin-Free Distilled Water 0.25 mg/ml Amphotericin B Stock Solution	Dissolve gelatin in 100 ml of hot water. Cool to room temperature and remaining ingredients. Add water to the 1 liter mark. filter sterilize using a 0.2 μm filter. Dispense 3 ml into sterile screw-cap vials. Store at 2-8°C for up to 2 months.
Viral Transport Medium (Veal Infusion Broth plus Antibiotics)	25.0 g 0.4 ml 2.0 ml 1.0 ml 1.0 L	Veal Infusion Broth (0344-17-6, Difco) 0.25 mg/ml Amphotericin B Stock Solution 10 mg/ml Gentamicin Solution Penicillin/Streptomycin Solution Endotoxin-Free Distilled Water	Add water to the 1 liter mark. Heat to boiling to dissolve. Dispense 3 ml into screw-cap vials. Autoclave at 121°C for 15 minutes. Store at 2-8°C for up to 1 year.

Subject Index

A

Abbreviation list, 241-247
Acquired immunodeficiency syndrome, 198-199
Adenovirus, 54-63
   centrifugation culture (shell vial) method, 59-60
      coverslip staining, 59
      procedure, 59
      quality control procedures, 60
      results interpretation, 59-60
   characteristics, 54
   direct smear
      control slide preparation, 62-63
      quality control procedures, 62
      results interpretation, 62
      specimen collection, 60-61
      specimen processing, 61
      test procedure, 61
   disinfectants, 55
   epidemiology, 55
   fluorescent antibody staining, 60, 62
   inactivators, 55
   isolation, 56-60
      specimen collection, 56
      specimen preparation, 56-57
      specimen storage, 56
   specimen source, 55
   time to result, 55
   transmission/incubation period, 55
   tube culture, 57
      culture confirmation, 57-58
      positive control inocula preparation, 58-59
      quality control procedures, 58
      results interpretation, 57, 58
   virus detection methods, 55
Aerosol
   biological safety, 6-7
   control, 6-7
Ambient air incubator, 4
Amniotic fluid
   antigen detection, 21
   culture, 21
   potential virus, 21
   specimen type, 21
   symptoms, 21
   turnaround time, 21
   virus testing protocol, 21
Amphotericin B, 35
Antibiotic, 34-35
Anticomplement immunofluorescence procedure, Epstein-Barr virus, 102-103
Autoclave, 5
   quality assurance, 52

B

B19 infection, 181
Back injury, laboratory safety, 10
Biological safety, 6-8
   aerosol, 6-7
   cell culture, 8
   droplet, 6-7
   gloves, 7
   infectious agent, 7-8
   infectious waste, 8
   laboratory coat, 7
   needle, 7
   syringe, 7
Biological safety cabinet, 2-3
Biopsy specimen
   antigen detection, 21
   culture, 21
   potential virus, 21
   specimen type, 21
   symptoms, 21
   turnaround time, 21
   virus testing protocol, 21
Blood
   antigen detection, 21
   culture, 21
   potential virus, 21
   specimen processing, 29
   specimen type, 21
   symptoms, 21
   turnaround time, 21
   virus testing protocol, 21
Body fluid, specimen collection, 22-23
Bone marrow
   antigen detection, 21
   culture, 21
   potential virus, 21
   specimen type, 21
   symptoms, 21
   turnaround time, 21
   virus testing protocol, 21
Brain biopsy
   antigen detection, 19
   culture, 19
   potential virus, 19
   symptoms, 19
   turnaround time, 19
   virus testing protocols, 19
Broncheoalveolar lavage, specimen processing, 31
Buffy coat, specimen processing, 29

C

Calf serum, 33
   inactivation, 35-36
Carbon dioxide incubator, 4
Cell count, 37, 38
   calculations, 38
Cell culture
   biological safety, 8
   cell rounding, 43
   contamination, 43
   CPE in uninoculated cultures, 44
   granular cytoplasm, 43
   immunization, 8
   normal-shaped cells sloughing from vessel, 43
   piling, 43
   procedure, 35-43
   propagation, 36-37
   quality assurance, 47-51
   troubleshooting guide, 43-44
   water for, 35
Cell culture isolation
   chlamydiae, 68-72
      cell culture procedure, 70-71
      coverslip staining, 71
      positive control culture preparation, 72
      positive control inocula, 72

Subject Index

quality control procedures, 72
results interpretation, 71-72
specimen collection, 68-70
specimen preparation, 68-70
traditional, 11
Cell culture medium, 33
pH control, 34
Cell culture neutralization assay,
Clostridium difficile
microtiter plate procedure, 79
quality control procedures, 80-81
results interpretation, 80
specimen collection, 79
specimen preparation, 79
specimen storage, 79
tube/vial procedure, 80
Cell dissociation, 35
Cell line
cell yield, 40
inoculation, 11
seeding density, 40
selection, 11
sensitivity
centrifugation-enhanced
(shell vial) isolation
methods, 13
standard virus isolation methods,
11, 12
Cell rounding, cell culture, 43
Cell smear
direct fluorescent antibody methods, 16
preparation, 30-32
Cell yield, cell line, 40
Centrifugation culture (shell vial)
method, 39
adenovirus, 59-60
coverslip staining, 59
procedure, 59
quality control procedures, 60
results interpretation, 59-60
cytomegalovirus, 87-88
procedure, 87
quality control procedures, 88
results interpretation, 88
herpes simplex virus, 115-117
coverslip staining, 116
procedure, 116
quality control procedures, 117
results interpretation, 116-117
measles virus, 146-148
coverslip staining, 147
procedure, 146-147
quality control procedures, 147-148
results interpretation, 147
mumps virus, 158-160
coverslip staining, 159

procedure, 159
quality control procedures, 160
results interpretation, 159-160
parainfluenza virus, 167-168
coverslip staining, 167-168
procedure, 167
quality control procedures, 168
results interpretation, 168
polyomavirus, 178-179
coverslip staining, 179
procedure, 179
quality control procedures, 179
results interpretation, 179
respiratory syncytial virus, 190-191
coverslip staining, 190
procedure, 190
quality control procedures, 191
results interpretation, 190-191
rotavirus, 219-221
coverslip staining, 219-220
procedure, 219
quality control procedures, 220-221
results interpretation, 220
varicella-zoster virus, 235-236
coverslip staining, 235-236
quality control procedures, 236
results interpretation, 235, 236
Centrifuge, 4
quality assurance, 52
Cerebrospinal fluid
antigen detection, 19
culture, 19
potential virus, 19
symptoms, 19
turnaround time, 19
virus testing protocols, 19
Cervical specimen, specimen collection, 23
Chemical safety, 8-10
Chickenpox, varicella-zoster virus, 229
Chlamydia pneumoniae, 66-68, 69
disinfectants, 69
epidemiology, 69
inactivators, 69
serologies, 69
specimen source, 69
time to result, 69
transmission/incubation period, 69
virus detection methods, 69
Chlamydia psittaci, 66, 67
disinfectants, 67
epidemiology, 67
inactivators, 67
serologies, 67
specimen source, 67
time to result, 67
transmission/incubation period, 67

virus detection methods, 67
Chlamydia slide preparation
cytobrush specimen, 23-24
swab, 23-24
Chlamydia smear, urine, 32
Chlamydia (sucrose-containing)
transport medium swab,
specimen processing, 31-32
Chlamydia trachomatis, 64-66, 74
disinfectants, 65
epidemiology, 65
inactivators, 65
serologies, 65
specimen source, 65
time to result, 65
transmission/incubation period, 65
virus detection methods, 65
Chlamydiae, 64-75
cell culture isolation, 68-72
cell culture procedure, 70-71
coverslip staining, 71
positive control culture preparation, 72
positive control inocula, 72
quality control procedures, 72
results interpretation, 71-72
specimen collection, 68-70
specimen preparation, 68-70
direct smear, 72-75
preparation of positive and
negative controls, 75
quality control procedures, 75
results interpretation, 74-75
specimen collection, 73-74
specimen preparation, 73-74
test procedure, 74
Clean area, 1-2
Clostridium difficile, 77-81
cell culture neutralization assay
microtiter plate procedure, 79
quality control procedures, 80-81
results interpretation, 80
specimen collection, 79
specimen preparation, 79
specimen storage, 79
tube/vial procedure, 80
disinfectants, 78
epidemiology, 78
inactivators, 78
specimen source, 78
virus detection methods, 78
Commercial p(24) antigen assay,
human immunodeficiency
virus, 206
Conjunctiva, specimen collection, 24-25
Contamination, cell culture, 43
Countertop, quality assurance, 52

Coverslip staining, 87-88
  shell vial centrifugation culture, 15
Culture confirmation procedure, 14
Cytobrush specimen, chlamydia slide preparation, 23-24
Cytomegalovirus, 82-91
  centrifugal enhanced (shell vial) method, 87-88
    procedure, 87
    quality control procedures, 88
    results interpretation, 88
  characteristics, 82
  direct smear
    BAL sample test procedure, 89
    control slide preparation, 90-91
    quality control procedures, 90
    results interpretation, 89-90
    specimen preparation, 89
  disinfectants, 83
  epidemiology, 83
  inactivators, 83
  serology, 83
  specimen source, 83
  time to result, 83
  transmission/incubation period, 83
  virus detection methods, 83
  virus isolation, 84-88
    culture confirmation procedure, 86
    positive control inocula preparation, 86-87
    quality control procedures, 86
    results interpretation, 85-86
    specimen collection, 84
    specimen preparation, 84-85
    specimen storage, 84
    standard tube culture, 85-87
Cytopathic effect, observation, 14

D

Departmental procedure manual, 46
Dermal lesion
  antigen detection, 20
  culture, 19-20
  potential virus, 19
  specimen, 19
  symptoms, 19
  turnaround time, 19
  virus testing protocols, 20
Diarrhea, rotavirus, 214
Dimethylsulfoxide, 9
Direct fluorescent antibody methods, 15-16
  cell smear, 16
  negative, 16
  positive, 16

Direct smear
  adenovirus
    control slide preparation, 62-63
    quality control procedures, 62
    results interpretation, 62
    specimen collection, 60-61
    specimen processing, 61
    test procedure, 61
  chlamydiae, 72-75
    preparation of positive and negative controls, 75
    quality control procedures, 75
    results interpretation, 74-75
    specimen collection, 73-74
    specimen preparation, 73-74
    test procedure, 74
  cytomegalovirus
    BAL sample test procedure, 89
    control slide preparation, 90-91
    quality control procedures, 90
    results interpretation, 89-90
    specimen preparation, 89
  herpes simplex virus, 117-119
    positive and negative control preparation, 119
    procedure, 117-118
    quality control procedures, 118-119
    results interpretation, 118
    specimen collection, 117
  influenza virus, 136-139
    control slides preparation, 139
    procedure, 137-138
    quality control procedures, 139
    results interpretation, 138-139
    specimen collection, 137
    specimen processing, 137
  measles virus, 148-150
    control slide preparation, 150
    procedure, 148-149
    quality control procedures, 149-150
    results interpretation, 149
    specimen collection, 148
    specimen processing, 148
  parainfluenza virus, 168-171
    control slide preparation, 170-171
    procedure, 169-170
    quality control procedures, 170
    results interpretation, 170
    specimen collection, 168
    specimen processing, 169
  respiratory syncytial virus, 191-194
    control slide preparation, 193-194
    procedure, 192-193
    quality control procedures, 193

    results interpretation, 193
    specimen collection, 192
    specimen processing, 192
  varicella-zoster virus, 236-239
    control slide preparation, 238-239
    quality control procedures, 238
    results interpretation, 238
    specimen collection, 237
    specimen preparation, 237
    test procedure, 237-238
Droplet
  biological safety, 6-7
  control, 6-7

E

Eagle's minimum essential medium, 33
ECHO 11 virus challenge, rubella virus, 226-227
Enterovirus, 92-97
  characteristics, 92
  disinfectants, 93
  epidemiology, 93
  inactivators, 93
  Lim and Benyesch-Melnick pool, 96-97
    interpretation, 97
    procedure, 96-97
  serologies, 93
  specimen sources, 93
  time to result, 93
  transmission/incubation period, 93
  tube culture, 95-96
    positive control culture preparation, 96
    quality control procedures, 95
    results interpretation, 95
  virus detection methods, 93
  virus isolation, 94-97
    specimen collection, 94-95
    specimen preparation, 95
Enterovirus 72. *See* Hepatitis A
Enzyme immunoassay test, 17
Epstein-Barr virus
  anticomplement immunofluorescence procedure, 102-103
  characterized, 98
  diagnosis, 98-100
  disinfectants, 99
  epidemiology, 99
  inactivators, 99
  serology, 99
  specimen source, 99
  time to result, 99
  transmission/incubation period, 99
  virus detection methods, 99

virus isolation, 100-104
   cell confirmation, 102
   quality control procedures, 101, 103
   results interpretation, 101, 103
   specimen collection, 100
   specimen preparation, 100
   specimen storage, 100
   test procedure, 100-101
Equipment, quality assurance, 51-52
Erythema infectiosum, human parvovirus, 181
Eye, specimen collection, 24-25

F

Fecal specimen
   antigen detection, 19
   culture, 19
   potential agent, 18
   specimen type, 18
   symptoms, 18
   turnaround times, 19
   virus testing protocols, 18-19
Fetal calf serum, 33
   inactivation, 35-36
Fire safety, 9-10
Fixative, specimen processing, 29
Fluorescence microscope, 3
   light sources, 3
Fluorescent antibody staining
   adenovirus, 60, 62
   direct smears, 15-16
Freezer, 4-5
   quality assurance, 51
Freezing, mammalian cell, 39-41
   recovery, 41

G

Gas cylinder, laboratory safety, 9
Gentamicin, 35
Glassware, laboratory safety, 8
Gloves, biological safety, 7
Granular cytoplasm, cell culture, 43

H

Hemadsorption testing, 11-13
   negative, 12-13
   positive, 12
   procedure, 12
Hemagglutination, procedure, 13-14
HEPA filter, 1
Hepatitis A virus, 105-108
   characterized, 105
   disinfectants, 105
   epidemiology, 105
   inactivators, 105
   specimen source, 105
   time to result, 105
   transmission/incubation period, 105
   tube culture, 107-108
      culture confirmation, 107-108
      procedure, 107
      quality control procedures, 108
      results interpretation, 108
   virus detection methods, 105
   virus isolation, 107-108
      specimen collection, 107
      specimen preparation, 107
Herpes simplex virus, 109-119
   centrifugation culture (shell vial) method, 115-117
      coverslip staining, 116
      procedure, 116
      quality control procedures, 117
      results interpretation, 116-117
   characterized, 109
   diagnosis, 111
   direct smear, 117-119
      positive and negative control preparation, 119
      procedure, 117-118
      quality control procedures, 118-119
      results interpretation, 118
      specimen collection, 117
   disinfectants, 110
   epidemiology, 110
   inactivators, 110
   serology, 110
   specimen source, 110
   time to result, 110
   transmission/incubation period, 110
   tube culture, 113-115
      culture confirmation procedure, 114
      positive control cultures preparation, 114-115
      procedure, 113
      quality control procedures, 114
      results interpretation, 113-114
   type 1, 109-111
   type 2, 109-111
   virus detection methods, 110
   virus isolation, 111-117
      specimen collection, 112
      specimen preparation, 112-113
      specimen storage, 112
Herpes zoster, varicella-zoster virus, 229
Household bleach, 9

Human foreskin fibroblast culture, preparation, 42-43
Human herpesvirus 6, 121-125
   characterized, 121
   disinfectants, 122
   epidemiology, 122
   inactivators, 122
   roseola, 121
   serology, 122
   specimen source, 122
   time to result, 122
   transmission/incubation period, 122
   virus detection methods, 122
   virus isolation, 121-125
      culture confirmation, 124-125
      quality control procedures, 125
      results interpretation, 125
      specimen collection, 123-124
Human immunodeficiency virus
   commercial p(24) antigen assay, 206
   immunofluorescence assay, 205
   lymphocyte separation, 202-203
   PHA-stimulated peripheral blood leukocyte, 203-204
   reverse transcriptase assay, 204-205
   virus isolation, 200-206
      confirmation, 204
      quality control procedures, 206
      results interpretation, 205-206
      specimen collection, 200-202
      specimen storage, 202
Human immunodeficiency virus type 1, 198-200, 201
   disinfectants, 201
   epidemiology, 201
   inactivators, 201
   serology, 201
   specimen source, 201
   time to result, 201
   transmission/incubation period, 201
   virus detection methods, 201
Human immunodeficiency virus type 2, 202, 203
   epidemiology, 203
   serology, 203
   specimen source, 203
   time to result, 203
   transmission/incubation period, 203
   virus detection methods, 203
Human papillomavirus, 172-174
   disinfectants, 173
   epidemiology, 173
   inactivators, 173
   serology, 173
   specimen source, 173
   transmission, 173
   virus detection methods, 173
Human parvovirus, 181-183

characterized, 181
diagnosis, 181-182
disinfectants, 182
epidemiology, 182
erythema infectiosum, 181
inactivators, 182
serology, 182
specimen source, 182
transmission/incubation period, 182
virus detection methods, 182
Human retrovirus, 196-206
Human T-cell lymphotropic virus-I, 196-198
   disinfectants, 197
   epidemiology, 197
   inactivators, 197
   serology, 197
   specimen source, 197
   time to result, 197
   transmission/incubation period, 197
   virus detection methods, 197
Human T-cell lymphotropic virus-II, 198, 199
   disinfectants, 199
   epidemiology, 199
   inactivators, 199
   serology, 199
   specimen source, 199
   time to result, 199
   transmission/incubation period, 199
   virus detection methods, 199
Hypochlorite, 9

I

Immunization, cell culture, 8
Immunofluorescence assay, human immunodeficiency virus, 205
Incubator, 4
   quality assurance, 51
Infectious agent
   biological safety, 7-8
   spills, 7-8
Infectious waste, biological safety, 8
Influenza virus, 127-139
   characterized, 127
   diagnosis, 127-129
   direct smear, 136-139
      control slides preparation, 139
      procedure, 137-138
      quality control procedures, 139
      results interpretation, 138-139
      specimen collection, 137
      specimen processing, 137
   disinfectants, 128
   epidemiology, 128
   inactivators, 128
   serology, 128
   specimen source, 128
   time to result, 128
   transmission/incubation period, 128
   tube culture, 130-135
      confirmation, 133-134
      hemadsorption procedure, 130-131
      hemagglutination (HA) procedure, 132-133
      positive control culture preparation, 134-135
      procedure, 130
      quality control procedures, 132, 133, 134
      results interpretation, 130, 131-132, 133, 134
   virus detection methods, 128
   virus isolation, 129-136
      specimen collection, 130
      specimen preparation, 130
Interference test, rubella virus, 225-226
Inverted microscope, 3
Isolation, adenovirus, 56-60
   specimen collection, 56
   specimen preparation, 56-57
   specimen storage, 56

L

Laboratory
   airflow, 1
   clean area, 1-2
   design, 1-2
   equipment, 2-5
   quality assurance, 51-52
   washup area, 2
   work area, 2
Laboratory coat, biological safety, 7
Laboratory safety, 6-10
   back injury, 10
   chemicals, 8-9
   gas cylinder, 9
   glassware, 8
   liquid nitrogen, 9
   Occupational Safety and Health Administration rules, 6
   sharps, 8
   slipping, 10
Laminar airflow hood, 2-3
   quality assurance, 51
Latex agglutination test, 16-17
Lim and Benyesch-Melnick pool, enterovirus, 96-97
   interpretation, 97
   procedure, 96-97
Liquid nitrogen, laboratory safety, 9
Liquid nitrogen freezer, 5
Lymphocyte separation, human immunodeficiency virus, 202-203

M

Mammalian cell
   freezing, 39-41
   recovery, 41
   primary culture
      frozen storage, 42
      tissue collection, 41-42
      tissue dissociation, 42
      tissue preparation, 41-42
Measles virus, 141-150
   centrifugation enhanced (shell vial) method, 146-148
      coverslip staining, 147
      procedure, 146-147
      quality control procedures, 147-148
      results interpretation, 147
   characterized, 141
   diagnosis, 141-143
   direct smear, 148-150
      control slide preparation, 150
      procedure, 148-149
      quality control procedures, 149-150
      results interpretation, 149
      specimen collection, 148
      specimen processing, 148
   disinfectants, 142
   epidemiology, 142
   inactivators, 142
   serology, 142
   specimen source, 142
   time to result, 142
   transmission/incubation period, 142
   tube culture, 143-146
      culture confirmation procedure, 144-145
      positive control inocula preparation, 146
      procedure, 143-144
      quality control procedures, 146
      results interpretation, 144, 145-146
   virus detection methods, 142
   virus isolation, 143-148
      specimen collection, 143
      specimen preparation, 143
      specimen storage, 143
Mechanical pipetting device, quality assurance, 52
Microscope, 3

acromats, 3
apochromats, 3
fluorites, 3
quality assurance, 51
types, 3
Microscope slide, specimen collection, 22
Mumps, 152
  diagnosis, 152
Mumps virus, 152-160
  centrifugation-enhanced (shell vial) method, 158-160
    coverslip staining, 159
    procedure, 159
    quality control procedures, 160
    results interpretation, 159-160
  disinfectants, 153
  epidemiology, 153
  inactivators, 153
  serology, 153
  specimen source, 153
  time to result, 153
  transmission/incubation period, 153
  tube culture, 155-158
    confirmation, 157-158
    hemadsorption procedure, 155-157
    positive control inocula preparation, 158
    procedure, 155
    quality control procedures, 157
    results interpretation, 155
  virus detection methods, 153
  virus isolation, 154-160
    specimen collection, 154
    specimen processing, 154
    specimen storage, 154

N

Nasal aspirate, specimen processing, 31
Nasal wash, specimen processing, 31
Nasopharyngeal aspirate, specimen collection, 25
Nasopharyngeal smear, specimen collection, 25
Nasopharyngeal swab, specimen collection, 25
Nasopharyngeal wash, specimen collection, 25-26
Needle, biological safety, 7
Negative pressure, 1
Neubauer brightline hemocytometer, 37

O

Occupational Safety and Health Administration rules, laboratory safety, 6
Ocular specimen
  antigen detection, 20
  culture, 20
  potential agents, 20
  specimen type, 20
  symptoms, 20
  turnaround time, 20
  virus testing protocols, 20

P

Parainfluenza virus, 161-171
  centrifugation-enhanced (shell vial) method, 167-168
    coverslip staining, 167-168
    procedure, 167
    quality control procedures, 168
    results interpretation, 168
  characterized, 161
  direct smear, 168-171
    control slide preparation, 170-171
    procedure, 169-170
    quality control procedures, 170
    results interpretation, 170
    specimen collection, 168
    specimen processing, 169
  disinfectants, 162
  epidemiology, 162
  inactivators, 162
  serotypes, 161
  specimen source, 162
  time to result, 162
  transmission/incubation period, 162
  tube culture, 163-167
    confirmation, 165-166
    hemadsorption procedure, 164-165
    positive control culture preparation, 166-167
    procedure, 163
    results interpretation, 163-164
  virus detection methods, 162
  virus isolation, 162-168
    specimen collection, 162
    specimen preparation, 163
Penicillin G, 34-35
pH control, cell culture medium, 34
pH meter, quality assurance, 52
PHA-stimulated peripheral blood leukocyte, human immunodeficiency virus, 203-204
Physical safety, 8-10
Physiological hazards, 10
Piling, cell culture, 43

Polyomavirus, 175-179
  centrifugation culture (shell vial) method, 178-179
    coverslip staining, 179
    procedure, 179
    quality control procedures, 179
    results interpretation, 179
  disinfectants, 176
  epidemiology, 176
  inactivators, 176
  progressive multifocal leukoencephalopathy, 175
  serology, 176
  specimen source, 176
  time to result, 176
  transmission/incubation period, 176
  tube culture
    confirmation, 177-178
    positive control cultures preparation, 178
    quality control procedures, 178
    results interpretation, 177
  virus detection methods, 176
  virus isolation, 176-179
    specimen collection, 177
    specimen preparation, 177
    specimen storage, 177
Primary culture, mammalian cell
  frozen storage, 42
  tissue collection, 41-42
  tissue dissociation, 42
  tissue preparation, 41-42
Procedure manual, quality assurance, 45-46
Products of conception
  antigen detection, 21
  culture, 21
  potential virus, 21
  specimen type, 21
  symptoms, 21
  turnaround time, 21
  virus testing protocol, 21
Proficiency testing, quality assurance, 52
Progressive multifocal leukoencephalopathy, polyomavirus, 175
Pseudomembranous colitis, 77
  diagnosis, 77-78

Q

Quality assurance, 45-52
  autoclave, 52
  cell culture, 47-51
  centrifuge, 52
  countertop, 52

equipment, 51-52
freezer, 51
incubator, 51
laboratory, 51-52
laminar airflow hood, 51
mechanical pipetting device, 52
microscope, 51
pH meter, 52
procedure manual, 45-46
proficiency testing, 52
reagent, 47
refrigerator, 51
specimen collection, 46-47
specimen transport, 46-47
spectrophotometer, 52
telephone, 52
thermometer, 52
tissue culture rotator, 52
water bath, 52
written procedures, 45-46

R

Reagent
   expiration dating, 48-50
   formulations, 255-266
   quality assurance, 47
   resources, 248-254
   storage, 48-50
Rectal swab
   antigen detection, 19
   culture, 19
   potential agent, 18
   specimen collection, 26
   specimen type, 18
   symptoms, 18
   turnaround times, 19
   virus testing protocols, 18-19
Refrigerator, 4-5
   quality assurance, 51
Respiratory specimen
   antigen detection, 18
   enzyme immunoassay, 18
   inoculation, 18
   potential agent, 18
   specimen type, 18
   symptoms, 18
   turnaround times, 18
   virus testing protocol, 18
Respiratory syncytial virus, 184-194
   centrifugation enhanced (shell vial) method, 190-191
      coverslip staining, 190
      procedure, 190
      quality control procedures, 191
      results interpretation, 190-191
   characterized, 184
   direct smear, 191-194
      control slide preparation, 193-194
      procedure, 192-193
      quality control procedures, 193
      results interpretation, 193
      specimen collection, 192
      specimen processing, 192
   disinfectants, 185
   epidemiology, 185
   inactivators, 185
   ribavirin, 186
   serologies, 185
   specimen sources, 185
   time to result, 185
   transmission/incubation period, 185
   tube culture, 188-190
      confirmation, 188-189
      positive control culture preparation, 189-190
      procedure, 188
      quality control procedures, 189
      results interpretation, 188, 189
   virus detection methods, 185
   virus isolation, 186-191
      specimen collection, 187
      specimen preparation, 187-188
      specimen storage, 187
Reverse transcriptase assay, human immunodeficiency virus, 204-205
Rhinovirus, 208-212
   disinfectants, 209
   epidemiology, 209
   inactivators, 209
   serology, 209
   specimen source, 209
   time to result, 209
   transmission/incubation period, 209
   tube culture, 210-212
      pH 3 inactivation test, 211-212
      positive control culture preparation, 212
      procedure, 210-211
      quality control procedures, 212
      results interpretation, 211
   virus detection methods, 209
   virus isolation, 209-212
      specimen collection, 210
      specimen preparation, 210
      specimen storage, 210
Ribavirin, respiratory syncytial virus, 186
Roller drum, 5
Roseola, human herpesvirus 6, 121
Rotavirus, 214-221
   centrifugation culture (shell vial) method, 219-221
      coverslip staining, 219-220
      procedure, 219
      quality control procedures, 220-221
      results interpretation, 220
   characterized, 214
   diarrhea, 214
   disinfectants, 215
   epidemiology, 215
   inactivators, 215
   serology, 215
   specimen source, 215
   time to result, 215
   transmission/incubation period, 215
   tube culture, 217-219
      culture confirmation procedure, 217-218
      positive control culture preparation, 219
      procedure, 217
      quality control procedures, 218-219
      results interpretation, 217, 218
   virus detection methods, 215
   virus isolation, 216-221
      sample preparation, 216
      specimen collection, 216
      specimen storage, 216
Rubella virus, 222-227
   characterized, 222
   disinfectants, 223
   ECHO 11 virus challenge, 226-227
   epidemiology, 223
   inactivators, 223
   interference test, 225-226
   serology, 223
   specimen source, 223
   time to result, 223
   transmission/incubation period, 223
   virus detection methods, 223
   virus isolation, 224-227
      procedure, 225
      specimen collection, 224-225
      specimen preparation, 225

S

Saliva, specimen collection, 26
Seeding density, cell line, 40
Semen, specimen processing, 29
Serum supplement, 33-34
Sharps, laboratory safety, 8
Shell vial centrifugation culture, 14-15
   coverslip staining, 15
   procedure, 15
Shingles, varicella-zoster virus, 229
Slipping, laboratory safety, 10
Specimen collection, 22-28
   body fluid, 22-23

cervical specimen, 23
conjunctiva, 24-25
eye, 24-25
microscope slide, 22
nasopharyngeal aspirate, 25
nasopharyngeal smear, 25
nasopharyngeal swab, 25
nasopharyngeal wash, 25-26
quality assurance, 46-47
rectal swab, 26
saliva, 26
sites, 23, 24
stool, 26
swab, 22
throat smear, 26-27
throat swab, 26
throat wash, 27
timing, 24
tissue, 27
urethral smear, 27
urethral swab, 27
urine, 27
vesicle fluid
    aspiration, 28
    swab, 28
vesicular cell smear, 28
Specimen processing, 29-32
blood, 29
broncheoalveolar lavage, 31
buffy coat, 29
chlamydia (sucrose-containing) transport medium swab, 31-32
fixative, 29
nasal aspirate, 31
nasal wash, 31
semen, 29
sputum, 29
stool, 30
swab, 30
throat wash, 31
throat washing, 29
tissue, 30
urine sediment, 30
viral transport medium swab, 31
wash and aspirate specimen, 30
Specimen transport, 28
quality assurance, 46-47
Spectrophotometer, quality assurance, 52
Sputum, specimen processing, 29
Stool
    specimen collection, 26
    specimen processing, 30
Streptomycin, 35
Swab
    chlamydia slide preparation, 23-24
    specimen collection, 22
    specimen processing, 30

Syringe, biological safety, 7

T

Telephone, quality assurance, 52
Thermometer, quality assurance, 52
Throat smear, specimen collection, 26-27
Throat swab, specimen collection, 26
Throat wash
    specimen collection, 27
    specimen processing, 29, 31
Tissue
    specimen collection, 27
    specimen processing, 30
Tissue culture rotator, quality assurance, 52
Tube culture, 37-39
adenovirus, 57
    culture confirmation, 57-58
    positive control inocula preparation, 58-59
    quality control procedures, 58
    results interpretation, 57, 58
enterovirus, 95-96
    positive control culture preparation, 96
    quality control procedures, 95
    results interpretation, 95
hepatitis A virus, 107-108
    culture confirmation, 107-108
    procedure, 107
    quality control procedures, 108
    results interpretation, 108
herpes simplex virus, 113-115
    culture confirmation procedure, 114
    positive control cultures preparation, 114-115
    procedure, 113
    quality control procedures, 114
    results interpretation, 113-114
influenza virus, 130-135
    confirmation, 133-134
    hemadsorption procedure, 130-131
    hemagglutination (HA) procedure, 132-133
    positive control culture preparation, 134-135
    procedure, 130
    quality control procedures, 132, 133, 134
    results interpretation, 130, 131-132, 133, 134
measles virus, 143-146
    culture confirmation procedure, 144-145
    positive control inocula preparation, 146
    procedure, 143-144
    quality control procedures, 146
    results interpretation, 144, 145-146
mumps virus, 155-158
    confirmation, 157-158
    hemadsorption procedure, 155-157
    positive control inocula preparation, 158
    procedure, 155
    quality control procedures, 157
    results interpretation, 155
parainfluenza virus, 163-167
    confirmation, 165-166
    hemadsorption procedure, 164-165
    positive control culture preparation, 166-167
    procedure, 163
    results interpretation, 163-164
polyomavirus
    confirmation, 177-178
    positive control cultures preparation, 178
    quality control procedures, 178
    results interpretation, 177
respiratory syncytial virus, 188-190
    confirmation, 188-189
    positive control culture preparation, 189-190
    procedure, 188
    quality control procedures, 189
    results interpretation, 188, 189
rhinovirus, 210-212
    pH 3 inactivation test, 211-212
    positive control culture preparation, 212
    procedure, 210-211
    quality control procedures, 212
    results interpretation, 211
rotavirus, 217-219
    culture confirmation procedure, 217-218
    positive control culture preparation, 219
    procedure, 217
    quality control procedures, 218-219
    results interpretation, 217, 218
varicella-zoster virus, 233-235
    culture confirmation procedure, 233-234
    results interpretation, 233

U

Urethral smear, specimen collection, 27
Urethral swab, specimen collection, 27
Urine
   antigen detection, 21
   chlamydia smear, 32
   culture, 21
   potential agent, 21
   specimen collection, 27
   specimens, 21
   symptoms, 21
   turnaround time, 21
   virus testing protocol, 21
Urine sediment, specimen processing, 30
Urogenital specimen
   antigen detection, 20-21
   culture, 20
   potential agent, 20
   specimen type, 20
   symptoms, 20
   turnaround time, 20
   virus testing protocol, 20-21

V

Varicella-zoster virus, 229-239
   centrifugation enhanced (shell vial) method procedure, 235-236
   coverslip staining, 235-236
   quality control procedures, 236
   results interpretation, 235, 236
   chickenpox, 229
   direct smear, 236-239
      control slide preparation, 238-239
      quality control procedures, 238
      results interpretation, 238
      specimen collection, 237
      specimen preparation, 237
      test procedure, 237-238
   disinfectants, 230
   epidemiology, 230
   herpes zoster, 229
   inactivators, 230
   serology, 230
   shingles, 229
   specimen source, 230
   time to result, 230
   transmission/incubation period, 230
   tube culture, 233-235
      culture confirmation procedure, 233-234
      results interpretation, 233
   virus detection methods, 230
   virus isolation, 232-236
      positive control inocula preparation, 234-235
      quality control procedures, 234
      specimen collection, 232-233
      specimen preparation, 233
      specimen storage, 232-233
Ventilation system, 1
Vesicle fluid, specimen collection
   aspiration, 28
   swab, 28
Vesicular cell smear, specimen collection, 28
Viral transport medium swab, specimen processing, 31
Virus isolation
   cytomegalovirus, 84-88
      culture confirmation procedure, 86
      positive control inocula preparation, 86-87
      quality control procedures, 86
      results interpretation, 85-86
      specimen collection, 84
      specimen preparation, 84-85
      specimen storage, 84
      standard tube culture, 85-87
   enterovirus, 94-97
      specimen collection, 94-95
      specimen preparation, 95
   Epstein-Barr virus, 100-104
      cell confirmation, 102
      quality control procedures, 101, 103
      results interpretation, 101, 103
      specimen collection, 100
      specimen preparation, 100
      specimen storage, 100
      test procedure, 100-101
   hepatitis A virus, 107-108
      specimen collection, 107
      specimen preparation, 107
   herpes simplex virus, 111-117
      specimen collection, 112
      specimen preparation, 112-113
      specimen storage, 112
   human herpesvirus 6, 121-125
      culture confirmation, 124-125
      quality control procedures, 125
      results interpretation, 125
      specimen collection, 123-124
   human immunodeficiency virus, 200-206
      confirmation, 204
      quality control procedures, 206
      results interpretation, 205-206
      specimen collection, 200-202
      specimen storage, 202
   influenza virus, 129-136
      specimen collection, 130
      specimen preparation, 130
   measles virus, 143-148
      specimen collection, 143
      specimen preparation, 143
      specimen storage, 143
   mumps virus, 154-160
      specimen collection, 154
      specimen processing, 154
      specimen storage, 154
   parainfluenza virus, 162-168
      specimen collection, 162
      specimen preparation, 163
   polyomavirus, 176-179
      specimen collection, 177
      specimen preparation, 177
      specimen storage, 177
   respiratory syncytial virus, 186-191
      specimen collection, 187
      specimen preparation, 187-188
      specimen storage, 187
   rhinovirus, 209-212
      specimen collection, 210
      specimen preparation, 210
      specimen storage, 210
   rotavirus, 216-221
      sample preparation, 216
      specimen collection, 216
      specimen storage, 216
   rubella virus, 224-227
      procedure, 225
      specimen collection, 224-225
      specimen preparation, 225
   varicella-zoster virus, 232-236
      positive control inocula preparation, 234-235
      quality control procedures, 234
      specimen collection, 232-233
      specimen preparation, 233
      specimen storage, 232-233

W

Wash and aspirate specimen, specimen processing, 30
Washup area, 2
Water bath, 4
   quality assurance, 52
Workbench, 2